$9.95 UCF

Thermodynamics of Energy Conversion and Transport

Springer
*New York
Berlin
Heidelberg
Barcelona
Hong Kong
London
Milan
Paris
Singapore
Tokyo*

Stanislaw Sieniutycz   Alexis De Vos
Editors

Thermodynamics of Energy Conversion and Transport

With 74 Figures

Springer

Stanislaw Sieniutycz
Institute of Chemical and
  Process Engineering
University of Warsaw
00-645 Waszawa
Poland

Alexis De Vos
Vakgroep voor Elektronika
  en Informatiesystemen
Universiteit Gent
Sint Pieternieuwstraat 41
B-9000 Gent
Belgium

Library of Congress Cataloging-in-Publication Data
Thermodynamics of energy conversion and transport / editors, Stanislaw Sieniutycz,
Alexis De Vos.
    p. cm.
    Includes bibliographical references and index.
    ISBN 0-387-98938-2 (hc. : alk. paper)
    1. Thermodynamics. 2. Direct energy conversion. 3. Energy transfer. I. Sieniutycz,
Stanislaw. II. Vos, Alexis de.
TJ265 .T4575 2000
621.402'1—dc21   99-045613

Printed on acid-free paper.

© 2000 Springer-Verlag New York, Inc.
All rights reserved. This work may not be translated or copied in whole or in part without the
written permission of the publisher (Springer-Verlag New York, Inc., 175 Fifth Avenue, New York,
NY 10010, USA), except for brief excerpts in connection with reviews or scholarly analysis. Use
in connection with any form of information storage and retrieval, electronic adaptation, computer
software, or by similar or dissimilar methodology now known or hereafter developed is forbidden.
The use of general descriptive names, trade names, trademarks, etc., in this publication, even if the
former are not especially identified, is not to be taken as a sign that such names, as understood by
the Trade Marks and Merchandise Marks Act, may accordingly be used freely by anyone.

Production managed by Allan Abrams; manufacturing supervised by Jacqui Ashri.
Photocomposed copy prepared from the authors' TeX files.
Printed and bound by Maple-Vail Book Manufacturing Group, York, PA.
Printed in the United States of America.

9 8 7 6 5 4 3 2 1

ISBN 0-387-98938-2 Springer-Verlag New York Berlin Heidelberg   SPIN 10745220

Preface

The challenge to humankind is to manage energy resources effectively, maximizing the satisfaction derived from them while minimizing their consumption. This challenge has either a time dimension or a spatial dimension, both of which are quite critical. In recent years more and more scientists and engineers have directed their research toward problems characterized by the need for a rational exploitation of energy in practical systems. This has been especially true for energy conversion, transmission, and storage, particulary when considered in the context of renewable energy resources such as solar energy.

It has been recognized that nonequilibrium thermodynamic approaches are capable of providing quite realistic performance criteria and bounds for real processes occurring in finite time and for practical systems of finite size. Finite-time thermodynamics and thermoeconomics, both related in spirit to an older field called Second Law Analysis, have been developed to aid in the search for the optimum ways to operate machines and processes, either by finding the best or optimum values of key parameters or by determining optimum pathways of operation associated with various thermal engines and unit operations in thermal and chemical plants. One of the reasons for the attractiveness of these newer branches of thermodynamics over older ones is that they admit explicitly that for given process requirements, operations must occur with finite intensities, and that inherent loss mechanisms need to be taken into account. Another reason is that the newer branches do not dissect the field on the basis of specific characteristics of individual processes as was done in some older approaches, but rather, they integrate the common characteristics of a number of processes, a feature that helps an engineer avoid being perplexed when shifting a design from one process to another.

Thanks to the generic and synthesizing nature of these approaches, as well as to their effectiveness, significant progress has been achieved in the design of new or improved thermal, separation, chemical, and radiative systems. In fact, progress in applications accompanies that in theory. With the help of thermodynamic approaches, a variety of practical and industrial systems can now be modeled and optimized, including thermal and solar-driven engines, heat exchangers, diffusional separators, semiconductor devices, etc. These applications, on the one hand, result in a deeper understanding of the theory and, on the other, lead to further improvements in the design of practical devices. Having developed formulations for evaluating system performance, design variables, and costs, researchers can now address various problems of system control and optimization. Dynamic bounds can be determined which are usually functions of operational constraints consistent with finite rates of operation for the process. These bounds can be contrasted with the static bounds of classical thermodynamics pertinent to infinitely slow processes.

The present book comprises the common efforts of the participants of the Copernicus program entitled *Thermodynamics and Thermoeconomics of Energy Conversion, Transmission, and Accumulation,* as well as that of a few of their external coworkers. The program has been coordinated by A. De Vos (Gent, Belgium) and conducted under the auspices of the European Union. Aimed at the state of the art in the thermodynamics of energy conversion and transmission, the book accomplishes its task through review chapters written by acknowledged researchers with a thermodynamic orientation. Thus, this book should be read by those scientists and engineers who work in the areas of applied physics and chemistry and various branches of engineering, especially solar, thermal, mechanical, chemical, and environmental engineering. The book consists of twelve topical chapters dealing with different, yet reasonably integrated, subjects in the field of energy conversion and transmission. The unifying factor for these subjects follows from the way most of them deal with thermodynamic systems which have some control that can be adjusted to achieve the best or extremal performance. Thus, in essence, this book focuses primarily on the optimization of energy conversion and transmission systems using thermodynamic methods. The structure of the Table of Contents as well as that of the Index should help the reader locate specific interests. The contributions indicate that the book should be of value to readers both in academia and in industry, thus, attracting researchers working either in fundamentals or in applications.

The first set of chapters provides a perspective on solar energy conversion. The second set focuses on the transfer and conversion of thermal and chemical energy. The third set deals with energy systems treated by thermodynamic geometry. The systems analyzed comprise solar cells, thermal and combustion engines, and separation units (distillation). For optimization purposes, system models are presented which contain, in general, control (decision) variables, state variables, and (usually) some uncontrolled (fixed) parameters. These models incorporate diverse process characteristics such as finite heat conductances, semiconductor bandgaps, diffusion and beam transfer channels, friction, heat loss, chemical resistances, and other factors which are essential in real energy conversion and transmission processes. Approaches from classical and finite-time thermodynamics, exergy analysis and thermodynamic geometry search for the best or optimum values of the principal parameters of various engines (thermal, solar, combustion, etc.) and unit operations or processes (distillation, chemical reactions, etc.)

as well as of some combined structures operating in thermal or chemical plants under specified operational constraints. Optimal paths and optimal controls (for instance, driving heat fluxes maximizing work produced) are determined. The role of mathematical programming and optimal control theory is essential when solving these problems. Optimization usually yields a few basic recommendations on how to run a practical system.

The distinction made in the Table of Contents between radiative and nonradiative systems, while not absolutely necessary, should prove useful. Even if the distinction is not always clear-cut and other distinguishing criteria could be used, one can take advantage of the fact that the transport laws in radiative systems contain inherent nonlinearities whereas those in nonradiative, i.e., matter-containing, systems may quite frequently be described by linear models. Fundamentals and applications of solar energy conversion are covered in Chapters 1–5. Chapter 1 gives an overview of microscopic fundamentals stemming from statistical mechanics and information theory. Chapters 2 and 3 deal, respectively, with the general macroscopic aspects of solar (radiative) energy conversion into mechanical and electrical work. Chapter 4 treats the theoretical and experimental efficiencies of solar cells. Chapter 5 analyzes solar buildings as thermodynamic automata. This chapter begins the group of reviews (Chapters 6–9) associated explicitly with optimal control and system theory. They refer, respectively, to multistage thermal machines (Chapter 6), internal combustion engines (Chapter 7), the qualitative properties of heat transfer systems (Chapter 8), and processes with large energy changes (Chapter 9). Chapters 10–12 deal with the theory and applications of thermodynamic geometry in energy systems. Chapter 10 reviews the general mathematical framework of differential geometry in thermodynamics; Chapter 11 illustrates the various contexts of thermodynamic distances; and Chapter 12 applies the principles of thermodynamic geometry to a distillation process as a chemical flow system.

Below we outline in some detail the contents of the chapters in this volume.

B. Månsson (Chapter 1) begins the group of reviews related to conversion of radiative and solar energy. He presents the statistical mechanical treatment of a class of processes in which solar energy is absorbed and then converted in multistage operations terminating at the energy storage. Examples include photoelectric (photovoltaic) and photochemical processes. Special attention is paid to loss mechanisms. Possibilities and limits for optimization arising from absolute limits and from irreducible losses are analyzed. The connection between statistical mechanics and information theory is investigated in the context of the solar energy conversion. Analyses which use both traditional and generalized (dissipative) exergy of incoherent photon beams are relevant. The conversion of black-body radiation is used as a benchmark case. It is shown that the actual exergy efficiencies are in some cases substantially less than those found for the ideal benchmark case.

V. Bădescu (Chapter 2) treats the thermodynamics of conversion of solar energy into mechanical work. He distinguishes two types of solar power systems. Those of the first type use photovoltaic cells in combination with electrochemical storage to convert the solar energy directly into electrical energy. Those of the second type are based on thermodynamic cycles. In a model system solar concentrators reflect the flux of solar radiation toward a collector where a working fluid is heated to drive conventional engines. Electrical energy can be generated by alternators coupled to these thermal engines. The main aim of the energy conversion model is to establish an upper limit for the conversion efficiency. The first part of this chapter refers to general models which take into account only thermal properties of two energy sources, the Sun and the ambient, and, possibly, the optical properties of the absorber. They are derived by using an original thermodynamic argument and usually allow easy (handy) computation of the upper bound formulas for the conversion efficiency of black-body radiation into work. However, all of these formulas predict upper bounds that are too high to be useful. The second part of the chapter deals therefore with more detailed models. They usually require the use of computers. They take into account both the nature of the incoming radiation and design parameters of the conversion system. These models are classified into two subcategories, as they refer to terrestrial and space applications. Developing a model of a terrestrial solar power plant requires taking into account that the available solar energy at the ground level has a direct (beam) and a diffuse component. Three different models of terrestrial power plants utilizing direct, diffuse, and global solar radiation are presented. The solar space power stations use solar beam radiation only. However, they differ from the terrestrial plants, as the main mechanism of heat rejection by the thermal engine is through radiation in space and by convection on Earth. Two different models of solar space power stations are presented. The first assumes the ideal case of a Carnot engine; the second exploits the concept of the irreversible engine of Novikov–Curzon–Ahlborn (NCA). This second model is made more realistic by taking into consideration irreversibilities caused by the heat transfer at the hot and cold parts of the engine. Irreversible engines are also treated in later chapters by De Vos (a single-stage system) and by Sieniutycz and Berry (multistage systems).

A. De Vos (Chapter 3) analyzes a solar cell as a nonlinear thermodynamic engine which works between the temperature of the Sun, and that of the Earth. For some value of the voltage the product of the voltage and current is the maximal electric power. Dividing this power by the incident power from the Sun the solar cell efficiency follows in terms of the semiconductor bandgap which is the ultimate optimized quantity. The resulting optimal efficiency refers to the irreversible case and is shown to be much lower than the one following from the Carnot (reversible) theory. However, by introducing multigap (tandem) solar cells (different materials with different bandgaps), one may achieve efficiencies quite close to the Carnot limit. This analysis shows the applicability of endoreversible thermodynamics in the photovoltaic context and its methodological similarity to the analysis of the endoreversible thermal engine of the Novikov–Curzon–Ahlborn type (NCA engine) containing the Carnot engine combined with thermal reservoirs by finite conductances.

P. T. Landsberg and V. Bădescu (Chapter 4) present a survey of experimental and theoretical cell efficiencies obtained in recent years and consider the problem of maximal conversion efficiency. High efficiencies are obtained if one bases oneself on various thermodynamic cycles or derives the results from the balance of the entropy and energy fluxes. Yet there are more realistic analyses involving the working of solar cells, using photovoltaics. Assuming perfect collection efficiencies for the carriers and neglecting their losses by recombination one still obtains ideally high efficiencies which can be worked out on the basis of various degrees of solar concentration. These theoretical efficiencies can be increased further by stipulating the use of tandem cells, by imagining pair production with various degrees of impact ionization, by taking into account multiple sources, and by stipulating hybrid systems. Later one tends to adjust these optimistic results by taking corrections into account, for example, by estimating the loss of carriers by recombination. The authors ask these questions: How do all these results compare? How different are actual efficiencies? To answer these questions they consider the physics of photovoltaics and thermovoltaic conversion in systems of mono- and multigap materials along with the survey of cell efficiencies obtained in recent years. Their efficiency table covers experimental and theoretical results and therefore covers a wide range from 11% to 88%. There is a tendency for the theoretical efficiencies to be lower the more detailed the model is on which they are based.

L. Sertorio and G. Tinetti (Chapter 5) treats solar buildings as thermodynamic automata. His review is still in the group of chapters related to solar energy problems but it begins several control-related chapters. A building that has the capacity to control the thermal flow interactions with the outside, the inlet and outlet flows, and the internal thermal flows between adjacent zones, formally acts as a thermodynamic automaton. The automaton rule minimizes the distance of the temperature of the core (living space) from a preassigned value T^*. The automaton rule must manage the chaotic variations of external inputs and also its own capacity to learn whether these inputs belong to a definite season of the year (summer, winter, etc.) as the automaton strategy changes each season. However, any backup injection of heating (cooling) energy perturbs the automaton logic; it may fool the automaton by making it believe that it is, say, summer when instead it is winter (or vice versa). The general issue addressed here is that given an automaton that performs according to a predetermined strategy of response to an external chaotic input, the addition of a correction is difficult. The smarter the automaton is the more difficult the correction is. An example may be an intelligent child who has some difficulty of behavior.

How would one try to correct him? A simple correction is elaborated by the child, and may enhance the trouble. If one is more intelligent than the child he may hope to find the strategy of correction, but that may not be the case. So, if at all possible, one has to invent a strategy of correction in the condition of incomplete knowledge. Thus the theory of correction for dynamical systems can be understood in terms of interacting machines.

S. Sieniutycz and R. S. Berry (Chapter 6) apply the general optimization theory of multistage systems to optimally controlled cascades composed of Novikov–Curzon–Ahlborn engines or heat pumps (NCA cascades). The optimal cascades are those which work in the engine mode (process approach to equilibrium) and maximize the work production, or those which work in the heat-pump mode (process departure from equilibrium) and minimize the work consumption. A unified mathematical description is proposed which deals with multistage NCA processes and the limiting continuous processes. A kinetic extension of the Carnot theory shows deviations of the stage efficiency from the Carnot formula, caused by the process irreversibility. A relatively unknown discrete theory with a Hamiltonian function constant along the optimal path is used for the purpose of work optimization in both discrete and continuous cases. Nonlinear difference equations which follow from the energy balance and kinetics of heat transfer are constraints in the work optimization. The optimal discrete set is canonical and preserves most properties of the classical Pontryagin algorithm of continuous optimization. A generalized exergy (available energy) is obtained from the discrete functionals of work for both discrete and continuous cases. This exergy refers to finite-time or finite-size systems and simplifies to the classical exergy in the case of infinite durations. The important issue is that the bounds provided by this generalized exergy are stronger than classical thermostatic bounds and hence they are closer to reality.

J. M. Burzler, P. Blaudeck and K.-H. Hoffmann (Chapter 7) consider optimization of internal combustion engines. As compared with an earlier work, the engine models are improved in several aspects thus assuring more realistic results and enabling an analysis of the effect of constraint modification. Their approach is focused on the thermodynamic process taking place inside the engine rather than on the technical realization of the engine as such. Very much in the tradition of early thermodynamics the idea is to abstract the engines enough to make them treatable, yet to include all major loss terms so that an analysis would indicate which loss terms could be most easily reduced. In this spirit the Otto engine is investigated as well as the Diesel engine. Both engines have a four-stroke cycle of successive intake, compression, power, and exhaust stroke. They operate at constant periods and with a fixed fuel consumption per cycle. The considered models are endoreversible in that the working fluid is in the internal thermal equilibrium. The models incorporate main loss mechanisms: friction, pressure drop, and heat leak. The optimal thermodynamic process has to assure a maximum work per cycle subject to coupled nonlinear differential equations (with boundary values) which are solved numerically with a computer. For exhaust and intake the changes in pressure and temperature are at least one order of magnitude smaller than during the compression and power stroke. But the compression results in an important increase of the gas temperature up to about 900–1500 K. Therefore the most important task is to additionally include the heat conduction and to optimize the compression and power stroke together. The authors have found an optimum of the compression and power stroke of the Diesel engine with full consideration of heat transfer also during the compression and constraints in piston position and acceleration. The heat production due to friction and the heat transport due to radiation have also been included in this new scheme.

H. Farkas, I. Faragó, and P. L. Simon (Chapter 8) discuss qualitative properties of unsteady heat transport by conduction. For the energy conversion processes the properties of the energy transport are quite essential, as known from the theory of endoreversible engines. First, the authors overview the classical theory formulated in thermodynamic terms (balance and state equations, Fourier law, transport equation, and a variational approach). The dissipative character of the heat conduction manifests itself through a basic inequality called the maximum principle and some related properties, such as the stability. These properties and their limits of validity are analyzed. They can be stated quite generally, far beyond the linear theory. However, the classical heat equation yields an infinite speed of propagation, and a hyperbolic heat equation is considered to overcome this paradoxical feature. The trouble is, however, that the maximum principle is not valid for hyperbolic equations, and a satisfactory theory involving the maximum principle as well as finite propagation is unknown. The required basic properties may still be used as postulates in searching for a new theory. Such an attempt is made by the authors via a system theory approach that uses only the temperature as a primary concept. It is demonstrated that a consistent theory for the heat current in a homogeneous one-dimensional medium can be derived from very general assumptions. Special characteristics of the continuous solution to the initial-boundary value problem for the classical linear heat-conduction are analytically proved without solving the differential equation. Numerical techniques of solving which define approximate discrete solutions at the discrete points of a mesh are also considered. The basic question is the convergence of these discrete solutions to the solution of the original continuous problem, when refining the mesh. The preservation of analogies with basic qualitative properties of the continuous solution is another important issue (see a similar issue of preservation of the Hamilton structure, for discrete optimal cascades, in S. Sieniutycz's and R. S. Berry's chapter). The authors formulate some exact conditions under which the discrete–continuous analogies hold. They also discuss damped traveling-wave solutions of the sourceless parabolic heat equation and list the "shape-preserving signal forms" which can propagate inside the body without distortion, after a transient period.

Z. Herman (Chapter 9) reviews the elementary collision processes in which large energy is exchanged between the colliding partners. These processes are of general interest in connection with energy conversion, storage, and transmission, amongst others for hybrid conversions with photothermal, photovoltaic, and photosynthesis effects. In particular, the question of the prevailing form of energy deposition in the products (electronic, vibrational, rotational, translational) is of importance. In elementary chemical reactions, the way of reaction heat disposal in exoergic processes, as well as the form of energy effective in overcoming the activation barrier of endoergic processes, is of prime importance. Both experimental and theoretical studies of these collision processes provide the information required. Relevant examples of collision processes of neutral particles as well as ions are discussed.

R. Mrugała's chapter (Chapter 10) begins focused on the theory and applications of differential geometry in thermodynamics and energy conversion. He reviews the general mathematical framework of modern differential geometry applicable in the thermodynamics of matter or radiation. Geometric structures can be obtained either phenomenologically or statistically. They usually refer to contact, symplectic, metric, and pseudometric geometries, although the more recent Poisson and Jacobi geometries can also be included. In thermodynamics, the underlying setting for all these geometries is the so-called thermodynamic phase space (TPS). For a system with n degrees of freedom, TPS is a $(2n+1)$-dimensional manifold endowed with a contact structure. The contact structure of TPS allows us to associate two types of flows, and hence two types of vector fields, X_f and $\overline{X}_f$, to any smooth function f on TPS. Especially fields X_f find numerous applications in thermodynamics. Formally they generate (one-parameter) continuous transformations of TPS, and, for some special choices of f's, integral curves of X_f's represent thermodynamic systems because they preserve equations of state. For other choices of f, X_f's allow us to find new equations of state from the known ones. The author presents novel examples of X_f's for various f's. An interesting problem is also that of finding f for a given X_f, i.e., finding a "contact Hamiltonian" f for a given vector field X_f. It is equivalent to solving a system of coupled partial differential equations ("contact Hamilton equations"). By means of X_f, one can also define three types of brackets on TPS; the Jacobi, Cartan, and Lagrange brackets which may be used for finding new invariants of thermodynamic systems.

The contact structure of TPS also leads to a pseudo-Riemannian metric G on TPS. Both the contact and the (pseudo)Riemannian structures allow one to compare not only different states of a given system, but also different states of different systems. They lead also to the well-known applications of the Riemannian metric on the Legendre submanifolds where the length of a curve can be related to the dissipation and the process efficiency. Of practical value are also local quantities. For instance, the Riemannian curvature is related to the stability of thermodynamic systems. This, in turn, is related to the fluctuations around the state. Indeed, the geometrical methods not only help understand thermodynamics but also lead to new results for complex thermodynamic systems via analyses of simple systems.

L. Diósi and P. Salamon (Chapter 11) discuss various contexts of thermodynamic distances, from statistics of distinguishability to minimally dissipative processes. The Riemannian metric structure of thermodynamics contains basic and relatively new information concerning a physical system. The presence of the structure can be felt at all levels of physical description. The metric is in fact a realization of R. Fisher's concept of statistical distinguishability, and is the basis for information geometry. The corresponding notion of statistical distance has since been introduced for various statistical systems. At the quantum level the distance measures the reliability of an experiment designed to optimally distinguish between the two states along a one-parameter family of density operators. At the statistical mechanical level, distance is the number of statistically distinguishable intermediate states as we transform one state into another. From this, Riemannian metricization of the state space emerges in the thermodynamic limit of Gibbs statistical ensembles. Numerous authors have assigned the curvature of this geometry to stability properties or interaction strength. The requirement of covariance with respect to this geometry can be used to give an important correction to thermodynamic fluctuation theory. Finally, at the very macroscopic level, the square of this same distance between two states of a thermodynamic equilibrium equals the minimum entropy produced in a process that transforms one state into the other, multiplied by the number of relaxations during the transformation. The authors recapitulate basic ideas and results concerning the Riemannian metric structure of thermodynamics while they attempt to shed light on the concept of statistical distance used in a much broader context.

B. Andresen and P. Salamon (Chapter 12) apply the principles of thermodynamic geometry to distillation processes as a chemical flow system. The analysis is not restricted to a distillation column but applies equally well to an arbitrary stagewise process, be it gas separation by diffusion, staged refrigeration, or chemical reactions. The geometrical methods have previously been used to find optimal paths for thermomechanical processes. Two basic results were proven:

(i) the entropy production or exergy loss of a given process is bounded from below by an expression proportional to the square of the thermodynamic distance between the initial and final states; and

(ii) the minimum dissipation occurs when the process is carried through at constant thermodynamic speed.

For a distillation process as an example, the approach analyzes the principle of constant thermodynamic speed for staged steady processes with thermochemical flows. It also shows how constant thermodynamic speed can be used as a general design principle.

For a given feed rate and fixed input and output concentrations the traditional distillation column really only has one free control, the heat flow through the column. Energy and mass balance equations from tray to tray then impose the corresponding temperatures and concentrations on each tray. As vapor and liquid constantly move up and down the column between thermodynamic surroundings which are not in equilibrium, such a traditional distillation column is inherently dissipative. Extensive modeling and optimization show substantial savings in the exergy required to separate a given mixture into its components. The optimal distillation column has external control of the temperature of each plate. This enables the designer to keep the trays a fixed thermodynamic distance apart, thus minimizing dissipation according to the general result of finite-time thermodynamics. The optimal policy is a gradual addition of heat at all trays below the feedpoint, and withdrawal of heat at all trays above the feedpoint with corresponding much smaller heat duties of the reboiler and the condenser. Although the overall amount of heat passed through the column is the same as for the traditional design, most of that heat is degraded over a much smaller temperature difference, thus being equivalent to a sharply reduced expenditure of exergy for separation. A test calculation for the separation of a 50/50 mixture of benzene and toluene into 99% pure products shows a large reduction of the driving exergy by a factor of 4.4.

Acknowledgments

Finally, acknowledgments constitute the last and most pleasant part of this Preface. The contributors who formed in 1993 the "Carnet", i.e., the Carnot Network, express their gratitude to the European Commission under the auspices of which our research was made possible in the framework of two subsequent programs, i.e., Copernicus contract no. CIPA-CT 92-4026 (1993–1996) and Inco-Copernicus contract no. ICOP-DISS-2168-96 (1997–1999).

In preparing this volume the editors received help and guidance from R. Stephen Berry, The University of Chicago, and Thomas von Foerster, Springer-Verlag, New York. A critical part of writing any book is the review process, and the authors and editors are very much obliged to the researchers who patiently helped them read through subsequent chapters and who made valuable suggestions: B. Andresen, L. Gradon, R. Mrugała, A. Radowicz, M. von Spakovsky, and Z. Szwast. The editors, furthermore, owe a debt of gratitude to all participants of the Carnet network. Working with these stimulating colleagues has been a privilege and a very satisfying experience.

Warsaw, 1 April 1999   S. Sieniutycz

List of Contributors

B. Andresen

Ørsted Laboratory,
Niels Bohr Institute
University of Copenhagen
Universitetsparken 5
DK-2100 København Ø
Denmark

V. Bădescu

Candida Oancea Institute of Solar Energy
Faculty of Mechanical Engineering
Polytechnic University of Bucharest
Splaiul Independentei 313
R-79590 Bucharest
Romania

R. S. Berry

Department of Chemistry
University of Chicago
5735 South Ellis Avenue
Chicago, IL 60637
U S A

P. Blaudeck

Department of Physics
Chemnitz University of Technology
Reichenhainer Strasse 70
D-09107 Chemnitz
Germany

J. M. Burzler

Department of Physics
Chemnitz University of Technology
Reichenhainer Strasse 70
D-09107 Chemnitz
Germany

A. De Vos

Department of Electronics and Information Systems
University of Gent
Sint Pietersnieuwstraat 41
B-9000 Gent
Belgium

L. Diósi

KFKI Research Institute for Particle and Nuclear Physics
Hungarian Academy of Sciences
Konkoly Thege út 29
H-1525 Budapest 114
Hungary

H. Farkas

Department of Chemical Physics
Technical University of Budapest
Budafoki út 8
H-1521 Budapest
Hungary

I. Faragó

Department of Applied Analysis
Eötvös Loránd University
Múzeum körút 6
H-1088 Budapest
Hungary

Z. Herman

V. Čermák Laboratory
J. Heyrovský Institute of Physical Chemistry
Academy of Sciences of the Czech Republic
Dolejškova 3
CZ-18223 Praha 8
Czech Republic

K. H. Hoffmann

Department of Physics
Chemnitz University of Technology
Reichenhainer Strasse 70
D-09107 Chemnitz
Germany

P. T. Landsberg

Faculty of Mathematical Studies
University of Southampton
Southampton SO17 1BJ
United Kingdom

B. Å. Månsson

Department of Engineering Science, Physics, and Mathematics
Karlstad University
S-65188 Karlstad
Sweden

R. Mrugała

Institute of Physics
N. Copernicus University
ulica Grudziądzka 5
PL-87100 Toruń
Poland

P. Salamon

Department of Mathematical Sciences
San Diego State University
San Diego, CA 92182
USA

L. Sertorio

Department of Theoretical Physics
University of Torino
via Pietro Giuria 1
I-10125 Torino
Italy

S. Sieniutycz

Faculty of Chemical Engineering
Warsaw University of Technology
Ulica Waryńskiego 1
PL-00645 Warszawa
Poland

P. L. Simon

Department of Applied Analysis
Eötvös Loránd University
Múzeum körút 6
H-1088 Budapest
Hungary

G. Tinetti

Department of Theoretical Physics
University of Torino
via Pietro Giuria 1
I-10125 Torino
Italy

Table of Contents

Preface   v

List of Contributors   xv

I   Conversion of Radiative Energy   1

1   Statistical Mechanics of Solar Energy Conversion   3
     1.1   Introduction 3
     1.2   Information Theory and Statistical Mechanics 4
         1.2.1   Relative Information 5
         1.2.2   Exergy 6
     1.3   Benchmark: Black-Body Radiation as a Free Photon Gas 6
     1.4   Problems with the Black-Body Radiation Model 8
         1.4.1   Isotropy 8
         1.4.2   Energy Extraction 9
         1.4.3   Distribution Function 9
     1.5   Solar Energy Absorption Devices 10
         1.5.1   Photochemical Solar Energy Conversion 10
     1.6   Further Conversion of the Photon Energy: Losses and Efficiency 11
         1.6.1   Dissipation Mechanisms 11
         1.6.2   Maximum Efficiency and Statistical Mechanics Models 12
     1.7   References 12

2   Thermodynamics of Solar Energy Conversion into Work   14
     2.1   Introduction 14
     2.2   Upper Bound Efficiencies 16
         2.2.1   Simple Upper Bounds for Black-Body Radiation Conversion 16
         2.2.2   Simple Upper Bound for Diluted Radiation Conversion 19
         2.2.3   More Accurate Simple Upper Bound Efficiency . . . 23

xviii   Table of Contents

    2.3 Terrestrial Applications 24
        2.3.1 Converting Direct Solar Radiation 24
        2.3.2 Converting Diffuse Solar Radiation 26
        2.3.3 Converting Global Solar Radiation 31
    2.4 Space Applications 36
        2.4.1 Solar Space Power System Model 37
        2.4.2 Classical Thermodynamic Model 39
        2.4.3 Finite-Time Thermodynamics Model 41
    2.5 Further Research and Studies 43
    2.6 References 43

3 Thermodynamics of Photovoltaics   49
    3.1 Introduction 50
    3.2 Endoreversible Thermal Engines 52
    3.3 Endoreversible Chemical Engines 57
    3.4 Endoreversible Thermochemical Engines 58
    3.5 Solar Cells 59
    3.6 Solar Cells with Larger-than-Unity
        Quantum Efficiency 62
    3.7 Tandem Solar Cells 63
    3.8 Conclusion 68
    3.9 References 69

4 Some Methods of Analyzing Solar Cell Efficiencies   72
    4.1 Introduction 72
    4.2 The Solar Cell Equation:
        Currents from Photon Fluxes 73
    4.3 Efficiencies in General 76
    4.4 Theoretical Efficiencies of a Simple Heterojunction 78
    4.5 Special Cases of the Simple Theory 79
        4.5.1 Homojunction with or without Impact Ionization .. 79
        4.5.2 Heterojunction without Impact Ionization 80
    4.6 Analysis of Heterojunction Cells
        Allowing for Impact Ionization 81
    4.7 The Graded Gap Solar Cell 83
        4.7.1 General 84
        4.7.2 Photon Absorption Coefficient 86
        4.7.3 Photon Emission Rates 89
        4.7.4 Solar Energy Conversion 90
    4.8 Thermophotovoltaic Conversion 93
        4.8.1 Definitions 93
        4.8.2 Theory of TPV Conversion 96
    4.9 Recent Results 100
    4.10 Conclusions 102
    4.11 References 103

5 Solar buildings — 106
- 5.1 Finalistic Systems. Introduction 106
- 5.2 The Geophysical Inputs 109
  - 5.2.1 The Incoming Solar Flux 109
  - 5.2.2 The Equation for T_E^{dry} 117
  - 5.2.3 The Equation for T_E^{wet} 118
- 5.3 The Model of the Solar House 125
  - 5.3.1 General Remarks on the Model with Fixed Controls 125
  - 5.3.2 The Annual Control 132
- 5.4 Backup and Adaptive Controls 134
- 5.5 References 139

II Conversion of Thermal and Chemical Energy — 141

6 Discrete Hamiltonian Analysis of Endoreversible Thermal Cascades — 143
- 6.1 Introduction: Multistage Novikov–Curzon–Ahlborn Process 143
- 6.2 A Single Stage with the Driving Heat Flux as a Control Variable 146
- 6.3 Applying Single-Stage Formulas to a Multistage Process 149
- 6.4 Pontryagin's Structure of Optimal Control 151
- 6.5 Work Maximizing in NCA Cascades by Discrete Maximum Principle 156
- 6.6 The Hamiltonian as the Lagrange Multiplier of a Time Constraint 162
- 6.7 Limiting Continuous Process 167
- 6.8 Concluding Remarks 168
- 6.9 References 170

7 Optimal Piston Paths for Diesel Engines — 173
- 7.1 Introduction 173
- 7.2 Model 175
  - 7.2.1 Combustion 177
  - 7.2.2 Frictional Losses 178
  - 7.2.3 Conductive and Convective Heat Leak 178
  - 7.2.4 Radiative Heat Leak 179
- 7.3 Optimization 180
  - 7.3.1 Control Theory 181
  - 7.3.2 Stochastic Optimization 183
- 7.4 Results 185

		7.4.1	Optimal Path	185
		7.4.2	Optimal Time of Ignition	190
	7.5	Conclusion		194
	7.6	References		195

8 Qualitative Properties of Conductive Heat Transfer — 199

- 8.1 Theoretical Background . . . 200
  - 8.1.1 Fourier's Differential Equation . . . 200
  - 8.1.2 Balance of Internal Energy . . . 200
  - 8.1.3 Material (Constitutive) Equations . . . 200
  - 8.1.4 Transport Equation. Initial and Boundary Conditions . . . 201
  - 8.1.5 Heat Conduction in Irreversible Thermodynamics . . . 201
  - 8.1.6 Variational Principles . . . 202
  - 8.1.7 Stationary Case . . . 203
  - 8.1.8 Temperature Scales: Pictures, Kelvin's Transformation . . . 204
- 8.2 Consequences of the Second Law . . . 204
  - 8.2.1 Heat Conductional Inequality . . . 204
  - 8.2.2 Maximum Principle . . . 205
- 8.3 The Velocity of Propagation . . . 206
- 8.4 System Theory Approach . . . 208
  - 8.4.1 Heat Conduction and Dynamical Systems Theory . 208
  - 8.4.2 Principle of Superposition . . . 208
  - 8.4.3 A Postulatory Approach to Stationary Heat Conduction . . . 209
- 8.5 Properties of the Solution of the Linear Heat Equation . . . 216
- 8.6 Numerical Solution of the Linear Heat Equation . . . 218
  - 8.6.1 Solution of the Problem by the Fourier Method . . . 218
  - 8.6.2 Finite Difference Method . . . 218
  - 8.6.3 Galerkin Finite Element Method . . . 222
- 8.7 Properties and Their Preservation for the Discretization . . . 224
  - 8.7.1 Qualitative Properties of the Numerical Solution . . 225
  - 8.7.2 Conditions for the Preservation of Qualitative Properties . . . 226
- 8.8 Temperature Waves . . . 230
  - 8.8.1 Shape Preserving Property . . . 230
  - 8.8.2 Classification of SPSFs . . . 231
  - 8.8.3 Asymptotic Behavior; Stability . . . 233
- 8.9 References . . . 234

9 Energy Transfer in Particle–Surface Collisions — 239

- 9.1 Introduction . . . 239

		9.1.1	Collision Energy Domains of Neutral and Ion Projectiles	240
	9.2	Neutral Particle–Surface Energy Transfer		241
		9.2.1	Translational Energy Transfer	241
		9.2.2	Rotational Energy Transfer	243
		9.2.3	Vibrational Energy Transfer	244
		9.2.4	Energy Exchange in Cluster–Surface Collisions	244
	9.3	Slow Ion–Surface Energy Exchange		244
		9.3.1	Neutralization of Ions at Surfaces	245
		9.3.2	Collisions of Atomic Ions with Surfaces	246
		9.3.3	Collisions of Simple Molecular Ions with Surfaces	246
		9.3.4	Collisions of Polyatomic Ions with Surfaces	247
		9.3.5	Collisions of Cluster Ions with Surfaces	252
	9.4	References		252

III Energy in Geometrical Thermodynamics 255

10 Geometrical Methods in Thermodynamics 257
  10.1 Introduction 258
  10.2 Contact Manifolds 259
  10.3 Contact Transformations and
       Contact Vector Fields 263
  10.4 Bracket Structures in Thermodynamics 266
  10.5 Thermodynamic Examples of Contact Flows 269
  10.6 Almost Contact and Contact Metric Structures 273
  10.7 Construction of a Contact Metric 275
  10.8 Statistical Derivation of G 278
  10.9 Relative Information and Riemannian Metric 280
  10.10References 284

11 From Statistical Distances to Minimally Dissipative Processes 286
  11.1 Introduction 286
  11.2 Empirical Statistical Distance 287
    11.2.1 Optimum Calibration 288
    11.2.2 Naive Optimum Control 289
    11.2.3 More Parameters 290
  11.3 Theory of Statistical Distance 292
    11.3.1 Classical Statistics 292
    11.3.2 Quantum Statistics 293
  11.4 Riemannian Geometry 295
    11.4.1 Parameterized Statistics 295
    11.4.2 From Gibbs Statistics to Thermodynamics 296

11.5 Relevance of Riemannian Geometry
   in Thermodynamics 299
   11.5.1 A Covariant Fluctuation Theory 300
   11.5.2 Entropy Production 301
   11.5.3 The Metric as a Symmetric Product 302
   11.5.4 The Group of Transformations 303
   11.5.5 Dissipation in a Small Equilibration 304
   11.5.6 The Discrete Horse–Carrot Theorem 304
   11.5.7 The Continuous Horse–Carrot Theorem 306
   11.5.8 Cooling Rates for Simulated Annealing 310
11.6 Staged Steady Flow Processes 312
   11.6.1 Dissipation in a Distillation Column 312
11.7 Conclusions 315
11.8 References 315

12 Distillation by Thermodynamic Geometry 319
12.1 Introduction 319
12.2 Thermodynamic Length 320
12.3 Optimization of a Step Process 321
12.4 A Classical Distillation Column 322
12.5 Optimal Temperature Profile 324
12.6 Example 329
12.7 References 330

Index   332

Part I

Conversion of Radiative Energy

1
Statistical Mechanics of Solar Energy Conversion

B. Å. Månsson

ABSTRACT. This chapter deals with the general class of processes in which solar energy is absorbed and in which the absorbed energy is converted in multistage processes until it is eventually in a storable form. Several such processes are discussed, as well as the difficulties of modeling the processes in the framework of statistical mechanics. The description pays special attention to loss mechanisms, including the temporal aspects. Possibilities and limits for optimization, arising from both absolute limits and from irreducible losses, are also discussed.

Furthermore, the connection between statistical mechanics and information theory is analyzed in the context of solar energy conversion. In this context, the usefulness of the exergy concept, as an efficiency measure in the analysis of solar energy conversion processes, is discussed. The results for the exergy, of general incoherent photon exergy as well as for a more general kind of radiation field, are reviewed.

The conversion of black-body radiation, in the form of free photons within a certain volume, is used as a benchmark case. Several problems with this widely used model are described.

It is clear that the actual achievable exergetic efficiency is in some cases considerably less than for the ideal benchmark case and also that the difference is not quantifiable with the existing models.

1.1 Introduction

There are three essential parts of a solar energy system:

(i) the radiation itself, i.e., photons with spectral distribution, directional and polarization properties according to solar radiation, possibly modified by passage through the Earth's atmosphere;

(ii) the absorption part, in which a photon interacts with matter and its energy produces an excited electron in either a molecule or a lattice; and

(i) the energy storage part, in which the electron excitation is converted into a useful energy form, in some cases through several energy conversion steps.

Here, the focus is on the first two parts. The third is mainly treated in later chapters. It is worth noting, however, that the second and third parts may actually be combined into one part in some devices. Furthermore, the interaction of the second and third parts may involve absorption of several photons.

Although sophisticated statistical mechanics models (as well as many useful models formulated within the framework of thermodynamics) do exist for all three parts separately, there are no combined and comprehensive statistical mechanics models embracing all three parts within a single framework. In fact, it is difficult to even find a model that accurately treats radiation with the particular spectral and spatial properties of solar radiation. Isotrophic black-body radiation in a cavity dominates the modeling. However, the corrections due to the particular solar radiation spectral properties are often small, but they may in certain "pathological" cases be significant. On the other hand, the spatial (directional) properties are essential for the physics of solar energy devices.

1.2 Information Theory and Statistical Mechanics

The microscopic foundation of thermodynamics is statistical mechanics. Statistical mechanics can be described as an application of information theory [1], [2], [3]. Information theory therefore provides a framework for the formulation of a complete and comprehensive model of the three parts described above.

The unit of information, one bit, is the information carried by one binary figure with unbiased a priori probability for being 0 or 1. The information content, I, in any event that has an a priori probability $p (0 < p \leq 1)$, is defined as

$$I = \log_2 \frac{1}{p}, \quad \text{bits} = \ln \frac{1}{p},$$

where the last equality is based on the convention that $1 \text{ bit} = \ln 2 \approx 0.693$.

Consider a system with a finite number n of available states, denoted by an index j $(j = 1, \ldots, n)$. Assume that there is an associated probability, p_j, such that the usual positivity and normalization constraints,

$$p_j \leq 0 \quad (j = 1, \ldots, n), \tag{1.1}$$

and

$$\sum_{j=1}^{n} p_j = 1, \tag{1.2}$$

are fulfilled. In statistical mechanics, the states j $(j = 1, \ldots, n)$ are *microstates*, and the entire probability distribution

$$P = \{p_j\}_1^n = \{p_1, \ldots, p_n\}, \tag{1.3}$$

1.2. Information Theory and Statistical Mechanics

satisfying (1.1) and (1.2), then defines a *macrostate* of the system.

The *maximum entropy principle* says that the probability distribution that is best suited to describe a system subject to certain constraints on the macrostates is the one that maximises the entropy under those constraints [2], [4].

In the context of utilizing solar radiation, there are at least two kinds of constraints involved. First, there are the constraints relating to the photons, e.g., concerning total energy and volume, isotropy, temperature, coherence, and polarization. Second, there are constraints pertaining to the absorbing device, e.g., concerning its energy gaps and levels, temperature, volume, pressure, mole numbers, and number of free electrons. As indicated above, there are practically no treatments working with both kinds of constraints simultaneously. A further limitation of the theory is that since the constraints are formulated in terms of macrostates, the theory cannot treat microscale phenomena, e.g., involving a single molecule; cf. the discussion on microscopic entropy [5]. Such microscale phenomena may play a key role in some of the three parts of a solar energy device, in particular when the time-dependence of the involved processes is essential; in those cases, the theory based on the maximum entropy principle is inadequate.

The bridge between statistical mechanics and information theory is built upon conventions on units and upon certain interpretations [6]. For entropy, the unit systems differ by Boltzmann's constant, k,

$$S \text{ (thermodynamics)} = kS \text{ (information theory)}.$$

Interpretations are made for extensive variables such as volume, V, the internal energy, U, and the numbers of molecules N_i of the M chemical species ($i = 1, \ldots, M$); correspondingly, interpretations are made for the pressure, p, temperature, T, and the chemical potentials, μ_i (per molecule).

1.2.1 Relative Information

Suppose that, for a system A with n possible microstates $j = 1, \ldots, n$ there is a probability assignment $P^{(0)} = \{p_j^{(0)}\}$, which defines a particular macrostate. If it is found that $P = \{p_j\}$ is a better probability assignment (macrostate) for A, then the expectation value of the information increase is [7]:

$$K[P^{(0)}; P] = \sum_{j=1}^{n} p_j \left(\ln \frac{1}{p_j^{(0)}} - \ln \frac{1}{p_j} \right) = \sum_{j=1}^{n} p_j \ln \frac{p_j}{p_j^{(0)}}. \quad (1.4)$$

Using the conventions and interpretations described above, it can be shown [6] that this information–theoretical quantity, sometimes called "relative information" or "contrast", plays an important role in statistical mechanics. It is then directly relatable to the thermodynamic "exergy" concept.

1.2.2 *Exergy*

One of the key concepts of thermodynamics is the maximal work yield of a system. For most systems, this is less than the energy within the system. One exception is a system with interior vacuum in a surrounding atmosphere. The main importance of the concept lies in that it provides a benchmark for thermodynamic efficiency.

The difference between maximal work yield and energy content is due to the kinds of allowed interactions and to constraints on the interactions and processes. Well-known special cases have been given certain names, e.g., the *Gibbs free energy* and the *Helmholtz free energy*. Lately, it has become common to refer to the general concept by the term *Exergy*, a term coined by Rant [8].

Using the connections between information theory and statistical mechanics outlined above, the exergy can be expressed as

$$\begin{aligned} E &= kT_0 K = \left(U + p_0 V - T_0 S - \sum_{i=1}^{M} \mu_{i0} N_i \right) \\ &= U - U_{\text{eq}} + p_0(V - V_{\text{eq}}) - T_0(S - S_{\text{eq}}) - \sum_{i=1}^{M} \mu_{i0}(N_i - N_{i\text{eq}}) \\ &= S(T - T_0) - V(p - p_0) + \sum_{i=1}^{M} N_i(\mu_i - \mu_{i0}), \end{aligned} \quad (1.5)$$

where T_0, p_0, and μ_{i0} are the intensive variables of the reference state, which is assumed to be in equilibrium, and where $U_{\text{eq}}, S_{\text{eq}}, V_{\text{eq}}$, and $N_{i\text{eq}}$ are the corresponding extensive quantities in the reference state.

The reference state may be given by an external (usually assumed infinite) reservoir. It may also be an internal "ground state" of the system, in which case it is the equilibrium state reached as the maximum work is extracted (i.e., extraction by reversible processes) from the system.

1.3 Benchmark: Black-Body Radiation as a Free Photon Gas

We shall now consider a gas of free photons within a certain volume V, a cavity with absorbtive/emissive (black) walls acting as a heat bath at temperature T [9], [10].

Since there is no restriction on the number of photons there is no chemical potential. The logarithm of the partition function for the system can be written as a spectral integral,

$$\alpha(\beta) = \ln Z(\beta) = V \int_0^\infty d\epsilon \, \rho(\epsilon) \ln z(\beta\epsilon).$$

1.3. Benchmark: Black-Body Radiation as a Free Photon Gas

Here $z(\beta\epsilon)$ is the partition function for photons in a state of energy ϵ at temperature $1/(k\beta)$,

$$z(y) = \sum_{n=0}^{\infty} e^{-ny} = \frac{1}{1-e^{-y}}, \qquad y = \beta\epsilon,$$

and

$$\rho(\epsilon) = \frac{8\pi}{c^3 h^3} \epsilon^2$$

(with c = the velocity of light and h = Planck's constant) is the number of photon states per unit energy interval and unit volume. The probability distribution over a photon number for a given photon energy is then at equilibrium

$$p_n = \frac{1}{z(y)} e^{-ny} = e^{-ny} - e^{-(n+1)y},$$

which gives the mean occupation number

$$\langle n \rangle = \sum_{n=0}^{\infty} n p_n = -\frac{z'(y)}{z(y)} = z(y) - 1 = \frac{1}{e^y - 1} = \frac{1}{e^{\beta\epsilon} - 1}.$$

The logarithm of the partition function is then

$$\alpha(\beta) = \frac{8\pi V}{c^3 h^3} \int_0^\infty d\epsilon\, \epsilon^2 \ln\frac{1}{1-e^{-\beta\epsilon}} = \frac{8\pi V}{(ch\beta)^3} \int_0^\infty dy\, y^2 \ln\frac{1}{1-e^{-y}}$$

$$= \frac{8\pi V}{3(ch\beta)^3} \int_0^\infty dy \frac{y^3}{1-e^{-y}} = \frac{8\pi V}{3(ch\beta)^3} \cdot \frac{\pi^4}{15} = \frac{\pi^2}{45(c\hbar)^3} \cdot \frac{V}{\beta^3}, \qquad \left(\hbar = \frac{h}{2\pi}\right).$$

For thermal equilibrium at temperature $T = 1/(k\beta)$, the energy density is

$$u(T) = \frac{1}{V} U(\beta) = -\frac{1}{V} \frac{\partial \alpha(\beta)}{\partial \beta} = 3\frac{kT}{V} \alpha(\beta) = \frac{\pi^2 k^4}{15(c\hbar)^3} T^4.$$

The entropy of black-body radiation is

$$\frac{S}{V} = \frac{k}{V}[\alpha(\beta) + \beta U(\beta)] = \frac{4}{3}\frac{u(T)}{T}, \qquad (1.6)$$

and the radiation pressure is

$$p = \frac{kT}{V} \alpha(\beta) = 1/3\, u(T).$$

The mean number of photons, N_γ, is then given by

$$\frac{N_\gamma}{V} = \int_0^\infty d\epsilon\, \frac{\rho(\epsilon)}{e^{\beta\epsilon} - 1} = \frac{2\zeta(3)}{\pi^2} \left(\frac{kT}{c\hbar}\right)^3,$$

where ζ is the Riemann zeta-function ($\zeta(3) \approx 1.202$).

Assuming now a reference state in which there is black-body radiation in the same volume at a different temperature T_0, the exergy density is

$$\begin{aligned}\frac{E}{V} &= u(T) - u(T_0) - T_0 \frac{4}{3}\left(\frac{u(T)}{T} - \frac{u(T_0)}{T_0}\right) \\ &= u(T)\left(1 - \frac{4}{3}\frac{T_0}{T} + \frac{1}{3}\left(\frac{T_0}{T}\right)^4\right).\end{aligned} \quad (1.7)$$

It is well worth noting that the expression in parentheses, which can be regarded as an efficiency factor, is smaller than a simple Carnot efficiency.

1.4 Problems with the Black-Body Radiation Model

In the perspective of solar energy utilization, there are three troublesome assumptions made in the black-body radiation model.

- First, it is assumed that the radiation is isotropic; for solar radiation this is not the case.
- Second, it is assumed that the cavity walls are at a single constant temperature; however, if energy is to be extracted and work produced, there must be a temperature difference or change.
- Third, it is assumed that the spectrum is black-body; for solar radiation outside the Earth's atmosphere there is a small discrepancy and for solar radiation at the Earth's surface the discrepancy is larger. For the radiation that has entered a solar energy device, e.g., through a glass sheet, the discrepancy may be quite significant.

For the calculation of exergy or the maximum efficiency, the particular choice of reference state is a very weak point.

1.4.1 *Isotropy*

The solar radiation impinging on the Earth is obviously nonisotropic. This allows, e.g., the use of focusing devices such as lenses and mirrors to acheive an increased energy density. In ideal solar energy devices, this is unimportant, but in real solar energy devices with time- (power-)dependent losses, focusing may have a strong effect on efficiency. Furthermore, focusing is necessary to achieve high temperatures, such as the ones used in device for direct thermal splitting of water (note that the efficiency of that particular kind of devices is very small, e.g., due to the losses caused by large temperature differences and the difficulties of separating the products at a high temperature).

1.4. Problems with the Black-Body Radiation Model

Looking closer at the benchmark model, it is clear that the presence of a device that absorbs photon energy and does not re-emit it, will produce nonisotropy within the cavity.

1.4.2 Energy Extraction

The solar photons may either be reflected or interact with matter to produce an electron excitation. In the first case, they are typically effectively lost by escaping to space. In the latter case, the excitation may be of a kind for which the energy remains in one or a few degrees of freedom and thus can be transformed into a useful form, or it may be quickly redistributed over many degrees of freedom, producing heat.

Looking again at the benchmark model, the second law tells us that if the temperature is everywhere equal, no work can be extracted. Here, a caveat is necessary: if a separate system at a different temperature is used, work may be extracted whilst the temperature of both the black-body radiation in the cavity and of the cavity walls decreases (but stays equal).

1.4.3 Distribution Function

Solar energy devices, except those intended for producing heat, involve a system-dependent energy threshold, such that photons with lower energy are not absorbed or are absorbed in a different manner, e.g., in vibrational and rotational bands. This is a selective mechanism, which will have an effect on the distribution function. The outgoing radiation will have less energy-rich photons than the incoming radiation. Again, this produces an effect that is incompatible with the assumptions of the benchmark model.

The effects on the exergy of a non-black-body spectral distribution were investigated by Karlsson [11]. For incoherent electromagnetic radiation, the probability distribution is of the form

$$p_n = \frac{\langle n \rangle^n}{(\langle n \rangle + 1)^{n+1}}$$

for each mode $(\mathbf{k}, \alpha)$, where $\mathbf{k}$ denotes the wave vector and α the polarization. The photon energy is

$$\epsilon = c\hbar|\mathbf{k}|.$$

Introducing

$$\beta(\mathbf{k}, \alpha) = \frac{1}{c\hbar|\mathbf{k}|} \ln\left(\frac{1}{n(\mathbf{k}, \alpha)} + 1\right),$$

yields an equivalent temperature $(k\beta(\mathbf{k}, \alpha))^{-1}$ for the mode $(\mathbf{k}, \alpha)$. The exergy can now be determined as a sum over exergies of the different degrees of freedom. With a black-body reference state $(\beta_0 = 1/(kT_0))$ of the kind used in the benchmark model, the exergy density is

$$\frac{E}{V} = (2\pi)^3 \sum_\alpha \int d^3\mathbf{k}\, n(\mathbf{k}, \alpha) c\hbar |\mathbf{k}| q(\mathbf{k}, \alpha).$$

Here $q(\mathbf{k}, \alpha)$ is an energy quality function [11]

$$q(\mathbf{k}, \alpha) = 1 - \frac{\beta}{\beta_0} + \frac{e^{\beta\epsilon} - 1}{\beta_0 \epsilon} \ln \frac{1 - e^{-\beta\epsilon}}{1 - e^{-\beta_0 \epsilon}}.$$

Again, the use of a black-body reference state should be noted as a limitation of the model.

1.5 Solar Energy Absorption Devices

This section briefly describes some of the less well-known possible pathways for useful absorption of solar energy. For these, statistical mechanics models giving ultimate limits to their efficiency do not exist. Microscale physical and chemical properties and the specificity of these for the systems as well as the different time-dependencies of the involved loss mechanisms are of crucial importance. Therefore, it is questionable whether it is at all possible to construct a *general* model that would yield the ultimate limit for the efficiency of such devices.

1.5.1 *Photochemical Solar Energy Conversion*

In addition to the commonly known photovoltaic solar cells, which are based on a p–n-junction in a semiconductor, there are several other kinds of possible solar energy devices. These are all able to absorb a photon in such a way that at least part of the energy can be delivered in chemical form or in the form of electricity.

Most photochemical devices aim at splitting water to produce hydrogen. For some devices, the output could be ammonia, methane, chlorine or hydrogen peroxide.

The primary step in a photochemical reaction is the absorption of one or more photons and the concomitant generation of excited species with redox properties different from the ground state.

The energy required for the dissociation of water depends on the number of electrons involved in the redox process. The most favorable reactions are the reduction with two electrons and the oxidation with four electrons:

$$2\mathrm{H}^+ + 2e^- \rightarrow \mathrm{H}_2, \qquad -0.41\,\mathrm{eV},$$
$$2\mathrm{H}_2\mathrm{O} \rightarrow \mathrm{O}_2 + 4\mathrm{H}^+ + 4e^-, \qquad +0.82\,\mathrm{eV},$$

in combination yielding the water dissociation reaction

$$\mathrm{H}_2\mathrm{O} \rightarrow \mathrm{H}_2 + 1/2\,\mathrm{O}_2, \qquad \Delta G = 1.23\,\mathrm{eV}.$$

Photoelectrochemical Cells

The basic idea of photoelectrochemical cells is to use an electrical field to influence the photon absorption levels.

An electrochemical cell is formed by connecting two half-cells, each consisting of an electrode and a redox couple in an electrolyte: The output is in the form of either electrical power (*regenerative cells*) or a chemical product, typically H_2 and O_2 (*photoelectrolysis cells*).

One or both electrodes may be semiconductors. The interaction of the electrolyte and the semiconductor produces a change in the band level close to the surface. At the interface, the electrostatic potentials in the semiconductor as well as in the electrolyte are different from the ones in the bulk.

Photochemical Reactions in Homogeneous Solutions

The most well-known photochemical reaction in homogeneous solution is the photosynthetic process of green plants. There are, however numerous other molecules that in solution can absorb photons and for which the energy can be converted into a useful form, e.g., hydrogen.

1.6 Further Conversion of the Photon Energy: Losses and Efficiency

By absorption of a photon, a molecule or an electron is put into an excited state. However, this excitation has to be converted into a form that precludes dissipation of the absorbed energy. In many cases, there is a conflict between achieving high efficiency in the absorption process and minimizing dissipation. The different *time constants* for all the different energy pathways and ways to influence them play key roles in the resulting optimization problem.

1.6.1 *Dissipation Mechanisms*

The dissipation mechanisms can be divided into two types: radiative mechanisms (fluorescence and phosphorescence) and nonradiative mechanisms. In the radiative mechanisms, the energy of the absorbed photons is re-emitted as photons. In the nonradiative mechanisms, the energy of the absorbed photons may either be converted into an unwanted chemical form or into heat. Since the loss processes in semiconductors are discussed elsewhere, only some features (with important temporal aspects) of photochemical systems will be noted here.

For photochemical cells, a critical point is the prevention of an immediate recombination of the primary photoproducts—the products should quickly be separated in space.

For a high efficiency of the conversion of the absorbed energy into storable chemical energy, fluorescence and phosphorescence must be avoided except when they can be used for sensitizing another molecule absorbing in a different wavelength region.

In a photoelectrochemical system, there are loss mechanisms involved both in the electrolyte and in the electrodes—but with different characters and different time-dependencies.

Fast and efficient redox reactions are one-electron transfer reactions, but the production of hydrogen and oxygen from water needs several electrons— for the formation of one oxygen molecule four electrons are used.

1.6.2 Maximum Efficiency and Statistical Mechanics Models

Statistical mechanics is most powerful in treating equilibrium systems. However, as indicated repeatedly above, a primary aim in the construction of solar energy devices is to prevent the equilibration processes. Thus, an efficient solar energy device will operate in nonequilibrium. This fact constitutes a challenge to statistical mechanics with similarities to the well-known ergodicity problem. Perhaps one way to proceed is to combine the approaches of finite-time thermodynamics with the framework of information theory?

1.7 References

[1] E. T. Jaynes: Information theory and statistical mechanics, *Phys. Rev.* **106**, 620, 1957.

[2] E. T. Jaynes: Information theory and statistical mechanics, in: (ed.) K. W. Ford, *Statistical Physics*, Brandeis Lectures, 1962, 1963.

[3] E. T. Jaynes: *Papers on Probability, Statistics and Statistical Physics*, (ed.) R. D. Rosenkrantz, Reidel, Dordrecht, 1983.

[4] R. D. Levine and M. Tribus (eds.): *The Maximum Entropy Principle*, MIT Press, Cambridge, MA, 1979.

[5] K. Lindgren: *Microscopic and macroscopic entropy*, *Phys. Rev.* **A38**, 4794, 1988.

[6] K.-E. Eriksson, K. Lindgren and B.Å. Månsson: *Structure, Context, Complexity, Organization*, World Scientific, Singapore, 1987.

[7] S. Kullback: *Information Theory and Statistics*, Wiley, New York, 1959.

[8] Z. Rant: *Exergie, ein neues Wort für "technische Arbeitsfähigkeit"*, Forschung auf dem Gebiete des Ingenieurwesens **22**(1), 36, 1956.

[9] F. Reif: *Fundamentals of Statistical and Thermal Physics*, McGraw-Hill, Tokyo, 1965.

[10] P. Landsberg: *Thermodynamics and Statistical Mechanics*, Oxford University Press, Oxford, 1978.

[11] S. Karlsson: *The exergy of incoherent electromagnetic radiation*, Phys. Scripta **26**, 329, 1982.

2
Thermodynamics of Solar Energy Conversion into Work

V. Bădescu

> ABSTRACT. The main aim of any energy conversion model is to establish upper limits for the conversion efficiency. This chapter deals with solar energy conversion into mechanical work. The first section refers to models which allow an easy (handy) computation of the conversion efficiency while the second section deals with models which usually require the use of computers.

2.1 Introduction

Systems that collect and convert solar energy into mechanical or/and electrical power appear to be attractive from many points of view. Presently, there are two major types of solar power systems. The first type uses photovoltaic cells to directly convert the energy of solar radiation into electrical energy, in combination with electrochemical storage. The second one is based on thermodynamic cycles. Solar concentrators reflect the flux of solar radiation toward a collector where a working fluid is heated up to drive conventional engines. Electrical energy could be generated by alternators coupled to these thermal engines. This last solar power system will be considered in this chapter.

The main aim of any energy conversion model is to establish upper limits for the conversion efficiency. The more detailed the thermodynamic model is, the more realistic upper bounds are obtained. However, the increase in the model's complexity is accompanied by more involved calculations. The body of this chapter is divided into three sections. Section 2.2 refers to models which allow an easy (handy) computation of the conversion efficiency while the Sections 2.3 and 2.4 deal with models which usually require the use of computers.

Section 2.2 one presents very general models, which take into account only the thermal properties of the two energy sources (i.e., the Sun and the ambient) and (eventually) the optical properties of the absorber. This approach allows us to obtain simple upper bound formulas for the conversion efficiency of black-body radiation into work such as those proposed

2.1. Introduction

by Jeter, by Spanner, or by Landsberg–Petela–Press. The last efficiency is derived here by using a thermodynamic treatment recently proposed by this author. This result is subsequently generalized to the case of diluted solar radiation. All the above formulas predict too high upper bounds to be of practical interest for the usual solar energy applications. The remainder of Section 2.2 presents the most accurate simple upper bound formula currently available. This formula was derived recently by this author and applies to the conversion of both black-body and diluted radiation.

Sections 2.3 and 2.4 deal with more detailed models. They take into account, on hand, the nature of the incoming solar radiation and, on the other one hand, different design parameters of the conversion system. The models are classified into two categories, as they refer to terrestrial and space applications (Sections 2.3 and 2.4, respectively).

Developing a model of a terrestrial solar power plant has to take into consideration that the available solar energy at ground level has a direct (beam) and a diffuse component. The power plants designed to use the direct solar radiation have a higher efficiency than a hypothetical device converting the diffuse component only. This is due to the fact that direct radiation can be concentrated. However, diffuse radiation is available all day long while the direct component sometimes vanishes because of cloudiness. Consequently, a thermal power plant utilizing the global solar radiation (i.e., both the direct and the diffuse components) is of practical interest in principle. Here we present three different models of power plants utilizing: (i) direct, (ii) diffuse, and (iii) global solar radiation, respectively. In case (i) one finds the atmospheric transmittance and the concentration ratio among the model parameters. Model (ii) is based on a thermodynamic approach of the multiply scattered solar radiation. It allows us to predict the maximum conversion efficiency of diffuse radiation as a function of the mean number of scatterings. Model (iii) is developed in two versions, as one deals with a Carnot thermal engine, in the first case, or with a Stirling engine in the second case.

The space solar power stations use beam solar radiation only. However, they essentially differ from terrestrial plants, as the main mechanism of heat rejection by the thermal engine is through radiation in space and by convection on Earth. Two different models of space solar power stations are presented here. The first one assumes the ideal case of a Carnot engine. The second model describes a more realistic situation, by taking into consideration the irreversibilities associated to the heat transfer at the hot and cold parts of the thermal engine. A Novikov–Curzon–Ahlborn model of a thermal engine is used in this last case.

2.2 Upper Bound Efficiencies

2.2.1 Simple Upper Bounds for Black-Body Radiation Conversion

Three different expressions for the maximum efficiency attainable with a device powered by solar radiation have been proposed by Landsberg–Petela–Press (LPP) [1]–[3], by Spanner (S) [4], and by Jeter (J) [5], respectively. They are given by

$$\eta_{\mathrm{LPP}} = 1 - \frac{4}{3}\frac{T_0}{T_s} + \frac{1}{3}\left(\frac{T_0}{T_s}\right)^4, \quad \eta_S = 1 - \frac{4}{3}\frac{T_0}{T_s}, \quad \eta_J = 1 - \frac{T_0}{T_s}, \qquad (2.1)$$

where T_0 and T_s are the ambient and Sun temperatures. Bejan presented a unification of the three previous published theories [6], [7]. Bejan showed that these theories "are individually correct, and they complement (rather than contradict) one another". They differ because of differences in the description of "investments" and in the way the radiation is used.

Many published papers deal with solar energy availability as calculated with the LPP efficiency (see, e.g., [8]–[12]). We shall derive here the LPP efficiency by using arguments presented for the first time in [13].

Let us consider a surface element Σ_1 (area dA_1) on the hypothetical external surface of the Sun's photosphere. The radiation emitted by the Sun doesn't depend on the place of Σ_1 on the photosphere, it is isotropic and has a spectral luminance $\mathcal{L}_{1\nu}$ (Js^{-1} m^{-2} sr^{-1} Hz^{-1}). Let us consider another surface element Σ_2 (area dA_2) placed at the level of the Earth's orbit (distance l_{12} as referred to the element Σ_1). The spectral energy flux $d^2\phi_{2\nu}$ (Js^{-1} Hz^{-1}) of solar radiation emitted by Σ_1 and received by Σ_2 is given by the Lambert law [14, p. 14]:

$$d^2\phi_{2\nu} = \frac{\mathcal{L}_{1\nu} \cos\alpha_1 \cos\alpha_2 \, dA_1 \, dA_2}{l_{12}^2}. \qquad (2.2)$$

In (2.2) α_1 and α_2 are the angles between the normals at Σ_1 and Σ_2, respectively, and the straight line joining the two elements. Σ_1 is seen from Σ_2 under a solid angle $d\Omega_2$ given by

$$d\Omega_2 = \frac{\cos\alpha_1 \, dA_1}{l_{12}^2}. \qquad (2.3)$$

By using (2.2) and (2.3) one finds

$$\frac{d^2\phi_{2\nu}}{dA_2 \, d\Omega_2} = \mathcal{L}_{1\nu} \cos\alpha_2. \qquad (2.4)$$

The spectral density $\varphi_{2\nu}$ of the solar energy flux incoming on the element

2.2. Upper Bound Efficiencies

Σ_2 from the whole Sun surface is

$$\varphi_{2\nu} = \frac{d\phi_{2\nu}}{dA_2} = \int \mathcal{L}_{1\nu} \cos\alpha_2 \, d\Omega_2 = \mathcal{L}_{1\nu} \int_0^{2\pi} d\beta \int_0^{\delta} \sin\alpha_2 \cos\alpha_2 \, d\alpha_2, \quad (2.5)$$

where δ is the half-angle of the Sun when viewed from Σ_2. One finds

$$\varphi_{2\nu} = \frac{\pi}{2}(1 - \cos 2\delta)\mathcal{L}_{1\nu}. \quad (2.6)$$

The spectral density of emitted solar radiation energy, per unit volume and unit solid angle, $W_{1\nu}$ (Jm^{-3} sr^{-1} Hz^{-1}), is given by [15, p. 206]:

$$W_{1\nu} = \frac{2h}{c^3} \frac{\nu^3}{e^{\beta_s \nu} - 1}, \quad (2.7)$$

where $\beta_s = h/(kT_s)$, and c and ν are the light speed and frequency, respectively, h – the Planck constant and k – the Boltzmann constant. The radiation diffusion in the photosphere is negligible. Consequently, the emitted electromagnetic field propagates by waves and the following simple relationship between $\mathcal{L}_{1\nu}$ and $W_{1\nu}$ is valid [16, p. 151]:

$$\mathcal{L}_{1\nu} = cW_{1\nu} = \frac{2h}{c^2} \frac{\nu^3}{e^{\beta_s \nu} - 1}. \quad (2.8)$$

By using (2.6) and (2.8) one finds the density φ_2 (Js^{-1} m^{-2}) of the total solar energy flux incident on Σ_2:

$$\varphi_2 = \int_0^\infty \varphi_{2\nu} \, d\nu = \frac{\pi^5 h}{15c^2 \beta_s^4} (1 - \cos 2\delta). \quad (2.9)$$

The internal energy U_2 of the solar radiation confined to the volume V_2 (placed at the level of Σ_2) and the total flux φ_2 are related through a relationship similar to (2.8):

$$\frac{U_2}{V_2} = \frac{\varphi_2}{c} = \frac{\pi^5 h}{15c^3 \beta_s^4} (1 - \cos 2\delta). \quad (2.10)$$

The entropy S_2 of the solar radiation confined to the volume V_2 is given by [15, p. 184]:

$$S_2 = k \int d^2 m_{2\nu} [(g_{2\nu} + 1)\ln(g_{2\nu} + 1) - g_{2\nu} \ln g_{2\nu}], \quad (2.11)$$

where $g_{2\nu}$ is the occupation factor and $d^2 m_{2\nu}$ is the number of quantic states of the photons of frequencies between ν and $\nu + d\nu$ incident on Σ_2 under the solid angle $d\Omega_2$, given by [15, p. 202]:

$$d^2 m_{2\nu} = \frac{2V_2}{c^3} \nu^2 \cos\alpha_2 \, d\nu \, d\Omega_2. \quad (2.12)$$

18   2. Thermodynamics of Solar Energy Conversion into Work

After emission the photons propagate without interaction. Consequently, the occupation factor remains unchanged and equals the occupation factor of the emitting black-body [15, p. 208]:

$$g_{2\nu} = g_{1\nu} = (e^{\beta_s \nu} - 1)^{-1}. \tag{2.13}$$

We replace (2.12) and (2.13) into (2.11) and integrate for all values of ν and Ω_2. We obtain

$$S_2 = \frac{4\pi^5 k V_2}{45 c^3 \beta_s^3} (1 - \cos 2\delta). \tag{2.14}$$

By using (2.10) and (2.14) we find

$$\beta_s = \frac{3h}{4k} \frac{S_2}{U_2}. \tag{2.15}$$

We replace (2.15) into (2.14) and obtain

$$S_2 = A V_2^{1/4} U_2^{3/4} (1 - \cos 2\delta)^{1/4}, \quad A \equiv \left(\frac{4\pi^5 k}{45 c^3}\right)^{1/4} \left(\frac{4k}{3h}\right)^{3/4} \tag{2.16}$$

From (2.16) we derive

$$U_2 = A^{-4/3} V_2^{-1/3} S_2^{4/3} (1 - \cos 2\delta)^{-1/3} \tag{2.17}$$

The solar radiation confined to volume V_2 is in nonequilibrium as the flux φ_2 is nonzero. However, all the conditions of statistical equilibrium are fulfilled as the macroscopic quantities for each part of the system equal the average values at the level of the whole system [15, p. 14]. Indeed, (2.10) and (2.14) prove that the volumetric densities of both the internal energy and entropy do not depend on the volume of the system. Consequently, we could extend the definition of temperature to this state of nonequilibrium:

$$T_2 = \left(\frac{\partial U_2}{\partial S_2}\right)_{V_2} = \frac{4}{3A} \left(\frac{U_2}{V_2}\right)^{1/4} (1 - \cos 2\delta)^{-1/4}. \tag{2.18}$$

By replacing (2.10) into (2.18) we find

$$T_2 = T_s. \tag{2.19}$$

The exergy of solar radiation, R, is defined as the maximum available work of a system placed into an environment characterized by pressure p_0 and temperature T_0. It can be evaluated by [17]:

$$R \equiv L - L_0, \quad L \equiv U - T_0 S + p_0 V. \tag{2.20}$$

We use (2.18) twice, for the black-body radiation source at temperatures

T_s and T_0, respectively. We obtain

$$U_0 = U_2 \left(\frac{T_0}{T_s}\right)^4. \qquad (2.21)$$

where U_0 is the internal energy of the environment. The pressure p_0 of the equilibrium environment radiation is given by [15, p. 205]:

$$p_0 = \frac{1}{3}\frac{U_0}{V_2}. \qquad (2.22)$$

By using (2.21) and (2.22) we obtain

$$p_0 V_2 = 1/3 \, U_2 \left(\frac{T_0}{T_2}\right)^4. \qquad (2.23)$$

We replace A from (2.18) into (2.16) and multiply by T_0. We derive

$$T_0 S_2 = 4/3 \, U_2 \frac{T_0}{T_2}. \qquad (2.24)$$

By replacing (2.23) and (2.24) into definition (2.20) we find

$$L = U_2 \left[1 - \frac{4}{3}\frac{T_0}{T_s} + \frac{1}{3}\left(\frac{T_0}{T_s}\right)^4\right]. \qquad (2.25)$$

In order to evaluate L_0 we use (2.25) under the assumption of thermal equilibrium (i.e., the elements Σ_1 and Σ_2 have the same temperature). We find $L_0 = U_0(1 - \frac{4}{3} + \frac{1}{3}) = 0$. Now, by using (2.20) and (2.25) we obtain

$$R = U_2 \left[1 - \frac{4}{3}\frac{T_0}{T_s} + \frac{1}{3}\left(\frac{T_0}{T_s}\right)^4\right]. \qquad (2.26)$$

The efficiency of converting solar energy into work can be defined as

$$\eta = \frac{R}{U_2}. \qquad (2.27)$$

By using (2.26) and (2.27) we can easily find the LPP efficiency (equation (2.1)).

2.2.2 *Simple Upper Bound for Diluted Radiation Conversion*

In this section we present a generalized upper-bound formula for the conversion efficiency, which can be used in case of diluted solar radiation and both black-body and selective converters [18], [19].

We assume a nonemissive ambient and a selective (non-black-body) converter with negligible conductive and convective thermal losses. Then, the steady-state balance equations per unit converter area are

$$\varphi_{s,\text{abs}} = \varphi + \dot{Q}', \quad \psi_{s,\text{abs}} + \dot{S}_g = \psi + \dot{Q}'/T. \qquad (2.28)$$

Here $\varphi_{s,\text{abs}}$ and $\psi_{s,\text{abs}}$ are the rates of energy and entropy *absorbed* from the source of radiation (diluted or undiluted), $\dot{Q}'$ is the unbalanced flux of thermal energy, φ and ψ are the rates of energy and entropy emission by the absorber to the ambient, $\dot{S}_g$ is the rate of entropy generation in the absorber, and T is the absorber temperature. Subsequently, the flux $\dot{Q}'$ is partially converted into work

$$\dot{Q}' = \varphi_{s,\text{abs}} - \varphi = \dot{Q} + \dot{W}, \qquad (2.29)$$

where $\dot{W}$ is the rate of mechanical, chemical, or other type of work performed, and $\dot{Q}$ is the flux of thermal energy finally reaching the surroundings. In the most favorable situation there is no further increase of entropy during work production. Then

$$\frac{\dot{Q}'}{T} = \frac{\varphi_{s,\text{abs}} - \varphi}{T} = \frac{\dot{Q}}{T_0}. \qquad (2.30)$$

Elimination of $\dot{Q}$ between (2.29) and (2.30) yields

$$\dot{W} = \left(1 - \frac{T_0}{T}\right)(\varphi_{s,\text{abs}} - \varphi). \qquad (2.31)$$

Let us call $\varphi_{s,\text{inc}}$ the energy flux of the (solar) radiation incident on the converter. The conversion efficiency can be defined in a way similar to (2.27) as

$$\eta = \frac{\dot{W}}{\varphi_{s,\text{inc}}} = \left(1 - \frac{T_0}{T}\right)\frac{\varphi_{s,\text{abs}} - \varphi}{\varphi_{s,\text{inc}}}. \qquad (2.32)$$

The derivation of (2.32) was performed without taking into consideration the nature of the incoming radiation. Consequently, this equation can be used for both diluted and undiluted black-body radiation.

Before applying (2.32) in the case of diluted solar radiation we must remember some useful results [8]. The effective temperature T_e of the diluted solar radiation depends on the Sun's temperature T_s by

$$T_e = \frac{T_s}{\chi(\varepsilon)}, \qquad (2.33)$$

where $\varepsilon \leq 1$ is the so-called dilution factor and $\chi(\varepsilon)$ is a monotonic function exactly calculated for the first time in [8], that can be approximated for small ε by $\chi(\varepsilon) = 0.9652 - 0.2777 \ln \varepsilon + 0.0511\,\varepsilon$, and such that $\chi(1) = 1$. The energy and entropy fluxes of diluted solar radiation may be written

$$\varphi_{s,\text{inc}} = \beta T_e^4, \quad \psi_{s,\text{inc}} = 4/3\beta T_e^3, \quad \beta = A(\Omega_s)\varepsilon\sigma\chi^4(\varepsilon). \qquad (2.34)$$

Here σ is Stefan's constant, and the function $A(\Omega_s)$ refers to a geometrical (view) factor given by

$$A(\Omega_s) = \frac{\Omega_s}{\pi}\left(1 - \frac{\Omega_s}{4\pi}\right), \qquad (2.35)$$

Ω_s being the solid angle subtended by the black-body radiation source (in our case the Sun). We can easily verify that in the case of a hemispherical source ($\Omega_s = 2\pi$) of undiluted radiation ($\varepsilon = 1$) (2.33) and (2.34) reduce to the usual (undiluted) black-body quantities.

The non-black-body (selective) converter we considered must have some special properties in order to give the best performance. Indeed, let us divide the whole domain of frequencies $(0, \infty)$ into two subdomains $\mathcal{D}_{T_e}$ and $\mathcal{D}_T$ in such a way that the difference between the energy of the radiation incoming from the source within $\mathcal{D}_{T_e}$ and the energy of the radiation emitted by the absorber within $\mathcal{D}_T$ can be maximized. For good performance, the reflectance of the absorber surface must be small for the domain $\mathcal{D}_{T_e}$ and high for $\mathcal{D}_T$ [20]. We shall make two hypotheses:

(1) the domains $\mathcal{D}_{T_e}$ and $\mathcal{D}_T$ contain the greatest part of the energy of the incoming and emitted radiation, respectively; and

(2) the absorber reflectance is independent of frequency within $\mathcal{D}_{T_e}$ and $\mathcal{D}_T$, respectively.

Under such circumstances the emitted radiation can be considered as diluted radiation and the theory from [8] applies. The two hypotheses are well verified in the case of selective solar converters, if the temperatures T_e and T are sufficiently far apart [20].

The spectrally averaged absorbtance and emittance of the selective converter will be noted with a and e, respectively. (For details of computation, see [20].)

The converter emits diluted radiation over the solid angle $\Omega^* = 2\pi$ with dilution factor e. We shall note T_e, T^* and β, β^* the effective temperatures and the corresponding quantities of incoming and emitted radiation, respectively. Then, from (2.33) and (2.34) we can write

$$\varphi_{s,\text{abs}} = a\beta T_e^4, \quad \psi_{s,\text{abs}} = 4/3 a\beta T_e^3, \qquad (2.36)$$
$$\varphi = \beta^* T^{*4} = A(\Omega^*)\sigma e \chi^4(e) T^{*4} = e\sigma T^4, \qquad (2.37)$$
$$\psi = 4/3\beta^* T^{*3} = 4/3 A(\Omega^*)\sigma e \chi^4(e) T^{*3} = 4/3 e\sigma \chi(e) T^3. \qquad (2.38)$$

If we replace (2.34), (2.36), and (2.37) into (2.32) we obtain

$$\eta = a\left(1 - \frac{T_0}{T}\right)\left(1 - \frac{e\sigma T^4}{a\beta T_e^4}\right). \qquad (2.39)$$

Let us introduce some notation

$$\theta \equiv \frac{T}{T_e}, \quad \theta_0 \equiv \frac{T_0}{T_e}, \quad \gamma \equiv \left(\frac{a\beta}{e\sigma}\right)^{1/4}. \qquad (2.40)$$

Now, (2.39) can be rewritten as

$$\eta = a\left(1 - \frac{\theta_0}{\theta}\right)\left[1 - \left(\frac{\theta}{\gamma}\right)^4\right]. \tag{2.41}$$

The maximum converter temperature $T_{\max} = T_e$ can be reached through effective equilibrium, for which $\varphi_{s,\text{abs}} = \varphi$ [8]. Consequently, from (2.36) and (2.37), we see that

$$\gamma^4 \leq 1. \tag{2.42}$$

Also, from (2.41), we see that η is positive only for $\theta_0 < \theta < \gamma$. By taking into account (2.42) we can write

$$0 < \theta_0 < \theta < \gamma \leq 1. \tag{2.43}$$

The second equation of balance (2.28) can be used to obtain the generation of entropy in the converter

$$\dot{S}_g = \frac{\varphi_{s,\text{abs}} - \varphi}{T} + \psi - \psi_{s,\text{abs}}. \tag{2.44}$$

From (2.31) and (2.44) we can write

$$\dot{W} = \varphi_{s,\text{abs}} - \varphi - T_0[\dot{S}_g - (\psi - \psi_{s,\text{abs}})]. \tag{2.45}$$

A new form of (2.41) can be derived if we use (2.40), (2.44), and (2.45):

$$\frac{\eta}{a} = 1 - 4/3\theta_0 + 1/3\theta_0^4 - \theta_0\dot{\sigma}, \quad \dot{\sigma} \equiv \frac{1}{\theta} - \frac{\theta^3}{\gamma^4} - \frac{4}{3} + \frac{\theta^4}{\gamma^4\theta_0} + \frac{\theta_0^3}{3}. \tag{2.46}$$

The function $\dot{\sigma}$ can be put in the form:

$$\dot{\sigma} = \frac{1}{3\gamma\theta_0\theta}\left\{(\theta - \theta_0)\theta_0^3\theta\left\{\left(\frac{\theta}{\theta_0}\right)^2\left[\frac{\theta}{\theta_0} - \gamma^4\right] + \frac{\theta}{\theta_0}\left[\left(\frac{\theta}{\theta_0}\right)^2 - \gamma^4\right] + \left(\frac{\theta}{\theta_0}\right)^3 - \gamma^4\right\} \right.$$
$$\left. + \gamma^4\theta_0(1 - \theta)(3 - \theta - \theta^2 - \theta^3)\right\}. \tag{2.47}$$

By taking into account (2.43) and (2.47) we see that $\dot{\sigma}$ is always nonnegative. Consequently, from (2.46) and (2.40) we obtain

$$\eta \leq \eta^{\text{sup}} \equiv a\left[1 - \frac{4}{3}\frac{T_0}{T_e} + \frac{1}{3}\left(\frac{T_0}{T_e}\right)^4\right]. \tag{2.48}$$

The new upper bound η^{sup} defined by (2.48) generalizes the result (2.1) to the case of diluted solar radiation and selective converters.

2.2.3 *More Accurate Simple Upper Bound Efficiency*

In many situations all three formulas (2.1) yield too high values to be useful in applications. In order to derive a more accurate upper bound for the conversion efficiency, we shall consider a hypothetic *black-body* converter at temperature T which receives radiation from a hypothetical hemispherical *black-body* radiation source at a temperature $T_{b,\mathrm{eq}}$ given by

$$T_{b,\mathrm{eq}} = \gamma T_e \leq T_e. \tag{2.49}$$

The absorbed and emitted fluxes of energy and entropy are now

$$\varphi_{s,\mathrm{abs,eq}} = \varphi_{s,\mathrm{inc}} = \sigma T_{b,\mathrm{eq}}^4, \quad \psi_{s,\mathrm{abs,eq}} = \psi_{s,\mathrm{inc}} = 4/3\,\sigma T_{b,\mathrm{eq}}^3, \tag{2.50}$$

$$\varphi_{\mathrm{eq}} = \sigma T^4, \quad \psi_{\mathrm{eq}} = 4/3\sigma T^3. \tag{2.51}$$

The conversion efficiency η_{eq} for this hypothetical case is defined by (2.32) again and is related to the real converter efficiency η:

$$\eta_{\mathrm{eq}} = \frac{\dot{W}_{\mathrm{eq}}}{\varphi_{s,\mathrm{inc}}} = \left(1 - \frac{T_0}{T}\right)\left(1 - \frac{\varphi_{\mathrm{eq}}}{\varphi_{s,\mathrm{inc}}}\right) = \frac{\eta}{a}. \tag{2.52}$$

A new form of η_{eq} can be obtained if we use (2.50)–(2.51) within the procedure described in Section 2.2.2:

$$\eta_{\mathrm{eq}} = 1 - \frac{4}{3}\frac{T_0}{T_{b,\mathrm{eq}}} + \frac{1}{3}\left(\frac{T_0}{T_{b,\mathrm{eq}}}\right)^4 - \theta_0 \dot{\sigma}_{\mathrm{eq}} = 1 - \frac{4}{3}\frac{\theta_0}{\gamma} + \frac{1}{3}\left(\frac{\theta_0}{\gamma}\right)^4 - \theta_0 \dot{\sigma}_{\mathrm{eq}}, \tag{2.53}$$

where the function $\dot{\sigma}_{\mathrm{eq}}$ is given by

$$\dot{\sigma}_{\mathrm{eq}} \equiv \frac{1}{\theta} - \frac{\theta^3}{\gamma^4} - \frac{4}{3\gamma} + \frac{\theta^4}{\gamma^4 \theta_0} + \frac{\theta_0^3}{3\gamma^4}. \tag{2.54}$$

The function $\dot{\sigma}_{\mathrm{eq}}$ can be put in the form

$$\dot{\sigma}_{\mathrm{eq}} = \frac{1}{3\gamma^4 \theta_0 \theta}\left[\theta^4(\theta - \theta_0)\left(3 - \frac{\theta_0}{\theta} - \frac{\theta_0^2}{\theta^2} - \frac{\theta_0^3}{\theta^3}\right)\right.$$
$$\left. + \gamma^3 \theta_0 (\gamma - \theta)\left(3 - \frac{\theta}{\gamma} - \frac{\theta^2}{\gamma^2} - \frac{\theta^3}{\gamma^3}\right)\right]. \tag{2.55}$$

By taking into account (2.43) we see that $\dot{\sigma}_{\mathrm{eq}}$ is always nonnegative. Then, from (2.53) we obtain the upper bound of η_{eq}:

$$\eta_{\mathrm{eq}} < 1 - \frac{4}{3}\frac{\theta_0}{\gamma} + \frac{1}{3}\left(\frac{\theta_0}{\gamma}\right)^4. \tag{2.56}$$

24   2. Thermodynamics of Solar Energy Conversion into Work

It is a simple matter to verify that $1 - \frac{4}{3}(\theta_0/\gamma) + \frac{1}{3}(\theta_0/\gamma)^4 \leq 1 - \frac{4}{3}\theta_0 + \frac{1}{3}\theta_0^4$. Consequently, by taking into account (2.48), (2.52), and (2.56), we can write

$$\frac{\eta}{a} = \eta_{\text{eq}} \leq 1 - \frac{4}{3}\frac{\theta_0}{\gamma} + \frac{1}{3}\left(\frac{\theta_0}{\gamma}\right)^4 \leq 1 - 4/3\theta_0 + 1/3\theta_0^4. \tag{2.57}$$

By using (2.33), (2.40), and (2.49) we obtain

$$T_{b,\text{eq}} = \gamma T_e = \left[\frac{a}{e} A(\Omega_s)\varepsilon\right]^{1/4} T_s. \tag{2.58}$$

From (2.40), (2.57), and (2.58) we see that

$$\eta \leq \tilde{\eta}^{\text{sup}} \equiv a\left\{1 - \frac{4}{3}\frac{1}{((a/eA)\varepsilon)^{1/4}}\frac{T_0}{T_s} + \frac{1}{3}\left[\frac{1}{((a/eA)\varepsilon)^{1/4}}\frac{T_0}{T_s}\right]^4\right\} < a \cdot \eta_{\text{LPP}}. \tag{2.59}$$

We conclude that $\tilde{\eta}^{\text{sup}}$ is a more accurate upper bound than $a \cdot \eta_{\text{LPP}}$. Consequently, it is recommended to use as a simple formula for the conversion efficiency of (diluted and undiluted) solar radiation. The quantity $\tilde{\eta}^{\text{sup}}$ was used for the first time in [13] and was rigorously derived in [18], [19].

As a numerical example we shall consider the conversion of undiluted solar energy ($\varepsilon = 1$) into work by using a black-body converter ($a = e = 1$) placed on Earth. In this case, one may assume $T_s = 5762$ K and $T_0 = 300$ K. The solid angle subtended by the Sun when viewed from Earth is $\Omega_s \approx 6.835 \times 10^{-5}$ sr. Consequently, the geometric factor given by (2.35) is $A(\Omega_s) = 2.17 \times 10^{-5}$. The upper limit efficiencies evaluated by using the formulas proposed by Jeter (J) [5], Spanner (S) [4], and Landsberg–Petela–Press (LPP) [1], [3] (see (2.1)) are: $\eta_J = 0.948$; $\eta_S \approx \eta_{\text{LPP}} = 0.931$. By using (2.59) we obtain $\tilde{\eta}^{\text{sup}} = 0.0965$. As one sees, the increase in accuracy when using $\tilde{\eta}^{\text{sup}}$ instead of the other formulas is significant.

2.3 Terrestrial Applications

2.3.1 *Converting Direct Solar Radiation*

The main advantage of using solar direct (beam) radiation is that it can be concentrated. The concentration ratio C ranges between 1 (in the case of unconcentrated solar radiation) and $\approx 46,200$ (in the case when, due to the solar concentrators, the receiver sees the Sun under a solid angle 2π). Let us consider a black-body converter which utilizes a flux of solar radiation that is affected by transmission through the atmosphere. In order to determine the maximum conversion efficiency, we neglect convective heat losses.

A slightly different form of the standard converter energy balance ((2.28)) will be used in the following [21], [22]:

$$\varphi_{s,\text{abs}} + \varphi_{0,\text{abs}} = \varphi + \dot{Q}'. \tag{2.60}$$

Here $\varphi_{0,\text{abs}}$ is the energy flux absorbed from the atmosphere (long wavelengths) while the other terms were already defined. Equation (2.29) will be used again to model the conversion of the unbalanced flux of thermal energy $\dot{Q}'$ into the rate of work $\dot{W}$. As usual, one assumes that no increase of entropy occurs during this conversion. Then

$$\frac{\dot{Q}'}{T} = \frac{\varphi_{s,\text{abs}} + \varphi_{0,\text{abs}} - \varphi}{T} = \frac{\dot{Q}}{T_0}. \tag{2.61}$$

Elimination of $\dot{Q}$ between (2.29), (2.60), and (2.61) leads to the following expression for the conversion efficiency

$$\eta = \frac{\dot{W}}{\varphi_{s,\text{inc}}} = \left(1 - \frac{T_0}{T}\right) \frac{\varphi_{s,\text{abs}} + \varphi_{0,\text{abs}} - \varphi}{\varphi_{s,\text{inc}}}. \tag{2.62}$$

One denotes by φ_s the flux of unconcentrated solar-beam radiation outside the Earth's atmosphere and by $\varphi_{0,inc}$ the flux of ambient radiation incident on the converter. Then the fluxes φ, φ_s, and $\varphi_{0,inc}$ are given by [12], [13]:

$$\varphi = \sigma T^4, \quad \varphi_{0,\text{inc}} = \sigma T_0^4, \quad \varphi_s = \frac{1 - \cos 2\delta}{2} \sigma T_s^4, \tag{2.63}$$

where δ is the half-angle of the cone subtended by the solar disk. One denotes by φ'_s the flux of unconcentrated solar radiation at the Earth's surface. Then the transmittance of the atmosphere is defined as

$$\tau \equiv \frac{\varphi'_s}{\varphi_s}. \tag{2.64}$$

The flux of solar radiation incident on a converter with concentration ratio C can be obtained by using (2.63) and (2.64)

$$\varphi_{s,\text{inc}} = C\varphi'_s = C\tau \frac{1 - \cos 2\delta}{2} \sigma T_s^4. \tag{2.65}$$

Note that both solar and atmospheric radiation incident on a black-body converter are entirely absorbed ($\varphi_{s,\text{abs}} = \varphi_{s,\text{inc}}$ and $\varphi_{0,\text{abs}} = \varphi_{0,\text{inc}}$). By using (2.62)–(2.65) we find

$$\eta = \left(1 - \frac{T_0}{T}\right)\left(1 - \alpha \frac{T^4 - T_0^4}{T_s^4}\right), \quad \alpha \equiv \frac{2}{C\tau(1 - \cos 2\delta)}. \tag{2.66}$$

The conversion efficiency equation (2.66) can be maximized with respect to the converter temperature T. The optimum value T_{opt} can be found by solving the equation

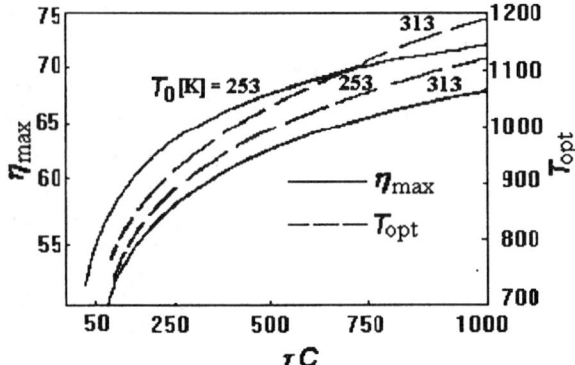

FIGURE 2.1. The maximum efficiency $\eta_{\max}$ and the optimum converter temperature T_{opt} as functions of the product of atmospheric transmittance τ and concentration ratio C. Parameter T_0 is the ambient temperature.

$$\left.\frac{\partial \eta}{\partial T}\right|_{T=T_{\text{opt}}} = 4T_{\text{opt}}^5 - 3T_0 T_{\text{opt}}^4 - \frac{1}{\alpha} T_0 T_s^5 = 0. \tag{2.67}$$

The maximum efficiency decreases abruptly when the concentration ratio C decreases (Figure 2.1). The influence of ambient temperature on the maximum efficiency is important and similar for any concentration ratio. $\eta_{\max}$ is considerably smaller than 0.931 (which is the value predicted by LPP efficiency). Thus, one confirms the usefulness of the upper bound efficiency $\tilde{\eta}^{\text{sup}}$ we proposed in Section 2.2.3. Also, the optimum converter temperature is much more strongly reduced than for fully concentrated radiation, for which $T_{\text{opt}} = 2464$ K [12].

2.3.2 Converting Diffuse Solar Radiation

A number of papers deal with the conversion of diffuse solar radiation into mechanical work [2], [12], [13]. The first optimization theory was presented in [12]. Bădescu [23] developed a different approach based on the concept of diluted solar radiation. The results confirmed previous conclusions derived in [13] by using a different method. The apparent contradiction between the theories from [12] and [23] was explained in [24] where it was shown that both of them are particular cases of a more general approach.

Thermodynamics of Diffuse Solar Radiation

Diffuse solar radiation may be treated as diluted black-body radiation. A full account on this subject is found in the work of Landsberg and Tonge [8], see also [15]. The most significant results are reviewed in [24]. The entropy and energy fluxes of diluted radiation are given by (2.34) as functions of the factor β and the effective temperature T_e ((2.33)). The factor β depends on the geometrical (view) factor A through (2.34).

In order to obtain the maximum efficiency of diffuse radiation one must envisage the most favorable situation [24]. Consequently, one considers the Sun at zenith. Also, one assumes that the Earth's atmosphere is not absorbent and emissive and that there is no back-scattered solar radiation. These assumptions allow us to define a "perfectly forward diffuser", i.e., a finite thin body, situated between the Sun and the observer whose surface elastically scatters into a 2π solid angle any narrow pencil of radiation incident on it. One assumes this diffuser as subtending a solid angle Ω_1 when viewed from the Earth. Consider first the direct sunlight (dilution factor ε_0) incident on a point M placed on the diffuser surface. Then, $\varepsilon_0 = 1$ and the geometrical factor (see (2.35)) is given by

$$A(\Omega_s) = \frac{\Omega_s}{\pi}\left(1 - \frac{\Omega_s}{4\pi}\right) \approx \frac{\Omega_s}{\pi}. \tag{2.68}$$

Consequently, the energy flux of solar beam radiation incident on the diffuser is given by (2.33), (2.34), and (2.68)

$$\varphi_s = \frac{\Omega_s}{\pi}\sigma T_s^4. \tag{2.69}$$

The energy flux φ_s is scattered over an $\Omega_1 = 2\pi$ solid angle around point M. After scattering, solar radiation has a dilution factor $\varepsilon < 1$ and the effective temperature is $T_{e,1}$. For an observer situated at point M the scattered radiation has a geometrical factor $A(\Omega_1) = 1$. The energy flux φ_1 of diluted radiation is equal to the incoming energy flux φ_s. By using (2.33), (2.34) and (2.35) one obtains

$$\varphi_1 = A(\Omega_1)\varepsilon_1 \sigma T_s^4 = \frac{\Omega_s}{\pi}\sigma T_s^4 = \varphi_s. \tag{2.70}$$

Consequently, the dilution factor is

$$\varepsilon_1 = \frac{\Omega_s}{\pi} \tag{2.71}$$

and the effective temperature of the scattered radiation can be derived by using (2.33)

$$T_{e.1} = \frac{T_s}{\chi(\varepsilon_1)} = 1459.5 \text{ K}. \tag{2.72}$$

After dilution the thermal radiation may be considered as undiluted with respect to the black-body radiation of temperature $T_{e.1}$.

Until now we have analyzed the scattered radiation from the point of view of an absorber situated on the surface of the diffuser. For an observer placed on the Earth's surface the source of singly scattered radiation may be formally described as a black-body of temperature $T_{e,1}$ (dilution factor $\varepsilon'_1 = 1$) which subtends the solid angle Ω_1. To determine Ω_1 we use (2.34) and the assumption that on the Earth's surface the energy flux φ_1 of scattered radiation equates the energy flux φ_s

$$\varphi_1 = A(\Omega_1)\varepsilon'_1\sigma T_{e,1}^4 = \frac{\Omega_s}{\pi}\sigma T_s^4 = \varphi_s. \tag{2.73}$$

By using (2.35) we obtain

$$\Omega_1 = 2\pi\left\{1 - \left[1 - \frac{\Omega_s}{\pi}\chi^4(\varepsilon_1)\right]^{1/2}\right\} = 0.01665 \text{ sr}. \tag{2.74}$$

The solid angle Ω_1 is enveloped by a cone symmetrically disposed around the nadir–zenith direction and whose half-angle can be derived from the relation

$$\Omega_1 = 2\pi(1 - \cos 2\delta_1). \tag{2.75}$$

We obtain $\delta_1 = 2.083°$. We conclude that after a single scattering solar radiation is still strongly anisotropic (compare δ_1 with the half-angle $\delta = 0.265°$ of the cone subtended by the Sun).

What happens if a second scattering occurs? This implies the existence of a second "perfectly forward diffuser". The incoming radiation consists of the energy flux φ_1 of single scattered solar radiation (dilution factor $\varepsilon_1 = \Omega_s/\pi$) or, which is equivalent, black-body radiation of temperature $T_{e,1}$ (dilution factor $\varepsilon'_1 = 1$). The energy flux φ_1 is again dispersed over a solid angle $\Omega_2 = 2\pi$, the scattered radiation having a geometrical factor $A(\Omega_2) = 1$ and a dilution factor $\varepsilon_2 < 1$. The energy flux φ_2 of doubly scattered radiation equates to the incoming energy flux φ_1. By using (2.70) we obtain

$$\varphi_2 = A(\Omega_2)\varepsilon_2\sigma T_{e,1}^4 = \varphi_1 = \varphi_s = \frac{\Omega_s}{\pi}\sigma T_s^4. \tag{2.76}$$

From (2.73) we find

$$\varepsilon_2 = \frac{\Omega_s}{\pi}\chi^4(\varepsilon_1) = 5.281 \times 10^{-3} \tag{2.77}$$

and the effective temperature of the doubly scattered radiation is given by (2.33)

$$T_{e,2} = \frac{T_{e,1}}{\chi(\varepsilon_2)} = \frac{T_s}{\chi(\varepsilon_1)\chi(\varepsilon_2)}. \tag{2.78}$$

For an observer placed on the Earth's surface the source of thermal radiation is a black-body of temperature $T_{e,2}$ (dilution factor $\varepsilon'_2 = 1$) which subtends a solid angle Ω_2. To determine Ω_2 we use (2.34) and the assumption that the energy flux φ_2 incident on the Earth's surface equates to the energy flux φ_s

$$\varphi_2 = A(\Omega_2)\varepsilon_2' \sigma T_{e,21}^4 = \varphi_s = \frac{\Omega_s}{\pi} \sigma T_s^4. \tag{2.79}$$

By using (2.35) we obtain

$$\Omega_2 = 2\pi \left\{ 1 - \left[1 - \frac{\Omega_s}{\pi} \chi^4(\varepsilon_1)\chi^4(\varepsilon_2) \right]^{1/2} \right\} = 0.599 \text{ sr.} \tag{2.80}$$

The half-angle δ_2 of the cone which envelops the solid angle Ω_2 can be determined with a relation similar to (2.75) and is $\delta_2 = 12.609°$. As one sees, the doubly scattered solar radiation is still anisotropic.

The above procedure can be repeatead for three and four scatterings, with the following results:

$$\varepsilon_i = \frac{\Omega_s}{\pi} \prod_{j=1}^{i-1} \chi^4(\varepsilon_j), \quad T_{e,i} = \frac{T_s}{\prod_{j=1}^{i} \chi(\varepsilon_j)}, \tag{2.81}$$

$$\Omega_i = 2\pi \left\{ 1 - \left[1 - \frac{\Omega_s}{\pi} \prod_{j=1}^{i-1} \chi^4(\varepsilon_j) \right]^{1/2} \right\}, \tag{2.82}$$

where we noted $T_s = T_{e,0}$.

TABLE 2.1. Thermodynamics of singly or multiply scattered solar radiation: $i = $ number of scatterings; $T_e = $ the effective temperature of scattered radiation; $\delta = $ the half-angle of the cone which subtends a black-body at the effective temperature T_e; $C_{\max} = $ the maximum concentration ratio; $T_{\text{opt}} = $ the optimum *black-body* converter temperature; $\eta_{\max} = $ the maximum conversion efficiency.

i	$T_{e,i}$ (K)	δ_i (deg)	$C_{\max,i}$	$T_{\text{opt},i}$ (K)	$\eta_{\max,i}$
0	5760.0	0.265	45963.0	2464	0.849
1	1459.5	2.083	189.5	837	0.573
2	602.8	12.609	5.5	397	0.198
3	418.9	30.961	1.3	363	0.114
4	393.4	90.000	1.0	348	0.053

Table 2.1 shows the results. After four scatterings solar radiation is completely isotropic and an observer on the ground would see a uniform brilliant sky. There are some indications that for a clear sky most of the diffuse solar radiation is received within a cone of half-angle $\delta_{\text{real}} = 20\text{--}30°$ [26], [27]. This implies a mean number of three scatterings.

Efficiency of Converting Diffuse Solar Radiation

Scattered solar radiation is generally anisotropic. Consequently, it can be concentrated. Under the already-made assumption of a nonabsorbent and nonemissive ambient, the energy flux φ_{inc} incoming to the receiver of a concentrator is (regardless of the number i of scatterings)

$$\varphi_{\text{inc}} = C\varphi_s = A(\Omega_i)\varepsilon_i \sigma T_{e,i}^4 = C\frac{\Omega_s}{\pi}\sigma T_s^4. \tag{2.83}$$

To determine the maximum concentration ratio of an i-scattered diffuse solar radiation we must observe that "fully concentrated scattered radiation" implies that the receiver sees over the hemispherical solid angle 2π a black-body at temperature $T_{e,i}$. Consequently, the maximum value of φ_{inc} is

$$\varphi_{\text{inc,max}} = \sigma T_{e,i}. \tag{2.84}$$

Now we can determine the maximum concentration ratio from (2.36), (2.84), and (2.85)

$$C_{\text{max},i} = \left[\frac{\Omega_i}{\pi}\left(1 - \frac{\Omega_i}{4\pi}\right)\right]^{-1}. \tag{2.85}$$

In the general case a selective converter will be considered. Consequently, the conversion efficiency is given by (2.39). If we take into consideration (2.83) we derive

$$\eta = \left(1 - \frac{T_0}{T}\right)\left(1 - \frac{e}{a}\frac{\pi}{C\Omega_s}\frac{T^4}{T_s^4}\right). \tag{2.86}$$

The optimum converter temperature is obtained by maximizing η, i.e., from the relation

$$\frac{\partial \eta}{\partial T} = 4T^5 - 3T_0 T^4 - \frac{a}{e}\frac{C\Omega_s}{\pi}T_0 T_s^4 = 0. \tag{2.87}$$

In the first case we consider *nonselective* (black-body) converters. Consequently, $a = e = 1$. Table 2.1 shows the results for $T_0 = 300$ K. In the case of unconcentrated scattered solar radiation ($C = 1$) the maximum efficiency is 0.053 for any number of scatterings. In the case of fully concentrated scattered radiation ($C = C_{\text{max},i}$) the maximum efficiency lies between 0.573 and 0.053 for one and four scatterings, respectively.

Better performance may be obtained by using *converters with selective surface* ($a > e$). Only the case of unconcentrated scattered radiation ($C = 1$) will be considered here. Then the maximum efficiency is 0.55 and corresponds to an ideal selective surface with $a/e = 127.5$ [28].

As a more realistic selective surface, we consider a tandem, infrared-reflecting, metalic substrate and an absorbing layer [23]. We see from (2.87) that the efficiency increases for greater a and smaller e. We shall use $a_{\text{max}} = 1$ and look for the minimum value of e. The Drude theory of conducting materials shows that the total hemispherical emittance e increases with temperature and electrical resistivity [20]. Of the usual materials, copper has the smaller resistivity $\rho_{Cu} = 1.7 \times 10^{-8}$ Ωm at 273.1 K. As the minimum value for e, we obtain for $T = T_0$ [20, p. 42] $e_{min} = 7.7\, T_0(\rho_{Cu}/275.15)^{1/2} = 0.017$. Using a_{max} and e_{min} in (2.87) and (2.86) we find $T_{opt} = 683.5$ K and $\eta_{max} = 0.474$. One should remember that these results were obtained for isotropic diffuse radiation (four scatterings). Better results will be obtained in the case of clear sky diffuse radiation (with two or three scatterings).

2.3.3 *Converting Global Solar Radiation*

A study dealing with devices converting global solar radiation into work is more complicated. First, because the converter optical properties generally depend on the radiation components (i.e., direct or diffuse). Also, the convective heat losses cannot always be neglected compared with the radiative ones. The dependence of solar converter performances on the convective heat losses was considered in [29]. Third, the radiation flux received by the converter from the atmosphere (longwave) cannot always be neglected compared with the solar radiation flux. Haught [21] studied the effect of the longwave radiation flux on converter performance.

Two models will be presented in the following. The first one treats the case of a solar converter coupled with an ideal Carnot engine. It is useful to obtain upper limits for the performance. The second model considers a Stirling engine-based system, which is an attractive solution for practical applications.

Solar Converter Coupled to a Carnot Engine

The steady-state balance equation per unit converter area is a modified form of (2.60) [30]:

$$\varphi_{s,abs} + \varphi_{0,abs} = \varphi + \dot{Q}_{conv} + \dot{Q}'. \tag{2.88}$$

The new term is the rate of thermal convective losses $\dot{Q}_{conv}$. Equation (2.29) is still valid while the expression of the second law of thermodynamics for a reversible engine is now given by

$$\frac{\dot{Q}'}{T} = \frac{\varphi_{s,abs} + \varphi_{0,abs} - \varphi - \dot{Q}_{conv}}{T} = \frac{\dot{Q}}{T_0}. \tag{2.89}$$

Elimination of $\dot{Q}$ between (2.29), (2.88), and (2.89) leads to the expression for the rate of mechanical work

$$\dot{W} = \left(1 - \frac{T_0}{T}\right)(\varphi_{s,abs} + \varphi_{0,abs} - \varphi - \dot{Q}_{conv}). \tag{2.90}$$

Usually the efficiency of converting *solar radiation* is defined by (2.32):

$$\eta_{sol} = \frac{\dot{W}}{\varphi_s} = \frac{\varphi_{s,abs}}{\varphi_s} \frac{\dot{W}}{\varphi_{s,abs}}. \tag{2.91}$$

32   2. Thermodynamics of Solar Energy Conversion into Work

One denotes by A_s the ratio $\varphi_{s,\text{abs}}/\varphi_s$, which we consider as depending on the optical properties of the converter only. Then, by using (2.90) and (2.91), we obtain

$$\eta_{\text{sol}} = A_s \left(1 - \frac{T_0}{T}\right)\left(1 - \frac{\varphi + \dot{Q}_{\text{conv}} - \varphi_{0,\text{abs}}}{\varphi_{s,\text{abs}}}\right), \tag{2.92}$$

η_{sol} gives pertinent information only if the flux $\varphi_{0,\text{abs}}$ is negligible. Also, η_{sol} can become overunitary if $\varphi_{0,\text{abs}} > \varphi_{s,\text{abs}}$. Consequently, we introduce the efficiency of converting both *solar and atmospheric radiation* as

$$\eta_{\text{sol+amb}} = \frac{\dot{W}}{\varphi_s + \varphi_0} = \frac{\varphi_{s,\text{abs}} + \varphi_{0,\text{abs}}}{\varphi_s + \varphi_0} \frac{\dot{W}}{\varphi_{s,\text{abs}} + \varphi_{0,\text{abs}}}, \tag{2.93}$$

where φ_0 is the flux of radiation incoming from the atmosphere. Again, we denote by A_{s+a} the ratio $(\varphi_{s,\text{abs}} + \varphi_{0,\text{abs}})/(\varphi_s + \varphi_0)$, assumed independent of temperature. By using (2.90) and (2.93) we obtain

$$\eta_{\text{sol+amb}} = A_{s+a}\left(1 - \frac{T_0}{T}\right)\left(1 - \frac{\varphi + \dot{Q}_{\text{conv}}}{\varphi_{s,\text{abs}} + \varphi_{0,\text{abs}}}\right). \tag{2.94}$$

For simplicity, in what follows we report on the efficiency of converting the *absorbed thermal energy*, defined as

$$\eta'_{\text{sol}} = \frac{\eta_{\text{sol}}}{A_s}, \quad \eta'_{\text{sol+amb}} = \frac{\eta_{\text{sol+amb}}}{A_{s+a}}. \tag{2.95}$$

The flux φ_s of unconcentrated solar beam radiation out of the atmosphere is given by (2.63). We define the following ratios

$$r_{\text{dir}} \equiv \frac{\varphi_{\text{dir}}}{\varphi_s}, \quad r_{\text{dif}} \equiv \frac{\varphi_{\text{dif}}}{\varphi_s}, \tag{2.96}$$

where φ_{dir} and φ_{dif} are the fluxes of direct and diffuse solar radiation at the Earth's surface, respectively. The flux $\varphi_{s,\text{abs}}$ consists of both direct ($\varphi_{\text{dir,abs}}$) and diffuse ($\varphi_{\text{dif,abs}}$) absorbed solar radiation

$$\varphi_{s,\text{abs}} = \varphi_{\text{dir,abs}} + \varphi_{\text{dif,abs}}. \tag{2.97}$$

Generally, a solar converter consists of a concentrator (concentration ratio C), a receiver (absorbtance α), and a transparent cover (transmittance τ). Then the above two fluxes are given by

$$\varphi_{\text{dir,abs}} = (\tau\alpha)_{\text{dir}} C r_{\text{dir}} \varphi_s, \quad \varphi_{\text{dif,abs}} = (\tau\alpha)_{\text{dif}} r_{\text{dif}} \varphi_s. \tag{2.98}$$

Here we have taken into account that the product transmittance–absorbtance generally differs for direct and diffuse radiation.

The fluxes $\varphi_{0,\text{abs}}$ and φ are given by

$$\varphi_{0,\text{abs}} = (\tau\alpha)_0 \varphi_0 = (\tau\alpha)_0 \sigma T_{\text{sky}}^4, \quad \varphi = e\sigma T^4, \tag{2.99}$$

where $(\tau\alpha)_0$ is the converter transmittance–absorbtance product for the atmospheric radiation, e is the converter emittance, and T_{sky} is the equivalent black-body temperature of the atmosphere given by [31]

$$T_{\text{sky}} = 0.0552 T_0^{1.5}. \qquad (2.100)$$

As usual, the rate of thermal convective losses is

$$\dot{Q}_{\text{conv}} = U(T - T_0), \qquad (2.101)$$

where U is the converter overall convective heat loss coefficient. We define the following dimensionless ratios:

$$\vartheta \equiv \frac{T}{T_s}, \quad \vartheta_0 \equiv \frac{T_0}{T_s}, \quad \vartheta_{\text{sky}} \equiv \frac{T_{\text{sky}}}{T_s} \qquad (2.102)$$

By using (2.92)–(2.95) and (2.97)–(2.102) we obtain

$$\eta'_{\text{sol}} = \left(1 - \frac{\vartheta_0}{\vartheta}\right)\left[1 - \frac{\frac{1}{4}\vartheta^4 + a(\vartheta - \vartheta_0) - b\vartheta_{\text{sky}}^4}{c}\right], \qquad (2.103)$$

$$\eta'_{\text{sol+amb}} = \left(1 - \frac{\vartheta_0}{\vartheta}\right)\left[1 - \frac{\frac{1}{4}\vartheta^4 + a(\vartheta - \vartheta_0)}{c + b\vartheta_{\text{sky}}^4}\right], \qquad (2.104)$$

where

$$a \equiv \frac{U}{4e\sigma T_s^3}, \quad b \equiv \frac{(\tau\alpha)_0}{4e}, \quad c \equiv [r_{\text{dir}}C(\tau\alpha)_{\text{dir}} + r_{\text{dif}}(\tau\alpha)_{\text{dif}}]\frac{1 - \cos 2\delta}{8e}. \qquad (2.105)$$

To determine the maximum values of η'_{sol} and $\eta'_{\text{sol+amb}}$ we apply the usual conditions

$$\frac{\partial \eta'_{\text{sol}}}{\partial \vartheta} = 0, \quad \frac{\partial \eta'_{\text{sol+amb}}}{\partial \vartheta} = 0. \qquad (2.106)$$

Simple calculations show that both η'_{sol} and $\eta'_{\text{sol+amb}}$ are maximized at the same temperature T_{opt}, obtained by solving the following equation for $\vartheta_{\text{opt}} \equiv T_{\text{opt}}/T_s$:

$$\vartheta_{\text{opt}}^5 + A_1 \vartheta_{\text{opt}}^4 + A_2 \vartheta_{\text{opt}}^2 + A_3 = 0, \qquad (2.107)$$

where

$$A_1 \equiv -3/4\vartheta_0, \quad A_2 \equiv a, \quad A_3 \equiv -b\vartheta_0\vartheta_{\text{sky}}^4 - a\vartheta_0^2 - c\vartheta_0. \qquad (2.108)$$

Figure 2.2 shows the values of η'_{sol} and $\eta'_{\text{sol+amb}}$ for some values of the parameters a, b, c, and T_0. The parameter c (related to the concentration ratio) has the most important influence on converter performance. Increasing c decreases both the influences induced by the other parameters and the differences between η'_{sol} and $\eta'_{\text{sol+amb}}$.

34   2. Thermodynamics of Solar Energy Conversion into Work

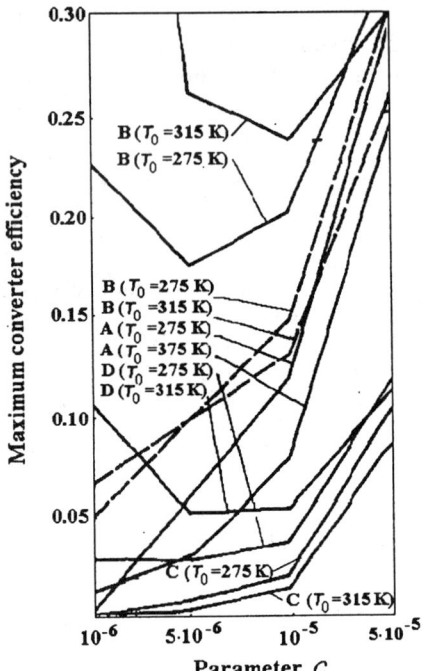

FIGURE 2.2. The dependence of η'_{sol} and $\eta'_{\text{sol+amb}}$ on the parameter c for some values of a, b, and ambient temperature T_0. (A) $a = 0.000092$, $b = 0.05$; (B) $a = 0.000092$, $b = 1$; (C) $a = 0.00184$, $b = 0.05$; (D) $a = 0.00184$, $b = 1$.

Solar Converter Coupled to a Stirling Engine

In this subsection we shall evaluate the performances of a solar converter in combination with a Stirling heat engine by taking into consideration both the convective losses and the ambient (longwave) radiation flux incident on the converter. This subject was approached by a number of authors within different theoretical frameworks [29], [32]–[35]. Here we follow [36].

A model similar to that of the preceding section will be used. However, the absorbed flux of solar radiation is more simply described here as

$$\varphi_{s,\text{abs}} = (\tau\alpha)_{\text{dif}} G_{\text{dif}} + (\tau\alpha)_{\text{dir}} C G_{\text{dir}}, \qquad (2.109)$$

where G_{dir} and G_{dif} are the incident energy fluxes of direct and diffuse solar radiation. The useful power P produced in the cycle can be found by multiplying $\dot{Q}'$ from (2.88) by the thermodynamic efficiency of the Stirling cycle

$$\eta_{\text{cycle}} = \frac{1 - T_0/T}{1 + D(1 - T_0/T)}, \quad D \equiv \frac{xc_v}{R\ln(v_1/v_2)}, \qquad (2.110)$$

where x is the fractional deviation from ideal regeneration, R is the gas constant, c_v is the heat capacity at constant volume of the working fluid, and v_1 and v_2 are the specific volumes of the constant volume regeneration process of the cycle (v_1/v_2 is the overall compression ratio). In (2.110), $x = 0$ means ideal regeneration and η_{cycle} reduces to the Carnot engine efficiency.

The notation (2.102) allows us to write the condition verified by the useful maximum power P:

$$\frac{\partial P}{\partial \vartheta} = \frac{d}{d\vartheta}(\dot{Q}' \cdot \eta_{\text{cycle}}) = 0. \tag{2.111}$$

By replacing (2.88), (2.99)–(2.102), and (2.109)–(2.110) into (2.111) we obtain

$$\vartheta^5 - \frac{8(D+3)}{4(D+1)}\vartheta^4 + \frac{D}{D+1}\vartheta^3 + \frac{b_1}{4a_1}\vartheta^2 - \frac{b_1}{2a_1}\frac{D}{D+1}\vartheta - \frac{1+c_1-b_1(D-1)}{4a_1(D+1)} = 0, \tag{2.112}$$

where

$$a_1 \equiv \frac{\varepsilon \sigma T_0^4}{\varphi_{s,\text{abs}}}, \quad b_1 \equiv \frac{UT_0}{\varphi_{s,\text{abs}}}, \quad c_1 \equiv \frac{(\tau\alpha)_0 \sigma T_{\text{sky}}^4}{\varphi_{s,\text{abs}}}. \tag{2.113}$$

By solving (2.112) we obtain a value (say ϑ_{opt}) which gives the optimum converter temperature via $T_{\text{opt}} = \vartheta_{\text{opt}} T_0$. The maximum power P_{max} and the maximum efficiency $\eta_{\text{sol,max}}$ of the solar converter–heat engine combination can simply be computed by

$$P_{\text{max}} = (\dot{Q}' \cdot \eta_{\text{cycle}})\big|_{T=T_{\text{opt}}}, \quad \eta_{\text{sol,max}} = \frac{P_{\text{max}}}{G}, \tag{2.114}$$

where the incident energy flux of (global) solar radiation G is given by $G = G_{\text{dir}} + G_{\text{dif}}$.

G is a quantity usually measured by most actinometric stations throughout the world. A relationship between the average values of G_{dif} and G is given by Page's formula

$$\frac{G_{\text{dif}}}{G} = 1.0 - 1.07 \frac{G}{G^0}, \tag{2.115}$$

where G^0 is the solar constant (1367 W/m^2).

The influence of design and climatological parameters on both the optimum ratio ϑ_{opt} and the maximum efficiency $\eta_{\text{sol,max}}$ was considered in [36]. The following design parameters were considered: concentration ratio C, receiver transmittance–absorptance product $(\tau\alpha)_{\text{dir}}$, overall heat loss coefficient U, and receiver emissivity e. Here only the dependence of the optimum ratio ϑ_{opt} on the solar global irradiance G is shown (Figure 2.3). A selective solar receiver (emissivity $e = 0.1$) and both a flat-plate solar

36   2. Thermodynamics of Solar Energy Conversion into Work

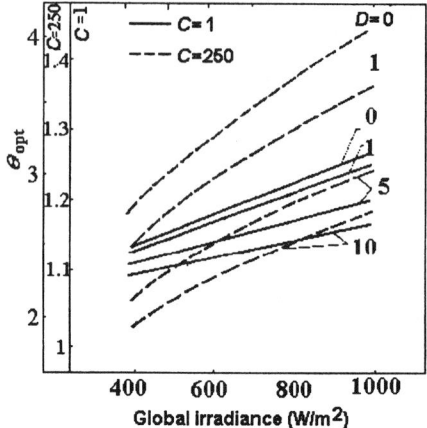

FIGURE 2.3. The dependence of the optimum ratio $\vartheta_{\rm opt} = T_{\rm opt}/T_0$ on the global irradiance G for some values of the concentration ratio C and parameter D [(2.110)]. The case of a selective converter ($e = 0.1$) was considered. The other parameters are: $U = 4$ W m^{-2} K^{-1}; $(\tau\alpha)_{\rm dir} = 0.8$; $(\tau\alpha)_{\rm dif} = 0.5$; $(\tau\alpha)_0 = 0.5$; and $T_0 = 293$ K.

collector (concentration ratio $C = 1$) and a concentrator ($C = 250$) are considered. In the first case, $\vartheta_{\rm opt}$ has a linear dependence on G. The dependence of $\vartheta_{\rm opt}$ on G is more pronounced and slightly nonlinear for solar concentrators.

Other conclusions from [36] are as follows. The maximum efficiency $\eta_{\rm sol,max}$ of the solar converter–heat engine combination has a strong dependence on the concentration ratio. Generally, flat-plate collectors are not recommended for power generation. A relatively small increase of $\eta_{\rm sol,max}$ requires a considerable augmentation of the concentration ratio and, consequently, of the receiver temperature. Performance improvements can be obtained by using selective solar receivers. During sunny days, the solar converter–heat engine combination is characterized not only by increased power output but also by an improvement in efficiency. The dependence of $\eta_{\rm sol,max}$ on the ambient temperature has a different shape for solar converters with or without concentration. In the first case, the maximum efficiency decreases at higher ambient temperatures.

2.4 Space Applications

The usefulness and life of a satellite or space vehicle is directly interrelated with the amount of energy available from the vehicle auxiliary power system. The use of systems based on nuclear or chemical energy sources implies limited life and, sometimes, excessive weight. Solar power systems that collect and convert solar radiation into electrical power appear to be attractive from many points of view. Presently, there are two major types of solar power systems. The first type uses photovoltaic cells to directly convert the solar radiation into electrical energy, in combination with electrochemical storage. The second one is the so-called "thermodynamic" system. Solar concentrators reflect the flux of solar radiation toward a collector where a working fluid is heated to drive conventional thermal engines. Electrical energy could be generated by alternators coupled to these thermal engines. This last solar power system will be considered in this section.

A number of studies can be found in [37]–[39]. A simple model of a solar space power station based on thermodynamic cycles was proposed by Bejan [7]. This model was developed in [40] and [41]. A review will be presented here.

2.4.1 Solar Space Power System Model

The main components of a solar space thermal power plant are the solar concentrator, the collector (receiver), the thermal engine, and the heat radiator (Figure 2.4). The other components of the power station, such as the living space, the energy accumulator, etc., are not relevant at this stage of our analysis. However, their weight will be taken into consideration subsequently. The solar concentrator has an opening area A_open. Its mirror has the optical reflectance ρ_m, the surface A_m, and a superficial mass density d_m. The collector area is denoted A_c. Its temperature, emittance, and absorptance, all of them supposed to be uniform, are denoted T_c, e_c, and a_c, respectively. Also, the radiator supperficial mass density is d_r.

There is general agreement that restrictions have to be imposed on the total mass of a solar space power station. We denote by M_conc, M_r, and M_tot the masses of the concentrator, radiator, and the total mass of the station, respectively. The following relation applies:

$$M_\text{conc} + M_r = \kappa M_\text{tot} \tag{2.116}$$

with $0 < \kappa < 1$.

The concentrator and radiator masses are given by

$$M_\text{conc} = A_m d_m, \quad M_r = A_r d_r, \tag{2.117}$$

while the geometric concentration ratio C_n is usually defined as

$$C_n = \frac{A_\text{open}}{A_c}. \tag{2.118}$$

By replacing (2.117)–(2.118) in (2.116) we obtain:

$$C_n f A_c + s A_r = A_e, \tag{2.119}$$

38   2. Thermodynamics of Solar Energy Conversion into Work

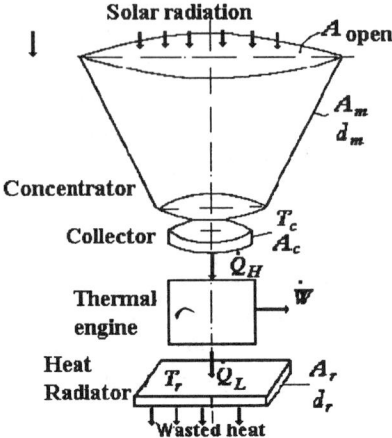

FIGURE 2.4. The main elements of a solar space power station.

where the following notations were used:

$$f \equiv \frac{A_m}{A_{\text{open}}}, \quad s \equiv \frac{d_r}{d_m}, \quad A \equiv \kappa \frac{M_{\text{tot}}}{d_m}. \qquad (2.120)$$

The shape factor f is a characteristic of the concentrating mirror. For the same value of the product $A_c C_n$ the mirrors with a smaller value of f are preferred because they are lighter.

TABLE 2.2. The shape function $f(k)$ for five mirrors usual in solar space applications. Here $k = h/r$, where h and r are the depth and characteristic entrance size of the mirror.

Mirror	Function $f(k)$
Conical	$(1+k^2)^{1/2}$
Sphere	$1+k^2$
Circular cylinder	$\dfrac{1}{2}\dfrac{k^2+1}{k}\arg\tan\dfrac{2k}{1-k^2}$
Parabolic cylinder	$\dfrac{1}{2}(1+4k^2)^{1/2} + \dfrac{1}{4k}\ln[2k+(1+4k^2)^{1/2}]$
Paraboloid	$\dfrac{2}{3}\left\{\left[(2k)^{2/3}+(\dfrac{1}{2k})^{4/3}\right]^{3/2} - \dfrac{1}{4k^2}\right\}$

The shape factor f can be generally expressed as a function of a parameter k defined as the ratio between a characteristic entrance size r and the depth h of the mirror (for more details, see [40]). The results obtained in the case of five mirrors usual in space power applications are shown in Table 2.2. The best results are provided by the conical and parabolic cylinder mirrors while the spherical concentrators could be recommended only at small values of the parameter k. The paraboloidal mirrors are not characterized by small values of f. Consequently, their use is not always justified for those concentration ratios which are accessible to the conical or parabolic cylinder mirrors ($C < 250$).

2.4.2 Classical Thermodynamic Model

First we assume the thermal engine is of Carnot type. In the next section the case of an endoreversible thermal engine will be considered. The energy balance equations for collector and radiator, respectively, are

$$\dot{Q}_H = A_c(\varphi_{s,\text{abs}} + \varphi_{0,\text{abs}} - \varphi_c), \quad \dot{Q}_L = A_r(\varphi_r - \varphi_{0,\text{abs}}). \quad (2.121)$$

Here $\dot{Q}_H$ and $\dot{Q}_L$ are the heat fluxes entering and leaving the thermal engine, respectively. Also, $\varphi_{s,\text{abs}}$ and φ_c are the energy fluxes of the radiation absorbed and emitted by the collector, respectively, while φ_r and $\varphi_{0,\text{abs}}$ are the energy fluxes of the radiation emitted by the radiator and absorbed (by collector or radiator) from the ambient, respectively.

The "Carnot" hypothesis is often used in the case of satellite power systems, where the best technologies are usually accessible (see [39]). Then two new equations can be written

$$\dot{W} = \dot{Q}_H\left(1 - \frac{T_r}{T_c}\right), \quad \frac{\dot{Q}_H}{T_c} + \frac{\dot{Q}_L}{T_r} = 0. \quad (2.122)$$

The former equation (2.122) gives the power supply while the latter is a consequence of the reversible thermal engine we used.

Some details follow. The energy flux of the concentrated radiation absorbed by the collector is given by

$$\varphi_{s,\text{abs}} = \rho_m a_c \frac{B(\Omega_c)}{\pi} \sigma T_s^4, \quad (2.123)$$

where a_c is the collector absorbtance while the geometric factor $B(\Omega_c) \equiv \pi A(\Omega_c)$ is a function of the enlarged solid angle Ω_c subtended by the Sun when viewed from the concentrator mirror. The quantity Ω_c is related to the geometrical concentration ratio C through

$$C = \frac{\Omega_c(4\pi - \Omega_c)}{\Omega_s(4\pi - \Omega_s)}. \quad (2.124)$$

40   2. Thermodynamics of Solar Energy Conversion into Work

The energy fluxes emitted by the collector and radiator are given by

$$\varphi_c = e_c \sigma T_c^4, \quad \varphi_r = e_r \sigma T_r^4, \tag{2.125}$$

where e_r is radiator emittance.

The ambient energy flux $\varphi_{0,\text{abs}}$ incident on both collector and radiator generally depends on the temperature and/or albedo of the neighboring celestial bodies. When the case of an interplanetary station is considered, this flux is mainly due to the ambient temperature (≈ 4.2 K). Consequently, it could be neglected. Moreover, the hypotheses $\varphi_{0,\text{abs}} \ll \varphi_{s,\text{abs}}$ and $\varphi_{0,\text{abs}} \ll \varphi_r$ are very good approximations in all practical cases.

The following notation will be used:

$$x = \frac{a_c \rho_m}{e_c} \frac{B(C_n)}{\pi} \left(\frac{T_s}{T_c}\right)^4. \tag{2.126}$$

By using (2.121) and (2.122)–(2.126) we obtain

$$\frac{T_r}{T_c} = \left(\frac{e_c A_c}{e_r A_r}\right)^{1/3} (x-1)^{1/3}. \tag{2.127}$$

Now the ratio T_r/T_c can be replaced in (2.122) and the power $\dot{W}$ becomes a function of the parameter x (i.e., of the temperature T_c). The overall thermal efficiency of the power station is defined as

$$\eta = \frac{\dot{W}}{A_c [B(C_n)/\pi] \sigma T_s^4}. \tag{2.128}$$

The maximum efficiency $\eta_{\max}$ is given by

$$\eta_{\max} = a_c \rho_m \frac{x_{\text{opt}} - 1}{x_{\text{opt}} + 3}, \tag{2.129}$$

where x_{opt} is the solution of the equation $d\eta/dx = 0$, i.e.,

$$p^3 (x_{\text{opt}} - 1)\left(1 + \frac{x_{\text{opt}}}{3}\right)^3 = 1, \quad p = \left(\frac{e_c A_c}{e_r A_r}\right)^{1/3}. \tag{2.130}$$

It is easy to see that $\eta_{\max}$ is a function of A_c if the restriction (2.119) is taken into account. Simple computations show that a second maximization of $\eta_{\max}$ with respect to A_c has no solution. However, if the maximum power $\dot{W}_{\max}$ ($= [B(C_n)/\pi] \sigma T_s^4 A_c \eta_{\max}$) is considered, such a second maximization is possible. With this aim we have to solve the equation $d\dot{W}_{\max}/dA_c = 0$ by taking into account the restriction (2.119). The optimum optimorum value of x is obtained from

$$(x_{\text{opt,opt}} - 1)^2 \left(1 + \frac{x_{\text{opt,opt}}}{3}\right)^3 = t \frac{e_r}{e_c}, \quad t \equiv \frac{C_n f}{s}. \tag{2.131}$$

The optimum values of the ratios A_r/A_c and A_e/A_c are given by

$$\left(\frac{A_r}{A_c}\right)_{\text{opt}} = \frac{t}{x_{\text{opt,opt}} - 1}, \quad \left(\frac{A_e}{A_c}\right)_{\text{opt}} = st \frac{x_{\text{opt,opt}}}{x_{\text{opt,opt}} - 1}. \quad (2.132)$$

The nondimensional maximum maximorum power $\dot{W}'_{\text{max,max}}$ and the maximum efficiency $\eta_{\text{max,max}}$ will be defined as

$$\dot{W}'_{\text{max,max}} \equiv \frac{\dot{W}_{\text{max,max}}}{a_c \rho_m A_{c,\text{opt}} [B(C_n)/\pi] \sigma T_s^4} = \frac{\eta_{\text{max,max}}}{a_c \rho_m} \equiv \frac{x_{\text{opt,opt}} - 1}{x_{\text{opt,opt}} + 3}, \quad (2.133)$$

where $A_{c,\text{opt}}$ can be evaluated with (2.132).

Finally, a usual performance indicator of space power stations could be obtained by using (2.120), (2.132), and (2.133) and taking into account that $B(C_n) = C_n B(C_n = 1)$:

$$\frac{M_{\text{tot}}}{\dot{W}_{\text{max,max}}} = \frac{d_m f}{\kappa a_c \rho_m} \frac{1}{[B(C_n = 1)/\pi] \sigma T_s^4} \frac{x_{\text{opt,opt}}(x_{\text{opt,opt}} + 3)}{(x_{\text{opt,opt}} - 1)^2}. \quad (2.134)$$

The value of this indicator may be used as criteria to choose among different design solutions.

2.4.3 Finite-Time Thermodynamics Model

Now we suppose the thermal engine is of Novikov–Curzon–Ahlborn (NCA) type (sometimes also called a Chambadal–Novikov–Curzon–Ahlborn engine or CNCA engine). This kind of engine is a particular case of the so-called "endoreversible engines" [43], which are supposed to consist of three parts:

(a) A reversible part (i.e., a Carnot part) working between two heat reservoirs (one at a high temperature, say t_1, and one at a low temperature, say t_2; usually, t_1 and t_2 are the temperatures of the working fluid during its isothermal expansion and compression, respectively).

(b) Two irreversible parts containing temperature drops (i.e., the temperature fall $T_c - t_1$ accompanying $\dot{Q}_H$ and the temperature fall $t_2 - T_r$ accompanying $\dot{Q}_L$).

The NCA engine is a special case of an endoreversible engine characterized by linear relationships between the heat flows and the temperature gradients (i.e., $\dot{Q}_H \sim (T_c - t_1)$ and $\dot{Q}_L \sim (t_2 - T_r)$). Details of the endoreversible and NCA engines can be found in the articles of Rubin [43] and Curzon and Ahlborn [44], respectively, or in the papers of De Vos [45], [46]. Under the above assumptions two new equations can be written [45]:

$$\dot{W} = \dot{Q}_H \left[1 - \left(\frac{T_r}{T_c}\right)^{1/2}\right], \quad \frac{\dot{Q}_H}{T_c^{1/2}} + \frac{\dot{Q}_L}{T_r^{1/2}} = 0. \quad (2.135)$$

TABLE 2.3. Results obtained in the case of a finite-time thermodynamic model. Numbers in parentheses denote the corresponding equations in Section 2.4.2.

$$x = \frac{a_c \rho_m}{e_c} \frac{B(C_n)}{\pi} \left(\frac{T_s}{T_c}\right)^4 \tag{2.126}$$

$$\frac{T_r}{T_c} = \left(\frac{e_c A_c}{e_r A_r}\right)^{2/7} (x-1)^{2/7} \tag{2.127}$$

$$\eta_{\max} = a_c \rho_m \frac{x_{\text{opt}} - 1}{x_{\text{opt}} + 7} \tag{2.129}$$

$$p^7 (x_{\text{opt}} - 1) \left(1 + \frac{x_{\text{opt}}}{7}\right)^7 = 1 \tag{2.130}$$

$$(x_{\text{opt,opt}} - 1)^2 \left(1 + \frac{x_{\text{opt,opt}}}{7}\right)^7 = t \frac{e_r}{e_c} \tag{2.131}$$

$$\left(\frac{A_r}{A_c}\right)_{\text{opt}} = \frac{t}{x_{\text{opt,opt}} - 1} \tag{2.132}$$

$$\left(\frac{A_e}{A_c}\right)_{\text{opt}} = st \frac{x_{\text{opt,opt}}}{x_{\text{opt,opt}} - 1} \tag{2.132}$$

$$\dot{W}'_{\max,\max} \equiv \frac{x_{\text{opt,opt}} - 1}{x_{\text{opt,opt}} + 7} \tag{2.133}$$

$$\frac{M_{\text{tot}}}{\dot{W}_{\max,\max}} = \frac{d_m f}{\kappa a_c \rho_m} \frac{1}{[B(C_n = 1)/\pi] \sigma T_s^4} \frac{x_{\text{opt,opt}}(x_{\text{opt,opt}} + 7)}{(x_{\text{opt,opt}} - 1)^2} \tag{2.134}$$

The former equation (2.135) gives the power supply while the latter is a consequence of the endoreversible thermal engine we used. After some computations similar to those given in Section 2.4.2 the results shown in Table 2.3 are obtained.

The theory was applied in the case of a space power station equipped with cylindrical parabolical concentrators. The station is situated at the mean distance Earth–Sun. The shape factor f has had two values, namely 1.03 and 1.47, which correspond to a ratio between the entrance size of the mirror and the mirror depth of 0.2 and 1.0, respectively. Two different values of the ratio $s = d_r/d_m$ were used, namely 1.5 and 10. They cover the usual range of the parameters d_r and d_m, which is 0.01...0.16 kg/m² [47]–[49] and 2.3...5 kg/m² [50], [51], respectively. Figure 2.5 shows the dependence of the maximum efficiency on the concentration ratio for the two sets of values of the factors f and s. The performance increases by increasing the concentration ratio, as expected. The mirrors with a more equilibrated ratio between the entrance size and depth give better results. Also, values of d_m and d_r close to each other are recommended.

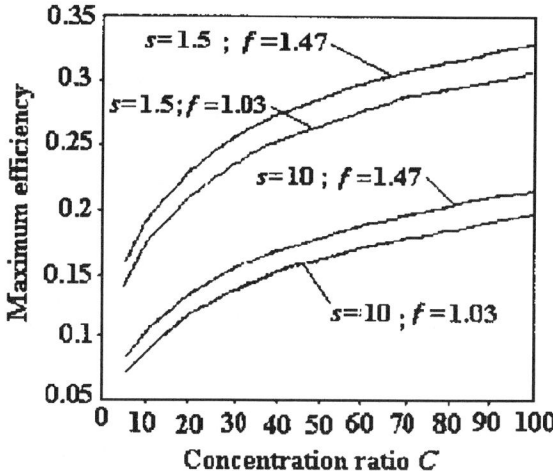

FIGURE 2.5. The dependence of the maximum efficiency $\eta_{max,max}$ on the concentration ratio C for two sets of values of the factors f and s (finite-time thermodynamics approach). The following numerical values were accepted: $a_c = 1$, $e_r = 0.8$, $e_c = 0.9$, and $\rho_m = 0.9$.

2.5 Further Research and Studies

There is now quite a literature on the conversion of solar energy into work. The case of heat engines when the primary mode of heat transfer between the engine and either or both of its heat reservoirs is radiative was studied in [52]. The energetically optimal operating temperature was found to be relatively insensitive to the engine design points [53]. Different optimization techniques for the solar collector–heat engine combination are presented in [54]–[59]. Time-dependent models were developed in [60] and [61] while nonisothermal collectors, providing heat to engines, were analyzed in [62]. Finally, conversion of solar energy into wind power is sometimes modeled by using concepts from heat engine theory [63]–[65]. A good review of the entropy generation minimization and other techniques used in the models of solar power plants is given in [66].

2.6 References

[1] R. Petela: Exergy of heat radiation, *J. Heat Transfer* **86**, 187–192, 1964.

[2] W. H. Press: Theoretical maximum for energy from direct and diffuse sunlight, *Nature* **264**, 734–735, 1976.

[3] P. T. Landsberg and J. R. Mallinson: Thermodynamic constraints, effective temperatures and solar cells, *Coll. Internat. sur l'Electricité Solaire*, Toulouse, CNES, pp. 27–35, 1976.

[4] D. C. Spanner: *Introduction to thermodynamics*, Academic Press, London, 1964, p. 218.

[5] S. M. Jeter: Maximum conversion efficiency for the utilization of direct solar radiation, *Sol. Energy* **26**, 231–236, 1981.

[6] A. Bejan: Unification of three different theories concerning the ideal conversion of enclosed radiation, *J. Sol. Energy Engng.* **109**, 46–51, 1987.

[7] A. Bejan: *Advanced engineering thermodynamics*, Wiley, New York, 1988.

[8] P. T. Landsberg and G. Tonge: Thermodynamics of the conversion of diluted radiation, *J. Phys. A* **12**, 551–562, 1979.

[9] J. A. Gribik and J. F. Osterle: The second law efficiency of solar energy conversion, *J. Sol. Energy Engng.* **106**, 16–21, 1984.

[10] A. De Vos and H. Pauwels: Letter to the editor, *J. Sol. Energy Engng.* **108**, 80–84, 1986.

[11] J. A. Gribik and J. F. Osterle: Authors' closure, *J. Sol. Energy Engng.* **108**, 83–84, 1986.

[12] M. Castañs, A. Soler and F. Soriano: Theoretical maximal efficiency of diffuse radiation, *Sol. Energy* **38**, 267, 1987.

[13] V. Bădescu: L'exergie de la radiation solaire directe et diffuse sur la surface de la Terre, *Entropie* **145**, 41–45, 1988.

[14] V. V. Meskhov: *Fundamentals of illumination engineering*, MIR, Moscow, 1981.

[15] L. Landau and E. Lifchitz: *Physique statistique*, Editions MIR, Moscow, 1967.

[16] L. Landau and E. Lifchitz: *Théorie des champs*, Editions MIR, Moscow, 1970.

[17] C. J. Adkins: *Equilibrium thermodynamics*, 2nd ed., McGraw Hill, New York, 1975.

[18] V. Bădescu: On the thermodynamics of the conversion of diluted radiation, *J. Phys. D: Appl. Phys.* **23**, 289–292, 1990.

2.6. References

[19] V. Bădescu: On the thermodynamics of the conversion of diluted and undiluted black-body radiation, *Space Power* **9**, 317–322, 1990.

[20] C. J. Hoogendorn: Optical properties of selective layers, in: (eds. H. Yuncu and B. Kilkis): *Solar energy utilisation: fundamentals and applications*, EIEI Printing Shop, Ankara, 1986.

[21] A. Haught: Physics considerations of solar energy conversion, *J. Sol. Energy Engng.* **106**, 3, 1984.

[22] V. Bădescu: The theoretical maximum efficiency of solar converters with and without concentration, *Internat. J. Energy* **14**, 237–239, 1989.

[23] V. Bădescu: Maximum conversion efficiency for the utilization of diffuse radiation, *Internat. J. Energy* **16**, 783–786, 1991.

[24] V. Bădescu: Maximum conversion efficiency for the utilization of multiply scattered solar radiation, *J. Phys. D* **24**, 1882, 1991.

[25] P. T. Landsberg and G. Tonge: Thermodynamic energy conversion efficiency, *J. Appl. Phys.* **51**, R1–R20, 1980.

[26] R. Perez, R. Seals, P. Ineichen, R. Stewart and D. Menicucci: A new simplified version of the Perez diffuse irradiance model for tilted surfaces, *Sol. Energy* **39**, 221–231, 1987.

[27] A. W. Harrison and C. A. Coombes: Angular distribution of clear sky short wavelength radiance, *Sol. Energy* **40**, 57–63, 1988.

[28] W. R. Menetrey : Space applications of solar energy, in: *Introduction to the utilization of solar energy*, (eds. A. M. Zarem and D. D. Erway), McGraw-Hill, New York, 1963.

[29] J. R. Howell and R. B. Bannerot: Optimum solar collector operation for maximizing cycle work output, *Sol. Energy* **19**, 149–153, 1977.

[30] V. Bădescu: Note concerning the maximal efficiency and the optimal operating temperature of solar converters with or without concentration, *Renewable Energy* **1**, 131–135, 1991.

[31] W. C. Swinbank: Long wave radiation from clear skies, *J. Roy. Meteorol. Soc.* **89**, 1963.

[32] M. K. Selcuk: Prediction of performance of paraboloid dish solar-power modules using graphical methods, in: *Solar energy utilisation: fundamentals and applications*, (eds. H. Yuncu and B. Kilkis), EIEI Printing Shop, Ankara, 1986, pp. 543–563.

[33] R. Constantinescu, M. Costea, C. Mladin, T. Brusalis and S. Petrescu: Metodologie de calcul pentru optimizarea temperaturii unui sistem de captare si conversie a energiei solare cu oglinda parabolica tip calota si receptor de radiatie concentrata disc (in Romanian), *Bull. Inst. Polytechnique de Bucharest* **50**, 59, 1988.

[34] S. Petrescu, T. Brusalis, R. Iordache, M. Costea, and V. Petrescu: Receptor de radiatie solara concentrata cu stocaj termic in topituri (in Romanian), *Energetica* **37**, 358, 1989.

[35] M. Costea: Augmentation des performances des échangeurs de chaleur en vue de l'optimisation thermodynamique du moteur de Stirling. Transfer de chaleur en régime instationnaire en milieu poreux, Ph.D. Thesis, Université 'H. Poincaré' de Nancy 1, 1997.

[36] V. Bădescu: Optimum operation of a solar converter in combination with a Stirling or Ericsson heat engine, *Internat. J. Energy* **17**, 601–607, 1992.

[37] K. Eguchi, S. Ogiwara and T. Fujiwara: An experimental Stirling engine for use in space dynamic power systems: preliminary tests, *Space Power* **9**, pp. 131–148, 1990.

[38] V. Bădescu: Optimization of Stirling and Ericsson cycles using solar radiation, *Space Power* **11**, pp. 99–106, 1992.

[39] V. F. Prisnjakov, I. N. Statsenko, A. I. Kondratjev, V. L. Markov, B. E. Petrov and V. A. Gabrinets: Developing space power Brayton system with solar heat input – Research of working process of high temperature latent heat storage system, *Proc. of SPS 91 Symposium*, Paris, 1991, pp. 465–470.

[40] V. Bădescu: Optimization of a solar space power system based on thermodynamic cycles, *Internat. J. Sol. Energy* **16**, 263–275, 1995.

[41] V. Bădescu: Optimum design and operation of a dynamic solar space power system, *Energy Conv. Mgmt.* **37**, 151–160, 1996.

[42] A. R. Martin and V. K. Thompson: The impact of technology advances upon satellite power systems, *Proc. of SPS 91 Symposium*, Paris, 1991, pp. 217–223.

[43] M. Rubin: Optimal configuration of a class of irreversible heat engines, *Phys. Rev.* **19**, pp. 1272–1276, 1979.

[44] F. Curzon and B. Ahlborn: Efficiency of a Carnot engine at maximum power output, *Am. J. Phys.* **43**, pp. 22–24, 1975.

[45] A. De Vos: Efficiency of some heat engines at maximum-power conditions, *Am. J. Phys.* **53**, pp. 570–573, 1985.

[46] A. De Vos: Reflections on the power delivered by endoreversible engines, *J. Phys. D* **20**, pp. 232–236, 1987.

[47] P. Compte, M. Ferronniere and C. Marchal: Application of the concept of inflatable and rigidifiable structures to large space power stations, *Proc. of SPS 91 Symposium*, Paris, 1991, pp. 253–259.

[48] W. R. Menetrey: Space applications of solar energy, in: *Introduction to the utilization of solar energy*, (eds. A. M. Zarem and D. D. Erway), Mc Graw-Hill, New York, 1963, pp. 326.

[49] T. Wallace and R. W. Bussard: A lightweight focusing reflector concept for space power applications, *Proc. of SPS 91 Symposium*, Paris, 1991, pp. 327–331.

[50] Y. A. Mozjorine, V. P. Senkevich, A. D. Koval and E. A. Narimanov: Small-scale space power stations: feasibility and usage prospects, *Proc. of SPS 91 Symposium*, Paris, 1991, pp. 381–392.

[51] E. H. Fay, M. Stavnes and R. C. Cull: Beam power options for the Moon, *Proc. of SPS 91 Symposium*, Paris, 1991, pp. 238–247.

[52] J. M. Gordon: Observations on efficiency of heat engines operating at maximum power, *Amer. J. Phys.* **58**, 370–375, 1990.

[53] J. M. Gordon: On optimized solar-driven heat engines, *Sol. Energy* **40**(5), 457–461, 1988.

[54] R. Borner, M. Feidt and P. Ramany Bala: *Proc. of the Internat. Conf. on Energy Systems and Ecology (ENSEC'93)*, Cracow, 1993, p. 113.

[55] Z. Yan and J. Chen: FLOWERS'94, *Proc. of the Florence World Energy Research Symposium*, (eds. E. Carnevalle, G. Manfrida and F. Martinelli), SGEditoriali, Padova (1994), pp. 1051–1057.

[56] A. De Vos, P. T. Landsberg, P. Baruch and J. E. Parrott: Entropy fluxes, endoreversibility, and solar energy conversion, *J. Appl. Phys.* **74**, 3631–3637, 1993.

[57] A. De Vos and P. van der Wel: Endoreversible models for the conversion of solar energy into wind energy, *J. Non-Equil. Thermodyn.* **17**, 77–89, 1992.

[58] S. Gotkun, S. Ozkaynak and H. Yavuz: Design parameters of a radiative engine, *Energy* **18**, 651–655, 1993.

[59] G. Popescu, V. Radcenco and M. Feidt: Finite time thermodynamics applied to the optimization of solar Stirling engines, *Proc. of Internat. Conf. FLOWERS'94*, Florence, 1994, pp. 13–20.

[60] A. Bejan: Extraction of exergy from solar collectors under time-varying conditions, *Internat. J. Heat Fluid Flow* **3**, 67–72, 1982.

[61] D. E. Chelghoum and A. Bejan: Second law analysis of solar collectors with energy storage capability, *J. Sol. Energy Engng.* **107**, 244–251, 1985.

[62] A. Bejan, D. W. Kearney and F. Kreith: Second law analysis and synthesis of solar collector systems, *J. Sol. Energy Engng.* **103**, 23–30, 1981.

[63] A. De Vos: *Endoreversible thermodynamics of solar energy conversion*, Oxford University Press, Oxford, 1992.

[64] J. M. Gordon and Y. Zarmi: Wind energy as a solar-driven heat engine: a thermodynamic approach, *Amer. J. Phys.* **57**, 995, 1989.

[65] A. De Vos and G. Flater: The maximum efficiency of the conversion of solar energy into wind energy, *Amer. J. Phys.* **59**, 751–754, 1991.

[66] A. Bejan: Entropy generation minimization: the new thermodynamics of finite-size devices and finite-time processes, *J. Appl. Phys.* **79**, 1191–1217, 1996.

3
Thermodynamics of Photovoltaics

A. De Vos

ABSTRACT. A solar cell is a thermodynamic engine working between two heat reservoirs, one at high temperature T_1 (= the temperature of the Sun = 5762 K) and one at low temperature T_2 (= the temperature of the Earth = 288 K). Its electric current consists of two parts: the light current, strongly dependent on T_1, and the dark current, strongly dependent both on T_2 and on material constants and technology parameters.

The maximum power, we can extract from the cell, is found by searching for the maximum rectangle inscribed in the current–voltage characteristic. This requests the solution of a transcendental equation. The numerical result (for a focused solar spectrum and for a "reasonable choice" of the material constants) is in the range 30% to 40%.

The above procedure stands in strong contrast to the Carnot theory of reversible heat engines. From the first and second law of thermodynamics, we immediately find the Carnot formula: $1 - T_2/T_1$, yielding 95%. Two questions arise:

(1) Why does photovoltaic theory, in contrast to Carnot theory, need a search for the maximum rectangle within a characteristic that is dependent on material constants?

(2) Why do photovoltaics yield so much lower efficiencies than the Carnot engine?

The first puzzle is illuminated by application of endoreversible thermodynamics. Novikov proposed a simple heat engine model, where an ideal Carnot engine is combined with a linear heat resistor. This leads to a maximum-power condition where some intermediate temperature T_3 plays the same role as the voltage V in the solar cell. By generalizing the Carnot engine by a reversible thermochemical engine and the finite heat conduction by heat radiation, one can model a solar cell as an endoreversible engine.

The second puzzle is solved by introducing multigap solar cells (or tandem solar cells), i.e., by applying different materials with different bandgaps E_g. This leads (in the limit) to an efficiency of 87%, rather close to the Carnot limit. The fact that we cannot get the Carnot efficiency itself, is caused by the fact that absorption of radiation without simultaneous emission of radiation with the same spectrum, is inevitably an irreversible process.

Endoreversible thermodynamics is not only useful for describing photovoltaics, but also has been successfully applied in numerous other fields, like chemistry, climatology, economics, computing, etc.

3. Thermodynamics of Photovoltaics

3.1 Introduction

A solar cell is a thermodynamic engine working between two heat reservoirs, one at high temperature T_1 and one at low temperature T_2:

$$T_1 = \text{temperature of the Sun} = 5762 \text{ K},$$
$$T_2 = \text{temperature of the Earth} = 288 \text{ K}.$$

Its current–voltage characteristic is given by

$$I(V) = I_l - I_s \left[\exp\left(\frac{qV}{kT_2}\right) - 1 \right], \tag{3.1}$$

- where q and k have their usual meaning: elementary charge and Boltzmann's constant, respectively;
- where the light current I_l is mainly a function of the incident solar spectrum (and thus of the temperature T_1); and
- where the dark saturation current I_s is a complicated expression of material constants (diffusion coefficients, lifetimes,...), material processing (doping levels,...), and the temperature T_2.

See, e.g., [1]. Figure 3.1 displays the $I(V)$ curve. The dark current is usually found by solving the diffusion equation of the electrons and holes within the semiconductor material. In its simplest form, the dark current is given by

$$I_s = q n_i^2 \left(\frac{1}{N_A} \sqrt{\frac{D_n}{\tau_n}} + \frac{1}{N_D} \sqrt{\frac{D_p}{\tau_p}} \right), \tag{3.2}$$

where n_i is the (strongly temperature- and bandgap-dependent) intrinsic charge carrier concentration. Further, N_A and N_D are the acceptor and donor doping concentrations, D_n and D_p are the electron and hole diffusion coefficients, and τ_n and τ_p the electron and hole lifetimes. The light current is given by

$$I_l = q \int_0^\infty Q(E) N(E) \, dE,$$

where E is the energy of the incident photon. Further, the incident solar spectrum is characterized by $N(E)\,dE$, i.e., the number of incident photons with energy in the range $(E, E + dE)$. The factor $Q(E)$ is called the quantum efficiency, i.e., the number of elementary charges q generated by a single incident photon. It is a function of material constants (diffusion coefficients, absorption coefficients,...) and material processing (doping profiles,...). For a "good" choice of materials and an appropriate technology, we are allowed to say that the quantum efficiency Q is an ideal step function:

$$Q(E) = \begin{cases} 0 & \text{for } 0 < E < E_g, \\ 1 & \text{for } E_g < E < \infty. \end{cases}$$

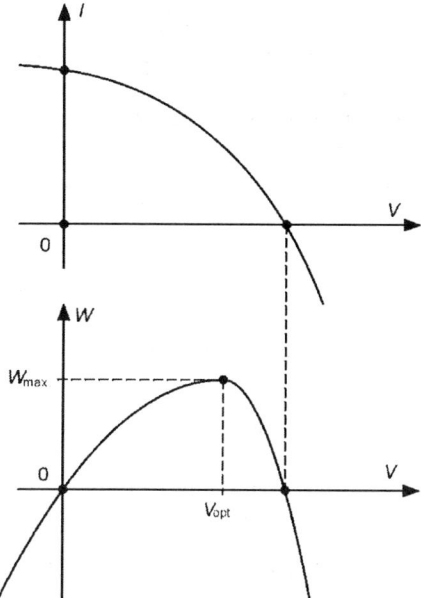

FIGURE 3.1. Current–voltage and power–voltage characteristics of a solar cell.

Here E_g is the semiconductor bandgap. Thus, the light current is

$$I_l = q \int_{E_g/h}^{\infty} N(E)\, dE . \qquad (3.3)$$

Figure 3.1 shows, besides the curve $I(V)$, also the $W(V)$ characteristic. Here W is the power produced by the cell: $W = V \times I$. The maximum power we can extract is given by $W_{\max} = V_{\text{opt}} \times I(V_{\text{opt}})$, where the optimal voltage V_{opt} is the solution of $(d/dV)[V \times I(V)] = 0$, a transcendental equation

$$\left(1 + \frac{qV}{kT_2}\right) \exp\left(\frac{qV}{kT_2}\right) = \frac{I_l}{I_s} + 1.$$

The resulting $W_{\max}$ is still a function of the bandgap E_g of the semiconductor material. For further optimization, we need

$$\frac{d}{dE_g} W_{\max} = 0,$$

eventually leading to the efficiency

$$w = W_{\max_{\max}} / Q_{\text{in}},$$

where Q_{in} is the incident energy:

$$Q_{\text{in}} = \int_0^{\infty} E N(E)\, dE.$$

The numerical result (for a concentrated solar spectrum and "reasonable" material constants) is in the range 30% to 40%, for a bandgap E_g of about 1.2 eV and a biasing voltage V of about 1.1 V.

The above reasoning stands in strong contrast to the Carnot theory. Indeed, let us for a moment assume we can connect a reversible heat engine simultaneously to a heat reservoir at the Sun's temperature and to a heat reservoir at the Earth's temperature: see Figure 3.2. From the first law of thermodynamics (i.e., conservation of energy):

$$W = Q_1 - Q_2,$$

the second law (i.e., conservation of entropy):

$$\frac{Q_1}{T_1} = \frac{Q_2}{T_2},$$

and the definition of efficiency $\eta = W/Q_1$, we immediately find the Carnot formula $\eta = 1 - T_2/T_1$, yielding 95%. Two questions arise here:

- Why does photovoltaic theory, in contrast to Carnot theory, need such complicated double optimization $\partial W/\partial V = 0$ and $\partial W/\partial E_g = 0$?

- Why do photovoltaics yield so much lower efficiencies than the Carnot engine ?

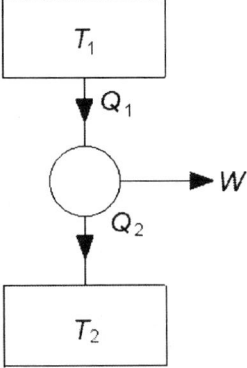

FIGURE 3.2. The Carnot engine.

3.2 Endoreversible Thermal Engines

The first puzzle is illuminated by application of endoreversible thermodynamics [2], [3]. Both Novikov [4] and Curzon and Ahlborn [5] proposed a simple model, where an ideal Carnot engine is combined with a linear heat resistor. The latter has a heat-temperature characteristic

$$Q_1 = g\,(T_1 - T_3). \tag{3.4}$$

See Figure 3.3(a). Here, g is a constant, called the thermal conductance.

Taking into account the Carnot theorem, we have

$$W = \left(1 - \frac{T_2}{T_3}\right) Q_1. \tag{3.5}$$

Thus

$$W = g\left(1 - \frac{T_2}{T_3}\right)(T_1 - T_3). \tag{3.6}$$

Now the crucial step is to consider T_3 in (3.4) and (3.6) as a parameter, we can vary freely. Figure 3.4(a) shows the functions $Q_1(T_3)$ and $W(T_3)$. Note that $Q_1(T_3)$ is a straight line, whereas $W(T_3)$ is a hyperbola. We will choose T_3 such that we obtain maximum power output

$$\frac{dW}{dT_3} = 0.$$

This leads to the maximum-power condition

$$T_3^2 - T_1 T_2 = 0,$$

a quadratic equation in T_3. We find immediately

$$(T_3)_{\mathrm{opt}} = \sqrt{T_1 T_2}.$$

Substitution subsequently into (3.5) and (3.4) immediately yields

$$\begin{aligned} W_{\max} &= \left(1 - \sqrt{\frac{T_2}{T_1}}\right) Q_1((T_3)_{\mathrm{opt}}) \\ &= g(\sqrt{T_1} - \sqrt{T_2})^2. \end{aligned}$$

The factor $1 - \sqrt{T_2/T_1}$ is known as the Novikov–Curzon–Ahlborn efficiency. One can easily demonstrate that (for any T_1 and T_2 obeying $0 < T_2 < T_1$) this number is smaller than the Carnot efficiency $1 - T_2/T_1$. This property is in agreement with the fact that part of the engine is irreversible, and with Carnot's law which states that any irreversible engine is less efficient than the reversible engine working between the same two temperatures.

We now relax condition (3.4). This means we do not restrict ourselves anymore to linear heat transport laws. We assume that the heat flux is given by

$$Q_1 = f(T_1) - f(T_3), \tag{3.7}$$

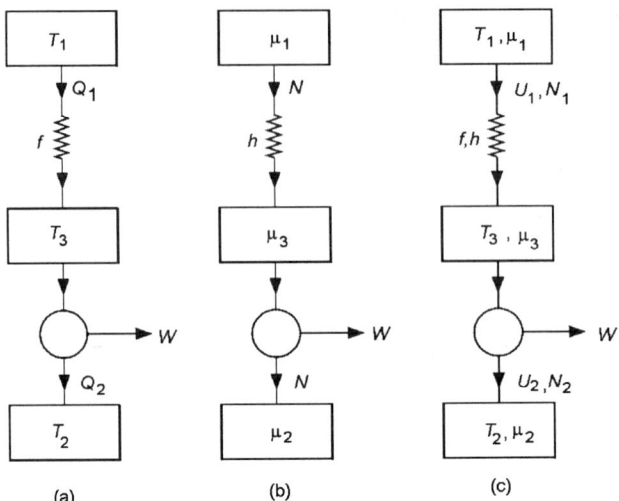

FIGURE 3.3. Endoreversible engines: (a) a thermal engine, (b) a chemical engine, and (c) a thermochemical engine.

with $f(x)$ some mathematical function, not yet specified. This expression automatically guarantees that no heat is exchanged if no temperature difference exists (i.e., $Q_1 = 0$ whenever $T_1 = T_3$). We assume $f(x)$ monotonously increasing (in order to guarantee that heat always flows from high to low temperature). The special case $f(x) = gx$, of course, recovers the Novikov–Curzon–Ahlborn case.

If we assume that heat is exchanged not by conduction but by radiation, we have to apply the appropriate laws. The simplest model is the case of black bodies. The heat radiated by a black body is proportional to the fourth power of its temperature, according to the Stefan–Boltzmann law. We therefore apply $f(x) = gx^4$, such that the net exchanged heat is

$$Q_1 = g(T_1^4 - T_3^4).$$

Thus the characteristic $Q_1(T_3)$ is a fourth-power curve, replacing the straight line of Figure 3.4(a). The corresponding $W(T_3)$ characteristic is a fifth-power curve:

$$W = g\left(1 - \frac{T_2}{T_3}\right)(T_1^4 - T_3^4).$$

The corresponding endoreversible model is the Müser model [6]. Maximum power is extracted if we satisfy

$$\frac{d}{dT_3}\left[\left(1 - \frac{T_2}{T_3}\right)(T_1^4 - T_3^4)\right] = 0,$$

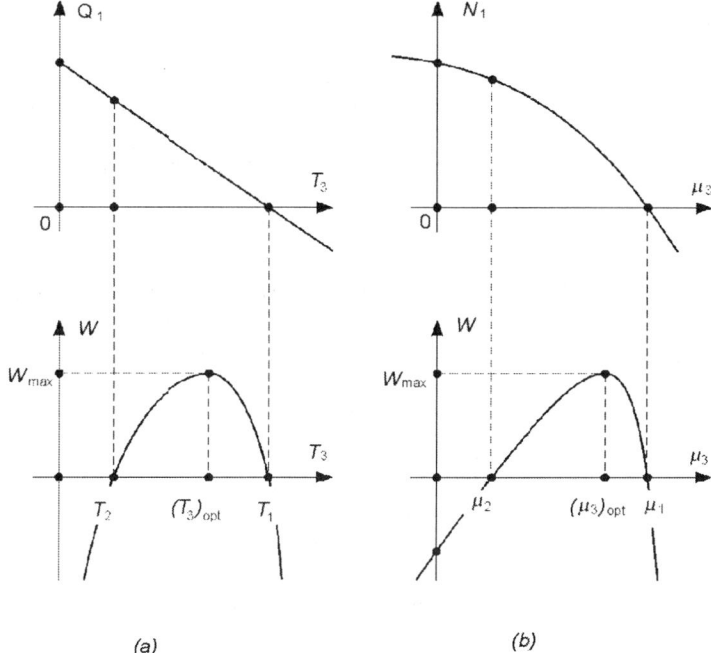

(a)   (b)

FIGURE 3.4. Characteristics of endoreversible engines: (a) heat-temperature and work-temperature characteristics of a thermal engine, and (b) particle-potential and work-potential characteristics of a chemical engine.

leading to the fifth-degree equation

$$4T_3^5 - 3T_2T_3^4 - T_1^4T_2 = 0.$$

Such a polynomial equation cannot be solved analytically. We thus have to recur to numerical calculations. For $T_1 = 5762$ K and $T_2 = 288$ K, we find $T_3 = 2443$ K. If we calculate the efficiency $\eta = W/Q_1 = 1 - T_2/T_3$, we find 88.2%. However, in solar energy conversion, the tradition exists to define the efficiency as the ratio of the produced power to the incoming solar radiation alone:

$$w = \frac{W}{Q_{in}}.$$

Whereas the denominator Q_1 of η is $g(T_1^4 - T_3^4)$, here the denominator of w is merely gT_1^4, i.e., a constant. Therefore w is like a dimensionless W. This solar conversion efficiency w amounts to

$$\frac{(1 - T_2/T_3)\, g(T_1^4 - T_3^4)}{gT_1^4} = \left(1 - \frac{T_2}{T_3}\right)\left(1 - \frac{T_3^4}{T_1^4}\right).$$

Substitution of the T_1, T_2, and $(T_3)_{opt}$ values leads to an efficiency of 85.4%.

56   3. Thermodynamics of Photovoltaics

We now go one step further: instead of black-body radiation, we assume selective black-body radiation:

$$f(T) = \int_{E_g/h}^{\infty} EN(E,T)dE,$$

where the spectral density $N(E,T)$ is assumed to obey Planck's law:

$$N(E,T) = \frac{2\pi}{c^2 h^3} \frac{E^2}{\exp(E/kT) - 1}, \quad (3.8)$$

with c the speed of light and h Planck's constant. This constitutes a generalization of the black-body case, as the Stefan–Boltzmann law corresponds to the special case $E_g = 0$. Indeed, because of

$$\int_0^{\infty} \frac{x^3 dx}{\exp(x) - 1} = \frac{\pi^4}{15},$$

we have

$$\frac{2\pi}{c^2 h^3} \int_0^{\infty} \frac{E^3}{\exp(E/kT) - 1} dE = \sigma T^4,$$

where $\sigma = 2\pi^5 k^4 / 15 c^2 h^3$ is called the Stefan–Boltzmann constant. For $E_g > 0$, we have

$$Q_1 = g \left[\int_{E_g/h}^{\infty} \frac{E^3}{\exp(E/kT_1) - 1} dE - \int_{E_g/h}^{\infty} \frac{E^3}{\exp(E/kT_3) - 1} dE \right]$$

and thus

$$W = g\left(1 - \frac{T_2}{T_3}\right) \left[\int_{E_g/h}^{\infty} \frac{E^3}{\exp(E/kT_1) - 1} dE - \int_{E_g/h}^{\infty} \frac{E^3}{\exp(E/kT_3) - 1} dE \right].$$

Again we consider T_3 (which appears twice in the right-hand side of the latter equation) as a free parameter. The maximum power condition

$$\frac{dW}{dT_3} = 0$$

now leads to a transcendental equation in T_3. Numerical calculations (for the above values of T_1 and T_2) lead to Figure 3.5(a), showing the efficiency w as a function of the material's bandgap E_g. We see that the optimum bandgap is zero. The maximum efficiency is thus the black-body efficiency we found earlier, i.e., 85.4%. A nonzero bandgap E_g is not advantageous for a photothermal solar energy converter. Nevertheless, the introduction of a bandgap E_g will turn out to be fruitful, in later sections.

To model a photovoltaic solar converter is somewhat more complicated. Indeed, a solar cell is not a thermal engine but a thermochemical engine, i.e., a combination of a thermal engine and a chemical engine. Thus we first have to introduce the endoreversible chemical engine.

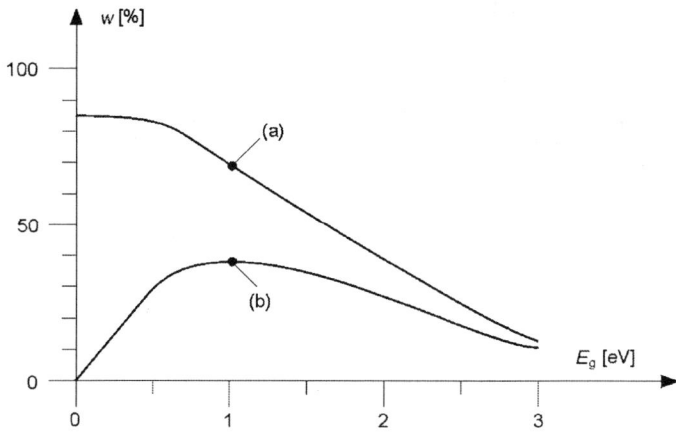

FIGURE 3.5. Efficiency of solar energy converters: (a) a photothermal converter, (b) a photovoltaic converter.

3.3 Endoreversible Chemical Engines

In a chemical engine [7] the intensive variable is the chemical potential μ, playing a role similar to temperature T in the thermal engine. The extensive variable exchanged between reservoirs is the particle flow N, playing a role similar to heat flow Q in the thermal engine. See Figure 3.3(b). We assume the following type of transport law holds:

$$N_1 = h(\mu_1) - h(\mu_3),$$

where $h(x)$ is some (monotonously increasing) mathematical function, not yet specified. This equation replaces (3.7), the Carnot formula (3.5) being replaced by the Gibbs law

$$W = (\mu_3 - \mu_2)N_1.$$

Thus:

$$W = (\mu_3 - \mu_2)[h(\mu_1) - h(\mu_3)].$$

In the same way as T_3 was the free parameter in the endoreversible thermal engine, now μ_3 is the parameter of the endoreversible chemical engine. See Figure 3.4(b). Maximum power output is found for

$$\frac{dW}{d\mu_3} = 0.$$

Interesting applications, are e.g., the linear law $h(x) = gx$ and the exponential law $h(x) = g\exp(x/kT)$. But we will not go into the details of these models. Instead, we will, in the next section, immediately combine the chemical engine with a thermal engine.

3.4 Endoreversible Thermochemical Engines

In a thermochemical system, both energy and particles are exchanged between the two reservoirs, #1 and #3. The energy current U_1 and the particle current N_1 are converted into a work flow W, according to the combined Carnot–Gibbs law, given by

$$W = \left(1 - \frac{T_2}{T_3}\right)U_1 + \left(\frac{T_2}{T_3}\mu_3 - \mu_2\right)N_1.$$

Here the input currents have to be replaced by their appropriate transport expressions

$$U_1 = f(T_1, \mu_1) - f(T_3, \mu_3),$$
$$N_1 = h(T_1, \mu_1) - h(T_3, \mu_3),$$

such that W is a function of two variables, i.e., T_3 and μ_3. See Figure 3.3(c).

Maximum power is guaranteed if

$$\frac{\partial W}{\partial T_3} = 0,$$
$$\frac{\partial W}{\partial \mu_3} = 0.$$

We now investigate the case where the exchanged particles are photons. For this purpose, we apply a generalization [8], [9], [10] of Planck's spectrum (3.8):

$$N(E, T, \mu) = \frac{2\pi}{c^2 h^3} \frac{E^2}{\exp\left((E-\mu)/kT\right) - 1},$$

such that

$$h(T, \mu) = g \int_{E_g/h}^{\infty} \frac{E^2}{\exp\left((E-\mu)/kT\right) - 1} dE.$$

For the exchanged energy, we need to multiply the number of photons by the photon energy E, such that the E^2 factor becomes an E^3 one:

$$f(T, \mu) = g \int_{E_g/h}^{\infty} \frac{E^3}{\exp\left((E-\mu)/kT\right) - 1} dE.$$

Thus

$$W = g\left(1 - \frac{T_2}{T_3}\right)\left[\int_{E_g/h}^{\infty} \frac{E^3 dE}{\exp\left((E-\mu_1)/kT_1\right)-1} - \int_{E_g/h}^{\infty} \frac{E^3 dE}{\exp\left((E-\mu_3)/kT_3\right)-1}\right]$$
$$+ g\left(\frac{T_2}{T_3}\mu_3 - \mu_2\right)\left[\int_{E_g/h}^{\infty} \frac{E^2 dE}{\exp\left((E-\mu_1)kT_1\right)-1} - \int_{E_g/h}^{\infty} \frac{E^2 dE}{\exp\left((E-\mu_3)kT_3\right)-1}\right].$$

The function $W(T_3, \mu_3)$ is a surface, displayed in Figure 3.6.

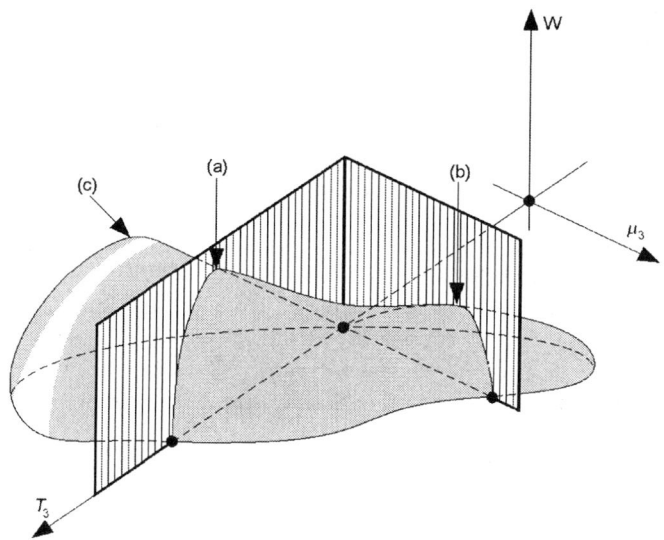

FIGURE 3.6. Solar energy conversion: (a) a photothermal maximum-power point, (b) a photovoltaic maximum-power point, (c) a hybrid photothermal–photovoltaic maximum-power point.

For $(T_1, \mu_1) = (5762 \text{ K}, 0)$ and $(T_2, \mu_2) = (288 \text{ K}, 0)$, this yields a maximum w of 86.6% at $(T_3, \mu_3) = (3890 \text{ K}, -0.75 \text{ eV})$. Because we have here two degrees of freedom (T_3 and μ_3), it is no surprise we attain higher efficiencies than the 85.4% we can obtain with a single degree of freedom (T_3), as in Section 2. The hybrid cell functions at a high temperature T_3, but thanks to the negative potential μ_3 the radiation of the semiconductor is limited [11] [12] and thus also the (negative) terms of W containing the factor $1/\exp\bigl((E - \mu_3)/kT_3\bigr) - 1$.

3.5 Solar Cells

In the previous section, we have investigated the optimum performance of the thermochemical solar energy converter, by imposing

$$\frac{\partial W}{\partial T_3} = 0,$$
$$\frac{\partial W}{\partial \mu_3} = 0. \qquad (3.9)$$

This resulted in a maximum efficiency w of 86.6% for this so-called hybrid device. We will now investigate the performance of purely thermal as well as purely photovoltaic solar energy converters.

60   3. Thermodynamics of Photovoltaics

In a purely photothermal converter, we do not use the μ degree of freedom. We choose μ_3 equal to the terrestrial value $\mu_2 = 0$. Thus the set of conditions (3.9) is replaced by the "less optimal" conditions

$$\frac{\partial W}{\partial T_3} = 0,$$
$$\mu_3 = \mu_2.$$

We thus intersect the $W(T_3, \mu_3)$-hill of Figure 3.6 with a plane $\mu_3 = \mu_2$, i.e., $\mu_3 = 0$, and search for the maximum on the resulting $W(T_3)$-curve. It turns out to amount to 85.4%, in accordance with Section 2.

In a purely photovoltaic converter, we do not use the T degree of freedom. We choose T_3 equal to the terrestrial value $T_2 = 288$ K. Thus the set of conditions (3.9) is replaced by the "less optimal" conditions

$$T_3 = T_2,$$
$$\frac{\partial W}{\partial \mu_3} = 0.$$

We thus intersect the $W(T_3, \mu_3)$-hill of Figure 3.6 with a plane $T_3 = T_2$ and search for the maximum on the resulting $W(\mu_3)$-curve:

$$W = g\mu_3 \left[\int_{E_g/h}^{\infty} \frac{E^2}{\exp(E/kT_1) - 1} dE - \int_{E_g/h}^{\infty} \frac{E^2}{\exp\left((E - \mu_3)/kT_2\right) - 1} dE \right] \tag{3.10}$$

with

$$\frac{dW}{d\mu_3} = 0.$$

The results of the calculations are displayed in Figure 3.5(b). We see that, in contrast to photothermal conversion, the photovoltaic conversion really needs a nonzero bandgap value in order to yield a nonzero power W. The optimal bandgap is 1.15 eV and the corresponding maximum efficiency is 40.8%, i.e., far less than the photothermal converter.

We now remark that μ_3 is changed by varying the bias voltage V of the semiconductor, according to the simple relation $\mu_3 = qV$. This means that by an electric voltage V we influence μ_3 and thus both the intensity and spectral distribution of the emitted radiation. This is what we call electroluminescence. Applying $\mu_3 = qV$ allows us to rewrite (3.10) as $W(V) = V \times I(V)$ with

$$I(V) = qg \left[\int_{E_g/h}^{\infty} \frac{E^2}{\exp(E/kT_1) - 1} dE - \int_{E_g/h}^{\infty} \frac{E^2}{\exp\left((E - qV)/kT_2\right) - 1} dE \right]. \tag{3.11}$$

Comparing this current–voltage characteristic with the classical characteristic (3.1), teaches us that the latter is an approximation, found by replacing "Bose statistics with chemical potential" by "Bose statistics without chemical potential multiplied by Boltzmann statistics":

$$\frac{1}{\exp\left((E-qV)/kT\right)-1} \approx \exp\left(qV/kT\right) \times \frac{1}{\exp(E/kT)-1}.$$

Indeed, this approximation converts (3.11) into

$$I(V) \approx qg\left[\int_{E_g/h}^{\infty} \frac{E^2}{\exp(E/kT_1)-1}dE - \exp(qV/kT_2)\int_{E_g/h}^{\infty} \frac{E^2}{\exp(E/kT_2)-1}dE\right],$$

such that identifying the coefficients of (3.1) leads to

$$I_s = qg \int_{E_g/h}^{\infty} \frac{E^2}{\exp(E/kT_2)-1}dE$$

$$I_l = qg\left[\int_{E_g/h}^{\infty} \frac{E^2}{\exp(E/kT_1)-1}dE - \int_{E_g/h}^{\infty} \frac{E^2}{\exp(E/kT_2)-1}dE\right].$$

The latter result slightly corrects expression (3.3), by substracting a small term. The former result replaces the "technological" expression (3.2) by its fundamental lower bound. This lower limit is reached when all possible electron–hole recombination processes are eliminated except for a single one, i.e., radiative recombination. In real semiconductors it is very difficult to quench completely the competing processes, especially the Auger recombination.

So, the circle is closed. First we introduced the Novikov–Curzon–Ahlborn model, consisting of a Carnot engine supplemented by a linear heat conductor. Then we generalized the Carnot engine to a reversible thermochemical engine and replaced the linear heat conduction by selective black-body radiation. As a result we demonstrated that a solar cell can be modeled as an endoreversible engine, thus revealing the underlying relationship between a solar cell and a Carnot engine. This constitutes the solution of the first puzzle, as it both demonstrates the relationship between the solar cell and the Carnot engine and explains why the photovoltaic efficiency is lower than the Carnot efficiency. Why the solar cell efficiency is so much below the Carnot efficiency we still have to explain. That constitutes the second puzzle of the Introduction. This second puzzle can be solved in two ways:

- Either we stick to a single material (i.e., a single bandgap), but introduce the possibility that a (sufficiently energetic) photon generates more than one electron–hole pair, e.g., benefiting by impact-ionization.
- Or we stick to the assumption that one photon creates only one electron–hole pair, but we introduce multi-gap solar cells (or tandem solar cells), by applying different materials with different bandgaps E_g.

The former approach will be treated in Section 3.6, the latter in Section 3.7.

3.6 Solar Cells with Larger-than-Unity Quantum Efficiency

In the previous sections, we tacitly assumed that each incident photon (with an energy E larger than the bandgap E_g) gives rise to one electron–hole pair and thus to a single charge q in the electric circuit. However, any sufficiently energetic photon (i.e., a photon with energy E larger than $2E_g$) can, in principle, generate more than one electron–hole pair [12], [13], [14], [15]. In practical cases, the phenomenon is not very probable, but it has nevertheless been observed in experiment. The most probable way is by first creating a single (energetic) electron–hole pair, which subsequently generates more pairs by means of impact ionization. We will assume that each photon with energy E between m times E_g and $(m+1)$ times E_g generates exactly m electron–hole pairs, with a maximum of M pairs.

Thus we choose for the quantum efficiency $\mathcal{Q}$ a staircase function:

$$\begin{aligned} \mathcal{Q}(E) = 0 & \quad \text{for} \quad 0 < E < E_g, \\ m & \quad \text{for} \quad mE_g < E < (m+1)E_g, \\ M & \quad \text{for} \quad ME_g < E < \infty, \end{aligned}$$

such that the electric current is

$$I = qg\bigg[\sum_{0}^{M-1} m \int_{mE_g}^{(m+1)E_g} \frac{E^2 dE}{\exp(E/kT_1)-1} - \sum_{0}^{M-1} m \int_{mE_g}^{(m+1)E_g} \frac{E^2 dE}{\exp\big((E-mqV)/kT_2\big)-1} \\ + M \int_{ME_g}^{\infty} \frac{E^2 dE}{\exp(E/kT_1)-1} - M \int_{ME_g}^{\infty} \frac{E^2 dE}{\exp\big((E-MqV)/kT_2\big)-1}\bigg].$$

Note that the outgoing spectrum has no single chemical potential. The chemical potential of the photons is dependent on the photon energy E, according to the staircase function

$$\begin{aligned} \mu_3 = mqV & \quad \text{for} \quad mE_g < E < (m+1)E_g, \\ MqV & \quad \text{for} \quad ME_g < E < \infty, \end{aligned}$$

resulting in a photoluminescent spectrum $N(E)$ with M peaks.

The numerical results are displayed in Figure 3.7. We see that the maximum efficiency increases strongly with increasing M:

- $w = 40.8\%$ for $M = 1$;
- $w = 55.6\%$ for $M = 2$;
- $w = 63.4\%$ for $M = 3$; and
- $w = 68.1\%$ for $M = 4$; etc.

We also see that impact-ionization favors small-bandgap materials.

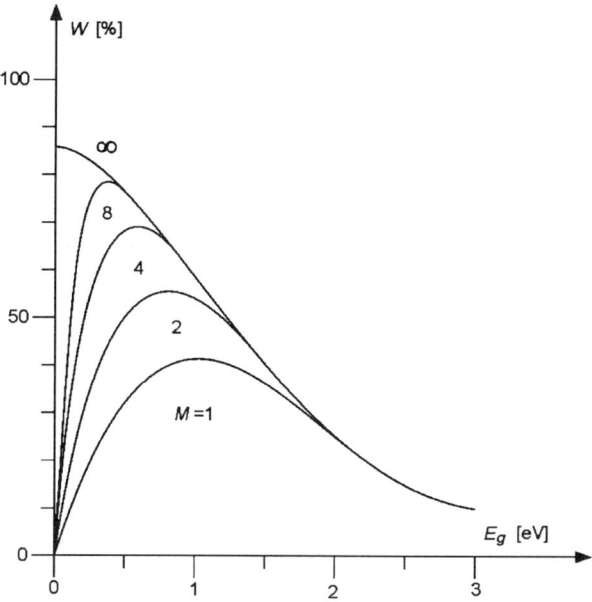

FIGURE 3.7. Efficiency of a solar cell with larger-than-unity quantum efficiency.

It is remarkable that the limit case $M \to +\infty$ together with $E_g \to 0$ leads to the same efficiency as the photothermal converter, i.e., 85.4%. The highest efficiency, however, is found in the limit case $M \to +\infty$, but for the finite bandgap of 48 meV, and amounts to 85.9%, as can be seen in Figure 3.8, a zoom-in of the curve $M = \infty$ in Figure 3.7.

3.7 Tandem Solar Cells

A second way of boosting the $w=40.8\%$ result of Section 3.5 toward higher efficiencies, is to stick to only one electron–hole pair per photon, but introducing more than one bandgap. This is realized by combining different semiconductors, each having its own bandgap. This leads to multi–gap solar cells or "tandem solar cells" [16], [17].

We assume that the n semiconductors are arranged in order of bandgap, the highest bandgap at the Sun-side:

$$E_{g1} > E_{g2} > \cdots > E_{g(i-1)} > E_{gi} > E_{g(i+1)} > \cdots > E_{gn}.$$

See Figure 3.9. The most energetic photons ($E > E_{g1}$) are absorbed in the uppermost semiconductor. Less energetic photons are absorbed in subse-

64   3. Thermodynamics of Photovoltaics

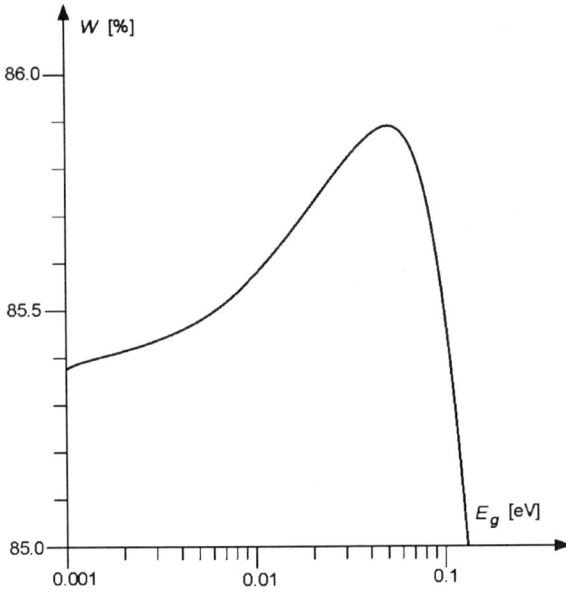

FIGURE 3.8. Efficiency of a solar cell with unlimited quantum efficiency.

quent materials underneath. The last semiconductor absorbs photons with little energy ($E_{gn} < E < E_{g(n-1)}$).

If each subcell of the tandem has its own voltage V_i, then the ith semiconductor emits a spectrum

$$g \frac{E^2}{\exp\left((E - qV_i)/kT_2\right) - 1} \quad \text{for} \quad E > E_{gi}.$$

Three spectra are incident on the cell:

- that part of the solar spectrum that transverses all the semiconductors 1 to $i - 1$ and is absorbed in semiconductor i:

$$g \frac{E^2}{\exp(E/kT_1) - 1} \quad \text{for} \quad E_{gi} < E < E_{g(i-1)};$$

- half of the spectrum emitted by the layer above:

$$g \frac{E^2}{\exp\left((E - qV_{i-1})/kT_2\right) - 1} \quad \text{for} \quad E > E_{g(i-1)};$$

- a part of half of the spectrum emitted by the layer beneath:

$$g \frac{E^2}{\exp\left((E - qV_{i+1})/kT_2\right) - 1} \quad \text{for} \quad E > E_{gi}.$$

3.7. Tandem Solar Cells

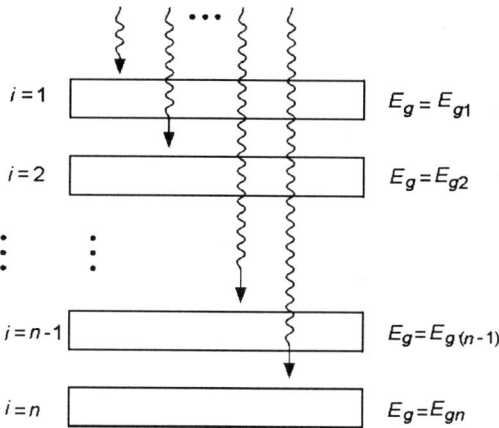

FIGURE 3.9. Tandem solar cell.

Each subcell not only has its own electric voltage V_i but also its own electric current I_i and thus produces its own electric power $W_i = V_i I_i$:

$$W_i = gqV_i \Bigg[\int_{E_{gi}}^{E_{g(i-1)}} \frac{E^2\, dE}{\exp(E/kT_1)-1} + \frac{1}{2}\int_{E_{g(i-1)}}^{\infty} \frac{E^2\, dE}{\exp\left((E-qV_{i-1})/kT_2\right)-1}$$

$$+\frac{1}{2}\int_{E_{gi}}^{\infty} \frac{E^2\, dE}{\exp\left((E-qV_{i+1})/kT_2\right)-1} - \int_{E_{gi}}^{\infty} \frac{E^2\, dE}{\exp\left((E-qV_i)/kT_2\right)-1} \Bigg].$$

Note that this expression is similar to (3.10) with two modifications:

- a different upper bound of the first integral; and
- two additional terms (i.e., the spectra incident from the two neighboring layers).

The total work delivered per unit time is $W = \sum_i W_i$. We will not go into the details of the calculation. Suffice it here to give the results [17]. Figure 3.10 shows the results for $n = 2$: the efficiency as a function of the two bandgaps E_{g1} and E_{g2}. We find a maximum of 55.7% for $E_{g1} = 1.7$ eV and $E_{g2} = 0.8$ eV. The maximum deliverable work is a strongly increasing function of n, the number of layers in the system:

- $w = 40.8\%$ for $n = 1$;
- $w = 55.7\%$ for $n = 2$;
- $w = 63.9\%$ for $n = 3$; and
- $w = 68.9\%$ for $n = 4$, etc.

66   3. Thermodynamics of Photovoltaics

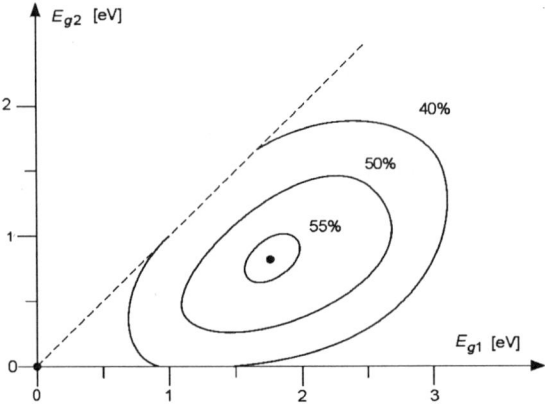

FIGURE 3.10. Efficiency of a tandem solar cell.

In the limit $n \to +\infty$, all bandgap increments $E_{g(i+1)} - E_{gi}$ become infinitesimally small. And so does the work contribution W_i. After some manipulations, we find that each subsystem acts as a monochromatic solar cell

$$dW = V\,dI = gqV\left[\frac{E_g^2}{\exp(E_g/kT_1) - 1} - \frac{E_g^2}{\exp\left((E_g - qV)/kT_2\right) - 1}\right]dE_g.$$

Eventually, the sum $\sum_i W_i$ becomes an integral

$$W = \int_{E_g=0}^{\infty} dW = qg \int_0^{\infty} V\left[\frac{E_g^2}{\exp(E_g/kT_1)-1} - \frac{E_g^2}{\exp\left((E_g-qV)/kT_2\right)-1}\right]dE_g. \tag{3.12}$$

The maximum of the functional W is obtained for the optimal choice of the function $V(E_g)$. One finds 86.8%. This limit is slightly in excess of the maximum of 85.9%, attained in the previous section. An extra advantage of the tandem solar cell with respect to the impact–ionization cell is the fact that we actually know how to build tandem devices, whereas we do not know well (yet) how to take technological advantage of impact ionization.

One can prove that the 86.8% result is not merely the upper limit of photovoltaic solar energy conversion, but actually constitutes the upper bound for any conceivable solar energy converter (working between the temperatures of 5762 K and 288 K). This result is rather close to the Carnot efficiency $1 - 288/5762 = 95.0\%$. The fact that we cannot get the Carnot efficiency itself, is caused by the fact that absorption of radiation without simultaneous emission of radiation with the same spectrum, is inevitably an irreversible process [18].

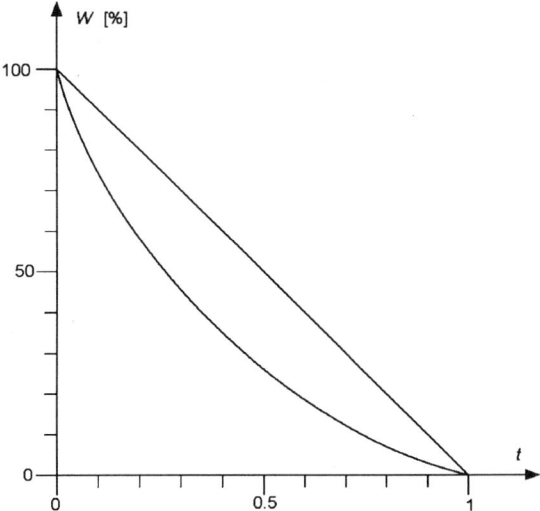

FIGURE 3.11. Efficiency of solar energy conversion.

We conclude this section with the formula for the upper bound of solar energy conversion for an arbitrary temperature ratio $t = T_2/T_1$. By merely introducing the dimensionless bandgap $u = E_g/kT_2$ and the dimensionless voltage $x = qV/kT_2$, formula (3.12) becomes

$$w = \frac{15t^4}{\pi^4} \int_0^\infty x \left[\frac{1}{\exp(tu) - 1} - \frac{1}{\exp(u - x) - 1} \right] u^2 \, du,$$

where $x(u)$ satisfies

$$\frac{d}{dx} \left\{ x \left[\frac{1}{\exp(tu) - 1} - \frac{1}{\exp(u - x) - 1} \right] \right\} = 0,$$

i.e., where $x(u)$ is the solution of

$$\frac{(1 + x)\exp(u - x) - 1}{[\exp(u - x) - 1]^2} = \frac{1}{\exp(tu) - 1}.$$

Figure 3.11 displays this function $w(t)$ and compares it with the Carnot function $1 - t$. The curve $w(t)$ is tangent to the ordinate axis for $t = 0$ and tangent to the abscissa for $t = 1$. A good approximation [19], [20] is given by

$$w \approx (1 - t)^2 [1 + 0.3t \log(t)].$$

3.8 Conclusion

In the present chapter, we have revealed the relationship between a solar cell and a Carnot engine. We have demonstrated that a solar cell can be modeled as an endoreversible thermochemical engine. The latter consists of two separate parts:

- an irreversible part taking care of the transport of particles (i.e., photons) and energy, by means of selective black-body radiation; and
- a reversible part taking care of the proper energy conversion.

We have found fundamental upper bounds for photovoltaic solar energy conversion efficiency:

- 40.8% if we use a single semiconductor material; and
- 86.8% if we use many semiconductor materials.

We compare here these numbers with present-day world records [21]:

- 27.6% for a GaAs ($E_g = 1.42$ eV) concentration solar cell; and
- 32.6% for a GaAs ($E_{g1} = 1.42$ eV)–GaSb ($E_{g2} = 0.72$ eV) tandem solar cell.

It is necessary to stress here that all theory in the present chapter was developed for sunlight combined with an ideal light concentrator, consisting either of ideal mirrors, ideal lenses, or a combination of both. For natural sunlight and for moderately concentrated sunlight, the mathematics is less transparent. For details the reader is referred to [2]. Suffice here to give the results for natural, i.e., unconcentrated, sunlight. The theoretical upper bounds are:

- 31.0% if we use a single semiconductor material; and
- 68.2% if we use many semiconductor materials.

We again compare these numbers with technological records [21]:

- 25.1% for a GaAs solar cell; and
- 30.3% for a GaInP–GaAs tandem solar cell.

For more details on record performances, we refer the reader to the next chapter of the book.

We would finally like to stress that endoreversible thermodynamics is not only useful for describing photovoltaics, but also has been successfully applied in numerous other fields, like chemistry [7], [22], climatology [23], [24], economics [25], [26], [27], computing [28], [29], [30], etc.

3.9 References

[1] P. Landsberg: An introduction to the theory of photovoltaic cells, *Solid-State Electron.* **18** 1043–1052, (1975).

[2] A. De Vos: Endoreversible Thermodynamics of Solar Energy Conversion, Oxford University Press, Oxford, 1992.

[3] A. De Vos: The endoreversible theory of solar energy conversion : A tutorial, *Solar Energy Mater. Solar Cells* **31**, 75–93, (1993).

[4] I. Novikov: Effektivyj koefficient poleznovo deystvia atomnoy energeticeskoj ustanovki, *Atomnaya Energiya* **3**, 409–412, (1957), in: English translation: The efficiency of atomic power stations (a review), *J. Nuclear Energy II* **7**, 125–128, (1958).

[5] F. Curzon and B. Ahlborn: Efficiency of a Carnot engine at maximum power output, *Amer. J. Phys.* **43**, 22–24, (1975).

[6] H. Müser: Behandlung von Elektronenprozessen in Halbleiter-Randschichten, *Z. Phys.* **148**, 380–390, (1957).

[7] A. De Vos: Endoreversible thermodynamics and chemical reactions, *J. Phys. Chem.* **95**, 4534–4540, (1991).

[8] A. De Vos and H. Pauwels: On the thermodynamic limit of photovoltaic energy conversion, *Appl. Phys.* **25**, 119–125, (1981).

[9] P. Landsberg: Photons at non-zero chemical potential, *J. Phys. C : Solid State Phys.* **14**, L 1025–1027, (1981).

[10] P. Würfel: The chemical potential of radiation, *J. Phys. C : Solid State Phys.* **15**, 3967–3985, (1982).

[11] A. De Vos and J. Landries: Endoreversible thermodynamics of the hybrid photothermal–photovoltaic converter, *11th European Photovoltaic Solar Energy Conference*, Montreux, 12–16 October 1992, pp. 363–366.

[12] W. Spirkl and H. Ries: Luminescence and efficiency of an ideal photovoltaic cell with charge carrier multiplication, *Phys. Review B* **52**, 11, 319–11,325, (1995).

[13] P. Landsberg, H. Nussbaumer, and G. Willeke: Band–band impact ionization and solar cell efficiency, *J. Appl. Phys.* **74**, 1451–1452, (1993).

[14] J. Werner, R. Brendel, and H. Queisser: New upper efficiency limits for semiconductor solar cells, *1st World Conference on Photovoltaic Energy Conversion*, Hawaii, 5–9 December 1994, pp. 1742–1745.

[15] A. De Vos and B. Desoete: On the ideal performance of solar cells with larger-than-unity quantum efficiency, *Sol. Energy Mater. Solar Cells* **51**, 413–424, (1998).

[16] E. Jackson: Areas for improvement of the semiconductor solar energy converter, *Conference on the Use of Solar Energy*, Tucson, 1–2 November 1955, pp. 122–126.

[17] A. De Vos: Detailed balance limit of the efficiency of tandem solar cells, *J. f Phys. D : Appl. Phys.* **13**, 839–846, (1980).

[18] A. De Vos and H. Pauwels: Comment on a thermodynamical paradox presented by Würfel, *J. Phys. C: Solid State Phys.* **16**, 6897–6909, (1983).

[19] C. Grosjean and A. De Vos: On the upper limit of the energy conversion efficiency in tandem solar cells, *J. Phys. D: Appl. Phys.* **14**, 883–894, (1981).

[20] A. De Vos, C. Grosjean and H. Pauwels: On the formula for the upper limit of photovoltaic solar energy conversion efficiency, *J. Phys. D: Appl. Phys.* **15**, 2003–2015, (1982).

[21] M. Green, K. Emery, K. Bücher, D. King and S. Igari: Solar cell efficiency tables (version 12), *Progr. in Photovoltaics* **6**, 265–270, (1998).

[22] A. De Vos: Thermodynamics of photochemical solar energy conversion, *Solar Energy Mater. Solar Cells* **38**, 11–22, (1995) and **40**, 1996 erratum.

[23] A. De Vos and G. Flater: The maximum efficiency of the conversion of solar energy into wind energy, *Amer. J. Phys.* **59**, 751–754, (1991).

[24] A. De Vos and P. van der Wel: The efficiency of the conversion of solar energy into wind energy by means of Hadley cells, *Theret. Appl. Climatol.* **46**, 193–202, (1993).

[25] A. De Vos: Endoreversible thermoeconomics, *Energy Conversion and Management* **36**, 1–5, (1995).

[26] A. De Vos: Endoreversible economics, *Energy Conversion and Management* **38**, 311–317, (1997).

[27] A. De Vos: Endoreversible thermodynamics versus economics, *Energy Conversion and Management* **40**, 1009–1019, (1999).

[28] A. De Vos: Reversible and endoreversible computing, *Internat. J. Theret. Phys.* **34**, 2251–2266, (1995).

[29] A. De Vos: Introduction to r-MOS systems, *4th Workshop on Phys. and Computation*, Boston, 22–24 November 1996, pp. 92–96.

[30] A. De Vos: Towards reversible digital computers, *European Conference on Circuit Theory and Design*, Budapest, 1–3 September 1997, pp. 923–931.

4
Some Methods of Analyzing Solar Cell Efficiencies

P. T. Landsberg
V. Bădescu

ABSTRACT. A survey is given of various theoretical approaches to estimating solar cell efficiencies. We start (Section 4.2) with a development of the usual solar cell equation which is widely used and assume the so-called shift theorem. It is itself an approximation as is shown again here. A theory of the heterojunction solar cell is then developed (Section 4.4), following a brief survey of properties of efficiencies in general (Section 4.3). In this section we also give an introduction to the problem of estimating the effects of impact ionization. This is done by introducing a probability that a current carrier which has enough energy to impact ionize will actually do so. Following simpler special cases (Section 4.5), a more detailed theory of heterojunction cells with impact ionization is then presented (Section 4.6).

As is well known, conversion efficiencies can be increased by connecting two or more cells in series, i.e., proceeding from a heterojunction or tandem cell to several cells, or even many cells. This problem is discussed in Section 4.7. It involves radiation theory, based on some elementary quantum mechanics and statistical mechanics. Thermophotovoltaic conversion (Section 4.8) has the benefit of yielding relatively high conversion efficiencies because the energy loss due to the thermalization of the current carriers which occurs in a normal solar cell is here reduced. This is due to the fact that the solar energy is first absorbed by a material that reemits radiation at a lower temperature.

4.1 Introduction

One of the usual methods of converting the energy of solar radiation into electrical energy is by means of solar photovoltaic cells (in short, solar cells). The conversion efficiency of these cells can be estimated in different ways. One can base oneself on various thermodynamic cycles and this treats the solar cell as a thermal engine. Then, by using the first and second laws of thermodynamics one can estimate the conversion efficiency and other indicators of performance. Results obtained by using this method usually lead to a significant overestimation of performances. Another method derives results from energy and entropy fluxes at the cell surface. Then, by taking into consideration the irreversibilities associated with the various processes occurring during the energy conversion, one can finally estimate the cell performance more realistically. More detailed analyses take into account the specificities of the solar cell operation. In this case, one usually treats the absorption of radiation as an ideal process, while neglecting the losses by carrier recombination. We again obtain ideally high efficiencies which can be worked out assuming various degrees of solar concentration. These theoretical efficiencies can be increased further by stipulating the use of tandem cells, by imagining pair production utilizing various degrees of impact ionization, by taking into account multiple sources, by stipulating hybrid systems, etc.

One can then adjust these optimistic results by taking corrections into account, for example, by estimating the loss of carriers by recombination.

How do all these results compare? How different are actual efficiencies? In answer to this question we follow two lines. First, we present a brief introduction to the physics of photovoltaics. Systems based on mono and multigap materials are considered here. Also, the basis of thermophotovoltaic conversion is briefly explained. Second, but in a separate publication [1], we give a survey of important solar cell efficiencies which have been obtained in recent years. Full references are given there to figures and curves in appropriate publications, and we also explain there if the result is experimental or theoretical. If the latter, we indicate the assumptions made. In this way we hope to supplement the tables published in [2]. The latter concentrate on confirmed experimental results and their method of confirmation. Our table of efficiencies covers experimental and theoretical results claimed in publications, but not necessarily independently confirmed. It therefore covers a range from 9% to 88%. There is, of course, a tendency for the theoretical efficiencies to be lower the more detailed the model is on which they are based.

The split of the present work into two parts is regretted, but it was forced on us by the need to keep the chapter down to a reasonable length. This is also a reason why we could not cover all the important recent work. We had to be satisfied with a discussion of the main lines of enquiry.

4.2 The Solar Cell Equation: Currents from Photon Fluxes

In a steady state the photons in a semiconductor at temperature T_c have a distribution function

$$n_\nu(T_c, V) = \frac{1}{\exp\left[(h\nu - qV)/kT_c\right] - 1}. \qquad (4.1)$$

Here q is the numerical value of the electric charge and V is the voltage across the semiconductor, while normally qV is the difference between the quasi-Fermi levels for conduction and the valence band. The number of photons per unit area per unit time (the photon *number* flux) emitted by a solar cell based on such a semiconductor is then approximately like the flux from a black-body at temperature T_c:

$$\gamma_e(T_c, V) = \int_{E_g/h}^{\infty} \frac{g(\nu) \, d\nu}{\exp[(h\nu - qV)/kT_c] - 1}. \tag{4.2}$$

Here E_g is the energy gap of the semiconductor and $g(\nu)$ is normally given by $2\pi\nu^2/c^2$ for unpolarized radiation and by $\pi\nu^2/c^2$ for polarized radiation. The pumping radiation, also modeled as black-body but at temperature T_p and normally due to the Sun, creates a photon *number* flux in the semiconductor

$$\gamma_p(T_p) \equiv B_p \int_{E_g/h}^{\infty} \frac{g(\nu) d\nu}{\exp(h\nu/kT_p) - 1}. \tag{4.3}$$

Here B_p is the view factor, often also referred to as the geometric factor, given by

$$B_p = \pi \sin^2 \delta, \tag{4.4}$$

where δ is the half-angle subtended by the Sun when viewed from Earth ($\delta \approx 2.2 \times 10^{-5}$).

Neglecting all nonradiative recombination, the current density in the cell is

$$J(E_g, V, T_c, T_p) = q[\gamma_e(T_c, V) - \gamma_p(T_p)]. \tag{4.5}$$

The light-induced density $q\gamma_p$ is taken in the negative direction as opposing the current density $q\gamma_e$ due to radiative recombination.

At open-circuit $V = V_{oc}$ (say) and (4.5) vanishes. Hence

$$\gamma_p(T_p) = \gamma_e(T_c, V_{oc}), \quad \text{and} \quad V_{oc} \text{ is a function of} T_c, T_p. \tag{4.6}$$

This assumes that the "shift theorem" holds, i.e., that γ_p does not depend on the voltage V.

Noting that the short-circuit current density J_{sc} is given by (4.5) with $V = 0$, and also using (4.6),

$$\frac{J}{J_{sc}} \equiv \frac{J(E_g, V, T_c, T_p)}{J(E_g, 0, T_c, T_p)} = \frac{\gamma_e(T_c, V) - \gamma_e(T_c, V_{oc})}{\gamma_e(T_c, 0) - \gamma_e(T_c, V_{oc})}$$
$$= \frac{w(T_c, V_{oc}) - w(T_c, V)}{w(T_c, V_{oc}) - 1}, \tag{4.7}$$

4.2. The Solar Cell Equation: Currents from Photon Fluxes

where

$$w(T_c, V) \equiv \gamma_e(T_c, V)/\gamma_e(T_c, 0). \tag{4.8}$$

Approximations for w are given below. But it is clear from (4.2) that for large negative V, corresponding to reverse voltages,

$$w(T_c, V) \sim \exp(qV/kT_c). \tag{4.9}$$

By considering first the case $J = 0$ (open circuit), and then $V = 0$ (short circuit) in (4.5), one finds

$$\gamma_p(T_p) = \gamma_e(T_c, V_{\text{oc}}) = \gamma_e(T_c, 0) - \frac{J_{\text{sc}}}{q}, \tag{4.10}$$

whence

$$J_{\text{sc}} = q[\gamma_e(T_c, 0) - \gamma_e(T_c, V_{\text{oc}})]. \tag{4.11}$$

Therefore (4.5) becomes, using (4.10),

$$\begin{aligned} J(E_g, V, T_c, T_p) &= J_{\text{sc}} - q[\gamma_e(T_c, 0) - \gamma_e(T_c, V)] \\ &= q\gamma_e(T_c, 0)[w(T_c, V) - 1] + J_{\text{sc}}. \end{aligned} \tag{4.12}$$

At short circuit without incident radiation there is no current. Thus if there is incident radiation and short circuit, then the current density can be denoted by $J_{\text{sc}} = -J_L$, where J_L is the light-induced current density. One thus obtains the standard "solar cell equation" from (4.12)

$$J(E_g, V, T_c, T_p) = J_0[w(T_c, V) - 1] - J_L, \tag{4.13}$$

where $J_0 \equiv q\gamma_e(T_c, 0)$ is the dark saturation current density. It also occurs in (4.7) in a rather general form. This formalism was introduced in equations (1)–(5) of [3] and generalized in equations (103) and (125) of [4]. An extended theory is presented in [5]–[8].

To estimate w, note that from (4.2) and (4.8)

$$w(T_c, V) \equiv \frac{\displaystyle\int_{E_g/h}^{\infty} \frac{g(\nu)d\nu}{\exp\frac{h\nu - qV}{kT_c} - 1}}{\displaystyle\int_{E_g/h}^{\infty} \frac{g(\nu)d\nu}{\exp\frac{h\nu}{kT_c} - 1}}.$$

The factors multiplying $g(\nu)\, d\nu$ in the integrals are of the form

$$\frac{1}{b^{-1} - 1} = \frac{b}{1 - b} = b(1 + b + b^2 + \ldots) = \sum_{i=1}^{\infty} b^i$$

76   4. Some Methods of Analyzing Solar Cell Efficiencies

for $b \equiv \exp[(qV - h\nu)/kT_c] < 1$, i.e., for $qV < E_g$. One finds

$$w = \frac{\sum_{s=1}^{\infty} J_s \exp(sqV/kT_c)}{\sum_{i=1}^{\infty} J_i} \approx \exp\frac{qV}{kT_c}\left[1 - \left(1 - \exp\frac{qV}{kT_c}\right)\frac{J_2}{J_1}\right],$$

$$J_s \equiv \int_{E_g/h}^{\infty} g(\nu) \exp\left(-\frac{sh\nu}{kT_c}\right). \tag{4.14}$$

Thus the approximate solar cell equation is from (4.13), see Figure 4.1,

$$J(E_g V, T_c, T_p) = J_0\left(\exp\frac{qV}{kT_c} - 1\right) - J_L. \tag{4.15}$$

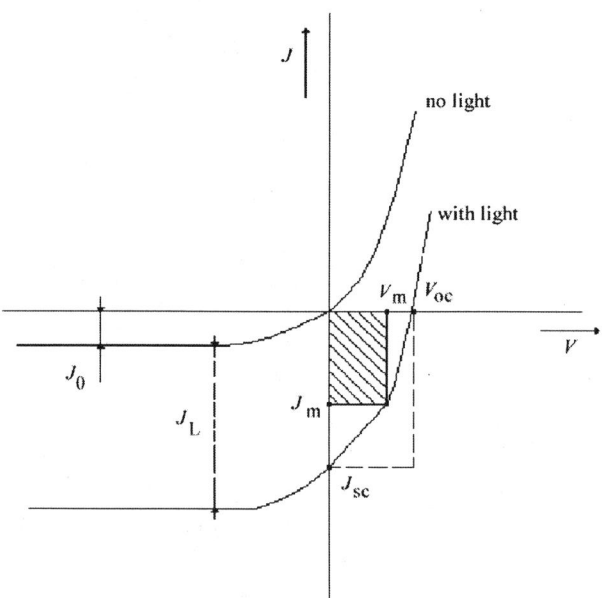

FIGURE 4.1. Schematic current–voltage characteristic of a p–n junction in the dark and under illumination.

4.3 Efficiencies in General

In the well-known "ultimate" efficiency calculation [9], [10] the cross-gap excitation due to solar radiation is assumed to produce an electron–hole pair. This can contribute an energy E_g to the work flux output. Energy above E_g per electron–hole pair is assumed dissipated by interaction with phonons and by nonradiative transitions. In the limit when the radiation from the cell can be neglected this is still a fair approximation [11]. Nevertheless, the results yield efficiencies in excess of what has been realized because of additional loss factors:

(a) incomplete collection of charge carriers due to recombination at the surface or in the space charge region;

(b) band gap shrinkage due to Coulomb interaction and other effects; and

(c) nonradiative Auger recombination.

Mechanisms (b) and (c) are important in the case of heavy doping [12]; (b) reduces the energy available per pair, while (a) and (c) reduce the available number of pairs.

Even higher efficiencies have been predicted. These theories envisage that nearly *all* the energy of nearly *all* the pairs can be used by utilizing an infinite sequence of materials of decreasing energy gaps (see Section 4.6). This yields the very high efficiencies of 68.2% for one sun and 86.8% for maximum concentration [13]. This is the true maximum efficiency which no solar cell structure can exceed. The Carnot efficiency for $T_c = 300$ K and $T_p = 6000$ K is at 95% quite unattainable.

A method of targeting high efficiencies does not go as far as the "infinite" tandem cells just described; instead it utilizes impact ionization. This is the process in which a high-energy electron of the conduction band makes a transition to a lower energy in the same band as a result of a collision with an electron in the valence band. The latter is promoted to the conduction band by gaining the energy lost by the first electron. (We shall neglect the corresponding impact ionization by energetic holes in Sections 4.3 and 4.4.) This new electron–hole pair can again contribute energy E_g to the work flux output. This model lies between the "ultimate" efficiency model and the "infinite" tandem cell model. The effect is that one has now three groups of photons:

(a) lower-energy photons which do not contribute to the photovoltaic energy output;

(b) intermediate energy photons which can contribute energy E_g; and

(c) higher-energy photons which can contribute $2E_g$ by impact ionization.

The remaining energy of electron–hole pairs is again assumed to be uselessly dissipated as heat. Higher-order impact ionizations are considered in Section 3.6.

4.4 Theoretical Efficiencies of a Simple Heterojunction

We now give a simplified treatment of the efficiency of a heterojunction solar cell which neglects recombination at the interface and also neglects spikes in the energy band [14]. The analysis utilizes: the reduced photon energy $x = h\nu/kT_p$ where T_p is again the "pump" or solar temperature; the Bose factor $g_1(x) \equiv [\exp x - 1]^{-1}$; and a function $g_2(x)\, dx$ (related to $g(\nu)$ from Section 4.1) which represents the number of photon states between x and $x + dx$ multiplied by a constant so that

$$\gamma_p(x)\, dx \equiv g_1(x)g_2(x)\, dx \tag{4.16}$$

is the number of photons in $(x, x+dx)$. We have normally

$$g_2(x) = \frac{2\pi(kT_p)^3}{h^3 c^2} x^2. \tag{4.17}$$

In the calculation of an efficiency, η, we have to divide the photon flux absorbed by the solar cell by the dimensionless energy flux, D, reaching the converter from the pump for maximum concentration, i.e., for the case when the solar cell behaves as if it were "surrounded" by the Sun, or, in another way of speaking, for a "2π geometry" [$\delta = \pi/2$ in (4.4)]. It is given by the input energy flux divided by

$$D = \int_0^\infty x\gamma_p(x)\, dx = \frac{2\pi^5 (kT_p)^3}{15 c^2 h^3}. \tag{4.18}$$

If the energy gap divided by kT_p is denoted by x_g the ultimate efficiency is then (see also [11])

$$\eta = \frac{x_g}{D} \int_{x_g}^\infty \gamma_p(x)\, dx. \tag{4.19}$$

What is the effect on (4.19) of impact ionization by electrons? Consider a simple heterojunction with the larger energy gap material (E_{g1}) exposed to the radiation. If $k_1(x)$ is the probability of absorption for photons with their energy subject to $x_{g1} < h\nu/kT_p < \infty$, these photons, as absorbed in material 1, contribute to ηD,

$$x_{g1}\left[\int_{x_{g1}}^\infty k_1(x)\gamma_p(x)\, dx + \int_{\theta_1 x_{g1}}^\infty P_1(x)k_1(x)\gamma_p(x)\, dx\right]. \tag{4.20}$$

The first term is essentially as in (4.19). The threshold photon energy to cause impact ionization is assumed to be $\theta_1 E_{g1}$ ($\theta_1 > 2$) and $P_1(x)$ is the probability that a photon of this energy will produce an electron in material 1 which will actually impact ionize. The second term in (4.20) is the new term.

Photons of the same energy range, and not absorbed in material 1, can be absorbed in material 2 with probability k_2, say. This contribution to ηD is

$$x_{g2}\left\{\int_{x_{g2}}^{\infty} k_2(x)[1-k_1(x)]\gamma_p(x)\,dx + \int_y^{\infty} k_2(x)[1-k_1(x)]P_2(x)\gamma_p(x)\,dx\right\}. \tag{4.21}$$

The first term is again essentially as in (4.19). The quantity y is the largest of x_{g1} and $\theta_2 x_{g2}$ where $\theta_2 x_{g2}$ specifies the threshold of impact ionization in material 2. Here $P_2(x)$ corresponds to $P_1(x)$ for material 2.

Lastly, photons in the energy range $x_{g2} < h\nu/kT_p < x_{g1}$ can be absorbed only in material 2 and so contribute

$$x_{g2}\left\{\int_{x_{g2}}^{x_{g1}} k_2(x)\gamma_p(x)\,dx + \Theta\langle x_{g1}-\theta_2 x_{g2}\rangle \int_{\theta_2 x_{g2}}^{x_{g1}} k_2(x)P_2(x)\gamma_p(x)\,dx\right\}. \tag{4.22}$$

The second term arises only if $x_{g1} > \theta_2 x_{g2}$ and this is expressed formally by the Heaviside function $\Theta\langle\ \rangle$. As an approximation we shall treat k_1, k_2, P_1, P_2 as constants, take them out of the integrals, and write

$$I(a,b) \equiv \int_a^b \gamma_p(x)\,dx. \tag{4.23}$$

One can then obtain the heterojunction efficiency in terms of k_1, k_2, P_1, P_2, x_{g1}, x_{g2}, from

$$\begin{aligned}\eta D =\ & k_1 x_{g1}[I(x_{g1},\infty)+P_1 I(\theta_1 x_{g1},\infty)] \\ & +k_2 x_{g2}[I(x_{g2},x_{g1})+(1-k_1)I(x_{g1},\infty)+(1-k_1)P_2 I(y,\infty) \\ & +\Theta\langle x_{g1}-\theta_2 x_{g2}\rangle k_2 x_{g2} P_2 I(\theta_2 x_{g2},x_{g1})].\end{aligned} \tag{4.24}$$

For a direct semiconductor with parabolic bands of effective masses m_e and m_h one has a rigorous expression for θ [15]:

$$\theta = 1 + \frac{2m_e + m_h}{m_e + m_h}. \tag{4.25}$$

These have been given as 2.035 (InSb) and 2.125 (GaAs) in [16, Table 1]. Figure 4.2 gives some typical values.

4.5 Special Cases of the Simple Theory

4.5.1 *Homojunction with or without Impact Ionization*

There is no absorption in material 2, so that it might as well be absent if $k_2 = 0$. In addition, $k_1 = 1$ so that one finds the earlier result [16] for a

80   4. Some Methods of Analyzing Solar Cell Efficiencies

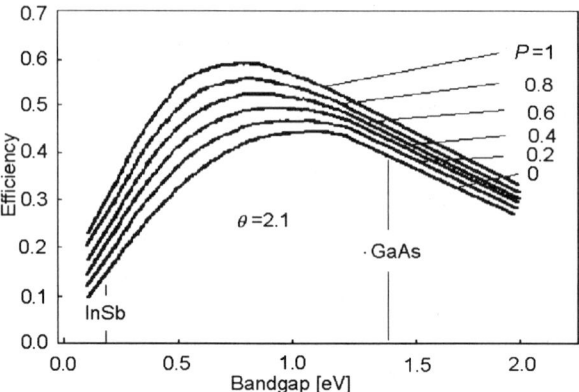

FIGURE 4.2. Theoretical efficiencies for homojunction, complete absorption assumed, for $T_p = 5785$ K, $\theta = 2.1$, and various values of $P \equiv P_1$. The value of θ applies approximately to InSb and GaAs.

homojunction with impact ionization:

$$\eta D = x_{g1}[I(x_{g1}, \infty) + P_1 I(\theta_1 x_{g1}, \infty)]. \qquad (4.26)$$

A similar result is found if $E_{g2} = 0$. If impact ionization has zero probability one recovers the ultimate efficiency (4.19).

The considerable increase in the theoretical efficiency that can result for $P_1 > 0$ depends, as shown in [16], on a value of P_1 in excess of 20%. This focuses attention on the need to reduce the electron–phonon and other interactions which compete with the impact ionization. It is not known as yet how one might achieve this.

4.5.2 *Heterojunction without Impact Ionization*

If $P_1 = P_2 = 0$ in (4.24) one recovers equations (11) and (12) of [14] from (4.24), viz.,

$$\eta D = k_1 x_{g1} I(x_{g1}, \infty) + k_2 x_{g2} I(x_{g2}, x_{g1}) + k_2(1 - k_1) I(x_{g1}, \infty). \qquad (4.27)$$

This gives the efficiency of a heterojunction in our simplified model if there is no contribution from impact ionization.

4.6 Analysis of Heterojunction Cells Allowing for Impact Ionization

In a more elaborate theory three energy ranges are considered (all divided again by kT_p):

(a) $0 \leq x < x_{g2}$;
(b) $x_{g2} \leq x < x_{g1}$; and
(c) $x_{g1} \leq x < \infty$;

where it has been assumed that $x_{g2} < x_{g1}$. Photon absorption with pair generation can occur only in regions (b) and (c).

The light-generated photon number flux is proportional to the light-induced current density and is

$$J_L = J_{L_1} + J_{L_2}. \tag{4.28}$$

Here J_{L_1} is due to semiconductor 1 and is generated in region (c):

$$\frac{J_{L_1}}{q} = k_1 \int_{x_{g1}}^{\infty} \gamma_p(x)[1 + P_{e1}\Theta\langle x - \theta_{e1}x_{g1}\rangle + P_{h1}\Theta\langle x - \theta_{h1}x_{g1}\rangle]\, dx. \tag{4.29}$$

The Heaviside step function Θ has again been used. The integral extends from the energy gap E_{g1} of the first semiconductor and the P's are again appropriate probabilities of impact ionization with threshold energies $\theta_{e1}x_{g1}$ and $\theta_{h1}x_{g1}$. Here e and h denote electrons and holes, respectively.

Semiconductor 2 contributes to the light-induced current by photons in regions (b) and (c):

$$\frac{J_{L2}}{q} = k_2 \int_{x_{g2}}^{x_{g1}} \gamma_p(x)[1 + P_{e2}\Theta\langle x - \theta_{e2}x_{g2}\rangle + P_{h2}\Theta\langle x - \theta_{h2}x_{g2}\rangle]\, dx$$

$$+ k_2(1 - k_1)\int_{x_{g1}}^{\infty} \gamma_p(x)[1 + P_{e2}\Theta\langle x - \theta_{e2}x_{g2}\rangle + P_{h2}\Theta\langle x - \theta_{h2}x_{g2}\rangle]\, dx. \tag{4.30}$$

For large reverse bias and no light the current density J_0 is called the saturation current density (Figure 4.1). For a simple p–n junction or heterojunction model it can be written as a sum over the two materials

$$\frac{J_0}{q} = \sum_{i=1}^{2} a_i \exp\left(-\frac{E_{gi}}{kT_c}\right) \tag{4.31}$$

with

$$a_i \equiv q \frac{N_{ci}N_{vi}}{N_i}\sqrt{\frac{kT_c\mu_i}{q\tau_i}}. \tag{4.32}$$

82   4. Some Methods of Analyzing Solar Cell Efficiencies

Here N_{ci}, N_{vi} are the effective densities of states at the band edges, N_i are the impurity concentrations (assumed uniform in each semiconductor), μ_i are the carrier mobilities, and τ_i are their lifetimes.

Results of the type (4.31) and (4.32) are obtained in most expositions of p–n junction theory. The current density is expressed as a sum of an electric field current density plus a diffusion current density. The latter involves a space derivative of electron and hole concentration, or, alternatively, of a chemical potential. Continuity equations for electron and hole concentrations then yield the results (4.31), (4.32) [17]. A simple form of a saturation current of this type occurs already in Shockley's book [18]

Several conversion efficiencies can be defined. The "spectrum factor", i.e., the portion of the Sun's energy used in pair generation is, using (4.18),

$$\eta_1 \equiv \frac{J_{L1}E_{g1} + J_{L2}E_{g2}}{kT_pD} \qquad (4.33)$$

in generalization of (4.24). The precise form of the integrals that occur in (4.33) depends on whether x_{g1} exceeds $\theta_2 x_{g2}$ or not.

Another efficiency, but not discussed in Section 4.4, is the voltage factor, η_2. The light-generated current components (4.28) require for their generation the energies E_{g1} and E_{g2}, respectively. The open-circuit voltage V_{oc} of (4.6) gives, however, at qV_{oc}, the upper limit to the energy available to the generated minority carriers. So one can define η_2 by

$$\eta_2 \equiv \frac{(J_{L1} + J_{L2})qV_{oc}}{J_{L1}E_{g1} + J_{L2}E_{g2}}. \qquad (4.34)$$

Using the solar cell equation (4.15),

$$\eta_2 = \frac{J_{L1} + J_{L2}}{J_{L1}E_{g1} + J_{L2}E_{g2}} \, kT_c \ln\left(1 + \frac{J_{L1} + J_{L2}}{J_0}\right). \qquad (4.35)$$

A third efficiency, also not discussed in Section 4.4, is obtained by noting from Figure 4.1 that the maximum available power, represented by the cross-hatched rectangle of area $J_m V_m$, is less than the power represented by the rectangle of area $J_{sc}V_{oc} = -J_L V_{oc}$. This gives rise to the "fill factor" (or the "impedance matching factor" or the "graph factor")

$$\eta_3 = -\frac{J_m V_m}{J_L V_{oc}}. \qquad (4.36)$$

This turns out to be given by

$$\eta_3 = \frac{v^2 e^v}{[(1+v)e^v - 1][v + \ln(1+v)]}, \qquad (4.37)$$

where $v \equiv qV_m/kT_c$ is given by

$$(1+v)e^v - 1 = \frac{J_L}{J_0}, \qquad (4.38)$$

which maximizes the power output.

We now derive (4.37) and (4.38) by dealing with J_m, V_m, J_{sc}, and V_{oc} in equation (4.36) in turn. By (4.15) the optimal J and V, i.e., J_m and V_m, are found from $\partial(JV)/\partial V = 0$, i.e., $J_m + vJ_0 e^v = 0$, i.e.,

$$J_0(e^v - 1) - J_L + vJ_0 e^v = 0. \tag{4.39}$$

Thus $J_m = J_0(e^v - 1) - J_L$. Next v_m is clearly $(kT_c/q)v$. The current J_L is obtained from (4.39) as

$$J_L = J_0 e^v (1 + v) - J_0. \tag{4.40}$$

Lastly, V_{oc} is from (4.15)

$$V_{oc} = \frac{kT_c}{q} \ln\left(\frac{J_L}{J_0} + 1\right) = \frac{kT_c}{q} [\ln(1 + v) + v], \tag{4.41}$$

where (4.39) has been used An empirical expression describing the dependence of η_3 on $v_{oc} (\equiv qV_{oc}/kT_c)$ for $v_{oc} > 10$ is [19]

$$\eta_3 = \frac{v_{oc} - \ln(v_{oc} + 0.72)}{v_{oc} + 1}. \tag{4.42}$$

The efficiencies plotted in Figure 4.3 are products of these three factors η_1, η_2, and η_3 [20]. Additional graphs can be found in [20] while an extended theory is given in [21].

4.7 The Graded Gap Solar Cell

The efficiency of photovoltaic conversion can be increased by using multi-cell systems formed by stacking n cells of varying bandgap in a linear or nonlinear fashion in order of decreasing energy gap. This arrangement allows for the preferential absorption of photons through the stack, with high-energy photons absorbed at the front of the system whilst photons of lower energy penetrate further into the stack before being absorbed. The gain from such a system is twofold. By suitable choice of the range of energy gaps one could ensure that only photons close in energy to the bandgap of a particular cell are absorbed in that cell, thus reducing excess energy loss and heating in any one cell. Second, the range of bandgaps employed could allow for a reduction in the low-frequency cut off of the entire system. The most favorable situation occurs for very large n. In the limit $n \to \infty$ this case corresponds to a infinite number of solar cells, each of them of infinitesimal thickness and having a bandgap infinitesimally different from that of the neighboring cells. This case has been considered in Chapter 3.

Instead of stacking cells on top of each other in a discrete manner one can gain efficiency alternatively by considering one cell, but grading its

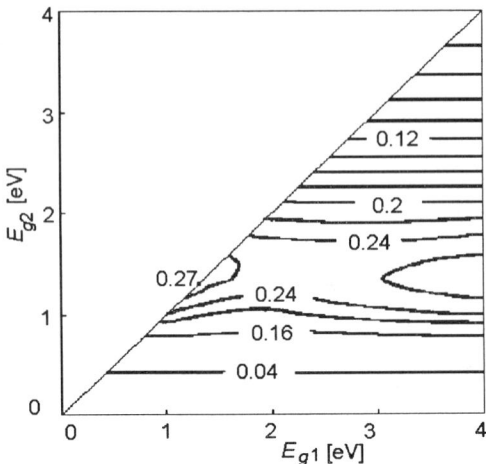

FIGURE 4.3. Heterojunction cell efficiency contours for $P_{e1} = P_{h1} = P_{e2} = P_{h2} = 0.4$, $\theta_{e1} = 2.20$, $\theta_{e2} = 2.00$, $\theta_{h1} = 2.36$, $\theta_{h2} = 3.00$. Maximum efficiency 27.5% (heterojunction) and 27.0% (if $E_{g1} = E_{g2}$). The latter case is not strictly a homojunction as material parameters other than E_g can differ from cell 1 to cell 2.

bandgap. Very roughly speaking, this corresponds to passing from an enumerable infinity of cells to a nonenumerable infinity. It has the advantage over the notion of stacking cells that the current generated in each unit can be drawn off more conveniently. We shall analyze the theory of such a structure (Section 4.7.4) after first deriving some essential preliminary results (Sections 4.7.1–4.7.3). The busy reader may proceed directly to Section 4.7.4.

The best way of utilizing such a cell is of course by edge-illumination as explained in Figure 4.4(b).

4.7.1 General

Consider the simplest case when a photon collides with an electron which jumps from some level in the valence band (energy state E_v) to some level in the conduction band (energy state E_c). One denotes

$$E_{cv} \equiv E_c - E_v. \tag{4.43}$$

The volumetric density rate of this upward electron transition [dimension: number of electrons per unit volume per unit time and per unit of absorbed photon energy] depends on:

4.7. The Graded Gap Solar Cell

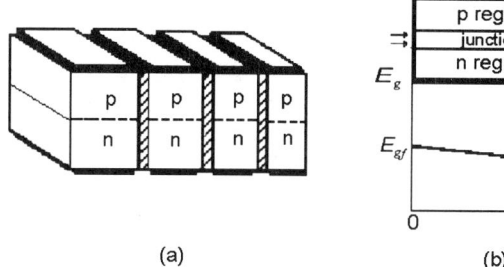

FIGURE 4.4. (a) Schematic arrangement of a four-segment edge-illuminated multigap cell. The metal electrodes are shown at top and bottom. (b) As in (a), but for a cell continuously graded from a front energy gap E_{gf} to a back energy gap E_{gb}.

(1) the probability f_v that the state E_v contains an electron;
(2) the probability $B_{v \to c}$ that the transition from state v to state c occurs in unit time;
(3) the probability $1 - f_c$ that the state E_c is empty; and
(4) the volumetric density $P(E_{cv})$ of photons of energy E_{cv} [dimension: number of photons per unit volume and per unit energy].

The occupation probabilities f_c and f_v are given as usual by Fermi–Dirac distributions

$$f_v = \frac{1}{\exp\left((E_v - F_v)/kT\right) + 1}; \quad f_c = \frac{1}{\exp\left((E_c - F_c)/kT\right) + 1}, \quad (4.44)$$

where F_v, F_c represent the quasi-Fermi levels for electrons at nonequilibrium in the valence and conduction bands, respectively. Here T denotes the temperature of the material.

The volumetric density of the upward electron transition rate [dimension: number of electrons per unit volume per unit time and per unit energy of absorbed photons] is then

$$\mathcal{R}_{v \to c} = f_v B_{v \to c} (1 - f_c) P(E_{cv}). \quad (4.45)$$

In addition to being absorbed, the photons can also stimulate the emission of similar photons by the transition of an electron from E_c to E_v. Thus, the total downward electron transition rate per unit volume per unit of emitted photon energy ($\mathcal{R}_{c \to v(\text{total})}$) is given by the sum of the spontaneous rate ($\mathcal{R}_{c \to v(\text{spon})}$) and the stimulated rate ($\mathcal{R}_{c \to v(\text{stim})}$) (both of them per unit volume and per unit of emitted photon energy):

86   4. Some Methods of Analyzing Solar Cell Efficiencies

$$\mathcal{R}_{c \to v}(\text{total}) = \mathcal{R}_{c \to v}(\text{spon}) + \mathcal{R}_{c \to v}(\text{stim}). \quad (4.46)$$

Probability arguments similar to those leading to (4.45) yield

$$\mathcal{R}_{c \to v}(\text{spon}) = f_c A_{c \to v}(1 - f_v), \quad (4.47)$$
$$\mathcal{R}_{c \to v}(\text{stim}) = f_c B_{c \to v}(1 - f_v) P(E_{cv}). \quad (4.48)$$

Here $A_{c \to v}$ and $B_{c \to v}$ are probability rates for spontaneous and stimulated electron transitions. Their dimensions are: probability per unit volume per unit time and per unit of emitted photon energy for $A_{c \to v}$ and probability per unit time for $B_{c \to v}$. (For a more detailed discussion of (4.45), (4.47), and (4.48) see Section 4.2 of [15]).

At thermal equilibrium $F_v = F_c$ and the upward electron transition rate must equal the total downward electron transition rate:

$$\mathcal{R}^{\text{eq}}_{v \to c} = \mathcal{R}^{\text{eq}}_{c \to v}(\text{total}) = \mathcal{R}^{\text{eq}}_{c \to v}(\text{spon}) + \mathcal{R}^{\text{eq}}_{c \to v}(\text{stim}). \quad (4.49)$$

Also, the equilibrium photon volumetric density is

$$P^{\text{eq}}(E_{cv}) = \frac{E_{cv}^2}{\pi^2 \hbar^3 (c/n)^3} \frac{1}{\exp(E_{cv}/kT) - 1}, \quad (4.50)$$

where the medium refraction index n was taken into account.

Then, (4.49) may be written by using (4.45), (4.47), and (4.50) as

$$\frac{n^3 E_{cv}^2}{\pi^2 \hbar^3 c^3} \left[B_{v \to c} \exp\left(\frac{E_{cv}}{kT}\right) - B_{c \to v} \right] = A_{c \to v} \exp\left(\frac{E_{cv}}{kT}\right) - A_{c \to v}. \quad (4.51)$$

Equating the temperature-independent terms of (4.51) gives

$$A_{c \to v} = \frac{n^3 E_{cv}^2}{\pi^2 \hbar^3 c^3} B_{v \to c}, \quad (4.52)$$

and then equating the temperature dependent parts of (4.51) with $A_{c \to v}$, given by (4.52), gives

$$B_{v \to c} = B_{c \to v}. \quad (4.53)$$

Equations (4.52) and (4.53) are sometimes called the Einstein relations. They strictly apply at equilibrium but also hold in nonequilibrium situations if quasi-Fermi levels exist (see [15, p. 317–318]).

4.7.2 *Photon Absorption Coefficient*

The photon absorption coefficienct α is usually related to the interaction between photons and electrons (the impact of photons on the nuclei is negligible in our cases).

4.7. The Graded Gap Solar Cell

First, stimulated emission is neglected. If one denotes by $\mathcal{A}$ the photon absorption rate per unit volume and per unit of absorbed photon energy, then $\mathcal{A} = \mathcal{R}_{v \to c}$. The photon absorption rate per unit volume is obtained after integration over *all* photon energies $\hbar\omega$:

$$\int \mathcal{A}d(\hbar\omega) = \int \mathcal{R}_{v \to c}d(\hbar\omega). \tag{4.54}$$

Now we shall analyze the right-hand side member of (4.45). Different models to evaluate the transition probability rate $B_{v \to c}$ are available [22]–[24]. For our purposes it is convenient to cite

$$B_{v \to c} = \frac{\pi q^2 \hbar}{m^2 \varepsilon_0 n^2 V \hbar\omega} |M|^2 \delta(E_{cv} - \hbar\omega), \tag{4.55}$$

where V, m, and ε_0 are the volume, the electron mass, and the dielectric constant of the medium, respectively, M is the electron momentum matrix element, and δ is Dirac's function. Equation (4.55) is similar to equation (3.3-34) of [23] which has to be amended twice: (i) the volume V has to be included (it should be used first on the left-hand side of equation (3.3-28) of [23]); and (ii) the δ-function has to be added (it was omitted in the definition of $B_{v \to c}$, (3.3-2) of [23]).

Now, by using (4.45) and (4.55) one evaluates the integral of (4.54) (keeping the notation $\hbar\omega$ for E_{cv}, as energy conservation is fulfilled):

$$\int \mathcal{A}d(\hbar\omega) = \frac{\pi q^2 \hbar}{m^2 \varepsilon_0 n^2 V \hbar\omega} |M|^2 f_v(1 - f_c)P(\hbar\omega). \tag{4.56}$$

After integration $E_c = E_v + \hbar\omega$. Thus, the product $f_v(1 - f_c)$ depends on E_v, F_v, F_c, and $\hbar\omega$; it is unity in the usual case of nondegenerate bands as $f_v \approx 1$ and $f_c \approx 0$.

The flux of photons per unit energy, $F(\hbar\omega)$, is related to the photon volumetric distribution per unit energy, $P(\hbar\omega)$, through the simple relation

$$\mathbf{F}(\hbar\omega) = \frac{c}{n} P(\hbar\omega). \tag{4.57}$$

Then the flux of photons of energy $\hbar\omega$ is

$$\mathbf{F}(\hbar\omega)\hbar\omega = \frac{c}{n} P(\hbar\omega)\hbar\omega. \tag{4.58}$$

The photon absorption coefficient is defined as

$$\tilde{\alpha}(\hbar\omega) \equiv \frac{\int \mathcal{A}d(\hbar\omega)}{\mathbf{F}(\hbar\omega)\hbar\omega} = \frac{\pi q^2 \hbar}{m^2 \varepsilon_0 n c V (\hbar\omega)^2} |M|^2 f_v(1 - f_c). \tag{4.59}$$

Here (4.56) and (4.58) were used. We used the tilde as a reminder that stimulated emission is neglected. Note that in (4.67), below, it is included.

88   4. Some Methods of Analyzing Solar Cell Efficiencies

One knows from band theory and the effective mass sum rule ([15, p. 532]) that it is sometimes a good approximation to write for the oscillator strength

$$\chi_{st}(\hbar\omega) \equiv \frac{2|M|^2}{m\hbar\omega} \approx \frac{E_g}{\hbar\omega}. \tag{4.60}$$

One denotes

$$\chi_f \equiv \frac{f_v(1-f_c)}{\hbar\omega}. \tag{4.61}$$

A simplification suggested by Kane [25] can be used: χ_f may be estimated near the band edge ($\hbar\omega = E_g$) and this behavior is extrapolated to all frequencies. The method is satisfactory for sources with low-radiation intensities. Let us put

$$\chi_{f,E_g} \equiv \left[\frac{f_v(1-f_c)}{\hbar\omega}\right]_{\hbar\omega=E_g} = \frac{f_v(1-f_c)}{E_g}. \tag{4.62}$$

When nondegenerate bands are considered, $\chi_{f,E_g} \approx 1/E_g$. By using (4.59)–(4.62) one obtains

$$\tilde{\alpha}(\hbar\omega) \approx \frac{\pi q^2 \hbar}{2m\epsilon_0 cnV} \chi_{st}(\hbar\omega)\chi_{f,E_g}. \tag{4.63}$$

The abrupt absorption model is the simplest one, and we shall use it here. It assumes the "strength" of the absorption process is given by a Heaviside step function $\Theta\langle E_g - \hbar\omega\rangle$ at the bandgap energy:

$$\text{absorption strength} = 0 \quad \text{for} \quad \hbar\omega < E_g \quad (\text{i.e, for } \chi_{st} > 1), \tag{4.64}$$
$$\text{absorption strength} = 1 \quad \text{for} \quad \hbar\omega \geq E_g \quad (\text{i.e., for } \chi_{st} \leq 1).$$

Replace $\chi_{st}(\hbar\omega)$ by $\Theta\langle E_g - \hbar\omega\rangle$ in (4.63), and one finds

$$\tilde{\alpha}(\hbar\omega) = \alpha_0 \Theta\langle E_g - \hbar\omega\rangle \quad \text{where} \quad \alpha_0 \equiv \frac{\pi q^2 \hbar}{2m\epsilon_0 cnV} \chi_{f,E_g}. \tag{4.65}$$

Now we shall treat the photon absorption coefficient including stimulated emission. The net absorption photon rate per unit volume and per unit photon energy, $\mathcal{A}_{net}$, is the difference between the upward electron transition rate $\mathcal{R}_{v\to c}$ and the downward stimulated electron transition rate $\mathcal{R}_{c\to v(stim)}$ (both of them per unit volume and per unit photon energy). By using (4.45), (4.48), and (4.53) one obtains

$$\mathcal{A}_{net} \equiv \mathcal{R}_{v\to c} - \mathcal{R}_{c\to v(stim)} = B_{v\to c}(f_v - f_c)P(E_{cv}). \tag{4.66}$$

The photon absorption coefficienct $\alpha(\hbar\omega)$ is defined similarly to (4.59). Then, use of (4.66) leads to

$$\alpha(\hbar\omega) \equiv \frac{\int \mathcal{A}_{\text{net}} d(\hbar\omega)}{\mathbf{F}(\hbar\omega)\hbar\omega} = \frac{\pi q^2 \hbar}{m^2 \varepsilon_0 n c V (\hbar\omega)^2} |M|^2 (f_v - f_c). \qquad (4.67)$$

One can write

$$f_v - f_c = f_v(1 - f_c)\left[1 - \frac{f_c(1 - f_v)}{f_v(1 - f_c)}\right]. \qquad (4.68)$$

Using the definitions of f_c and f_v ((4.44)), and the notation $qV \equiv F_c - F_v$, one obtains

$$\frac{f_c(1 - f_v)}{f_v(1 - f_c)} = \exp\left(\frac{qV - \hbar\omega}{kT_c}\right). \qquad (4.69)$$

Finally, by using (4.60)–(4.62), (4.65), and (4.67)–(4.69) one derives a simple approximate expression for the absorption coefficient

$$\alpha(\hbar\omega) = \alpha_0 \left[1 - \exp\left(\frac{qV - \hbar\omega}{kT_c}\right)\right] \Theta\langle E_g - \hbar\omega\rangle. \qquad (4.70)$$

The exponential term is the contribution of stimulated emission, an effect which was omitted in (4.65).

4.7.3 *Photon Emission Rates*

The spontaneous photon emission rate per unit volume and per unit photon energy, $\mathcal{E}_{\text{spon}}$, is defined as

$$\mathcal{E}_{\text{spon}} \equiv \mathcal{R}_{c \to v(\text{spon})} = \frac{n^3 E_{cv}^2}{\pi^2 \hbar^3 c^3} B_{v \to c} f_c (1 - f_v). \qquad (4.71)$$

Here we used (4.47), (4.52), and (4.53). The spontaneous photon emission rate per unit volume is obtained by using (4.71), (4.55), and (4.67):

$$\int \mathcal{E}_{\text{spon}} d(\hbar\omega) = \frac{n^2(\hbar\omega)^3}{\pi^2 \hbar^3 c^2} \frac{f_c(1 - f_v)}{f_v - f_c} \alpha(\hbar\omega). \qquad (4.72)$$

One now again uses the definitions of f_b and f_v ((4.44)) and the notation $qV \equiv F_c - F_v$. By the "play of probabilities" (see [15, p. 318]) one has

$$\frac{f_c(1 - f_v)}{f_v - f_c} = \frac{1}{\exp\left((\hbar\omega - qV)/kT\right) - 1}. \qquad (4.73)$$

Replacing (4.73) into (4.72), one obtains:

$$\int \mathcal{E}_{\text{spon}} d(\hbar\omega) = \frac{n^2(\hbar\omega)^3}{\pi^2 \hbar^3 c^2} \left[\exp\left(\frac{\hbar\omega - qV}{kT}\right) - 1\right]^{-1} \alpha(\hbar\omega). \qquad (4.74)$$

It has become common usage to define $A_{c \to v}(f_c - f_v)$ as the stimulated photon emission rate per unit volume and per unit photon energy, $\mathcal{E}_{\text{stim}}$, because of similarity in form to the spontaneous electron transition rate in (4.47). By using (4.47), (4.55), and (4.67) one finds the stimulated photon emission rate per unit volume

$$\int \mathcal{E}_{\text{stim}} \, d(\hbar\omega) \equiv \int A_{c \to v}(f_c - f_v) d(\hbar\omega) = -\frac{n^2(\hbar\omega)^3}{\pi^2 \hbar^3 c^2} \alpha(\hbar\omega). \quad (4.75)$$

The *net* stimulated photon emission rate per unit volume and per unit photon energy, $\mathcal{E}_{\text{stim(net)}}$, is defined as the difference between the downward electron transition rate $\mathcal{R}_{c \to v(\text{stim})}$ and the upward electron transition rate $\mathcal{R}_{v \to c}$ (both of them per unit volume and per unit photon energy):

$$\mathcal{E}_{\text{stim(net)}} \equiv \mathcal{R}_{c \to v(\text{stim})} - \mathcal{R}_{v \to c} = B_{v \to c}(f_c - f_v) P(\hbar\omega). \quad (4.76)$$

Here we used (4.48), (4.45), and (4.53). In fact, from (4.66) and (4.76) one can see that $\mathcal{E}_{\text{stim(net)}} = -\mathcal{A}_{\text{net}}$. The net stimulated photon emission rate per unit volume is obtained by using (4.76), (4.53), (4.67), and (4.50):

$$\int \mathcal{E}_{\text{stim(net)}} d(\hbar\omega) = \frac{n^2(\hbar\omega)^3}{\pi^2 \hbar^3 c^2} \left[1 - \exp\left(\frac{\hbar\omega}{kT}\right)\right]^{-1} \alpha(\hbar\omega). \quad (4.77)$$

The total photon absorption rate per unit volume, $\int \mathcal{E}_{\text{total}} d(\hbar\omega)$, is defined as

$$\int \mathcal{E}_{\text{total}} d(\hbar\omega) \equiv \int \mathcal{E}_{\text{spon}} d(\hbar\omega) + \int \mathcal{E}_{\text{stim(net)}} d(\hbar\omega). \quad (4.78)$$

This rate will be used when the total electron–hole recombination rate will be considered ((4.91)).

4.7.4 *Solar Energy Conversion*

A number flux $\gamma_p(\hbar\omega)$ of solar photons per unit energy is normally incident on a plane semiconductor surface $z = 0$, as shown in Figure 4.4 [26], [27]. The energy gap $E_g(z)$ of the semiconductor is arranged to be an arbitrary but monotonically decreasing function of z. One defines z_ω as that value of z at which $\hbar\omega = E_g(z)$. Under the assumptions of the abrupt absorption model, photons with energy $\hbar\omega$ can be absorbed only at $z > z_\omega$. The electron–hole pair generation rate per unit volume and unit photon energy $G(\hbar\omega, z)$ is due to photons with energy $\hbar\omega \geq E_g(z)$ and is related to the absorption coefficient $\alpha(\hbar\omega)$ through a simple relation similar to (4.59):

$$\alpha(\hbar\omega) = \frac{G(\hbar\omega, z)}{\gamma(\hbar\omega, z)}, \quad (4.79)$$

4.7. The Graded Gap Solar Cell

where $\gamma(\hbar\omega, z)$ is the number flux of solar photons per unit energy at z, given by Beer's (or Bouguer–Lambert) law:

$$\gamma(\hbar\omega, z) = \gamma_p(\hbar\omega) e^{-\alpha(\hbar\omega)(z-z_\omega)}. \tag{4.80}$$

For simplicity, in the photon absorption computation we shall use the approximation $\alpha(\hbar\omega) \approx \tilde{\alpha}(\hbar\omega)$. By using (4.79), (4.80), and (4.65) one obtains the electron–hole pair generation rate per unit volume $G(z)$:

$$G(z) \equiv \int_0^\infty G(\hbar\omega, z) d(\hbar\omega) = \alpha_0 \int_{E_g(z)}^\infty \gamma_p(\hbar\omega) e^{-\alpha_0(z-z_\omega)} d(\hbar\omega), \tag{4.81}$$

where α_0 is given by (4.65) and the properties of the Heaviside function were used.

Let us now consider a slab of semiconductor, of thickness dz, located at abscissa z and *electrically* isolated from the surrounding semiconductor with other values of z. If one denotes by η_c the pair collection efficiency, then the short circuit current density dJ_{sc} generated by that slab is

$$dJ_{sc} = q\eta_c G(z) \, dz. \tag{4.82}$$

Let us expand $E_g(z')$ about $E_g(z)$ in a Taylor series

$$E_g(z') = E_g(z) + (z' - z) \frac{dE_g}{dz}\bigg|_z + \ldots . \tag{4.83}$$

One neglects higher terms and notes that when $z' = z_\omega$ one has $E_g(z') = E_g$ (according to the definition of z_ω). Then, from (4.83), one obtains

$$z - z_\omega = \frac{\hbar\omega - E_g}{|dE_g/dz|}. \tag{4.84}$$

With our conventions dE_g/dz is negative.

If the solar cell is surrounded by air (refractive index $n_{air} = 1$), $\gamma_p(\hbar\omega)$ and the spectrally integrated solar energy flux φ_p are given by

$$\gamma_p(\hbar\omega) = \frac{B_p}{4\pi^3 \hbar^3 c^2} \frac{(\hbar\omega)^2}{\exp(\hbar\omega/kT_p) - 1}, \tag{4.85}$$

$$\varphi_p = \frac{B_p}{4\pi^3 \hbar^3 c^2} \int_0^\infty \frac{(\hbar\omega)^3 d(\hbar\omega)}{\exp(\hbar\omega/kT_p) - 1} = \frac{B_p \pi k^4}{60 \hbar^3 c^2} T_p^4. \tag{4.86}$$

One can eliminate the geometric factor B_p between (4.85) and (4.86) and one obtains

$$\gamma_p(\hbar\omega) = \varphi_p \frac{15(\hbar\omega)^2}{(\pi kT_p)^4} \left[\exp\frac{\hbar\omega}{kT_p} - 1\right]^{-1}. \tag{4.87}$$

By using (4.87), the integral in (4.81) becomes

$$\frac{15\varphi_p}{\pi^4 kT_p} \exp(K\varepsilon) \cdot F(\varepsilon, K), \qquad (4.88)$$

where

$$\varepsilon \equiv \varepsilon(z) \equiv \frac{E_g(z)}{kT_p}, \quad K \equiv K(z) \equiv \frac{\alpha_0}{|d\varepsilon/dz|}, \quad F(s, u) \equiv \int_s^\infty \frac{x^2 e^{-ux}}{e^x - 1} dx. \qquad (4.89)$$

By using (4.81), (4.82), and (4.88) one obtains

$$dJ_{\text{sc}} = \frac{15}{\pi^4} \frac{q\eta_c\alpha_0}{kT_p} \varphi_p \exp(K\varepsilon) F(\varepsilon, K) \, dz. \qquad (4.90)$$

In order to derive an appropriate expression for the reverse (dark) current density, dJ_{rev}, one should take into account that the total photon emission rate per unit volume, $\int \mathcal{E}_{\text{total}} d(\hbar\omega)$, equals the total electron–hole recombination rate per unit volume. By assuming a perfect pair collection ($\eta_c = 1$) the reverse current density dJ_{rev} associated with the infinitesimal slab is

$$dJ_{\text{rev}} = q \left[\int \mathcal{E}_{\text{total}} d(\hbar\omega) \right] dz. \qquad (4.91)$$

Use of (4.70), (4.74), (4.76), and (4.78) yields

$$\int \mathcal{E}_{\text{total}} d(\hbar\omega) = \frac{n^2 \alpha_0}{\pi^2 c^2 \hbar^3} (\hbar\omega)^2 \frac{\exp(qV/kT_c) - 1}{\exp(\hbar\omega/kT_c) - 1}. \qquad (4.92)$$

Using (4.91) and (4.92) one finds

$$dJ_{\text{rev}} = \frac{qn^2\alpha_0}{\pi^2 c^2 \hbar^3} (kT_c)^3 \left[\exp\left(\frac{qV}{kT_c}\right) - 1 \right] F(\varepsilon t, 0) \, dz, \qquad (4.93)$$

where $t \equiv T_p/T_c$. One recognizes in (4.93) the exponential term of the usual solar cell equation.

The open-circuit voltage V_{oc} across the slab is obtained in steady state (i.e., when $dJ_{\text{sc}} - dJ_{\text{rev}} = 0$). By using (4.90) and (4.93), and taking into account that $\varphi_p = (B_p \pi k^4/60\hbar^3 c^2) T_p^4$, one obtains

$$V_{\text{oc}}(z) = \frac{kT_c}{q} \ln\left[1 + \frac{\eta_c B_p t^3}{4\pi n^2} \frac{\exp(K\varepsilon) F(\varepsilon, K)}{F(\varepsilon t, 0)} \right]. \qquad (4.94)$$

The power output P_{el} from all slabs is

$$P_{el} = \int \eta_3(z) \frac{dJ_{\text{sc}}}{dz} V_{\text{oc}}(z) \, dz. \qquad (4.95)$$

where $\eta_3(z)$ is the fill factor defined in (4.36) whose approximation (4.42) is used in computations. It is convenient to change in (4.95) the variable z to ε (which is defined by (4.89)). Thus

$$P_{el} = \frac{1}{\alpha_0} \int_0^\infty \eta_3(z) \left(\frac{dJ_{sc}}{dz}\right) V_{oc}(\varepsilon) K \, d\varepsilon. \quad (4.96)$$

The efficiency can be obtained by substituting (4.90) and (4.94) into (4.96) followed by a division by φ_p. One obtains

$$\eta = \frac{15\eta_c}{\pi^4 t} \int_0^\infty \eta_3 \exp(K\varepsilon) F(\varepsilon, K) \ln\left[1 + \frac{\eta_c B_p t^3}{4\pi n^2} \exp(K\varepsilon) \frac{F(\varepsilon, K)}{F(\varepsilon t, 0)}\right] K \, d\varepsilon. \quad (4.97)$$

The increase in efficiency in the case of an "infinite" tandem cell is important: $\eta_{max} \approx 64\%$ for concentration ratio $C = 1$, $\eta_{max} \approx 77\%$ for $C = 1000$, and $\eta_{max} \approx 81\%$ for $C = 10,000$. At maximum concentration ($C = C_{max} \approx 46,300$) one obtains $\eta = 88\%$ [26] in the case of a continuously graded cell; only slightly in excess of the value of 86.8% noted in Section 3.7. Owing to small differences in approximations the two upper limits do not agree completely.

4.8 Thermophotovoltaic Conversion

Solar radiation is usually considered as black-body radiation at temperature $T_p \approx 5762$ K. For a typical semiconductor solar cell with bandgap $E_g \approx 1.0$ eV, a fraction 0.80 of the incident solar radiation energy is absorbed. The average energy of the absorbed solar photons with $\hbar\omega \geq E_g$ is ≈ 1.9 eV. The amount of $1.9 - 1.0 = 0.9$ eV is transferred to the cell lattice as heat upon thermalization of photo-generated electron–hole pairs. The efficiency of solar energy conversion into electricity is improved if incident solar radiation is first absorbed by an intermediate absorber at temperature T_A lower than T_p. The thermal radiation emitted by the absorber is directed onto the solar cell. Under the new circumstances the average energy of the photons absorbed by the cell is, of course, smaller than 1.9 eV. Consequently, the heat losses by thermalization decrease and the overall conversion efficiency increases. Photovoltaic conversion with the addition of an intermediate absorber/emitter is known as thermophotovoltaic energy conversion ("TPV").

4.8.1 Definitions

As the TPV conversion theory deals with *energy* and *number* fluxes in the case of both black-body and bandgap materials a brief presentation of concepts used in the thermodynamic theory of radiation is useful. Let us consider a volume V containing isotropic polarized radiation emitted by a body at temperature T.

Consider a general macroscopic property Ξ of the radiation (its energy $E(V,T)$ or the photon number $N(V,T)$ are examples). Its properties can be expressed by three different functions.

(i) The "spectral" property per unit frequency ω is then

$$\mathcal{P}_\Xi(\omega, V, T) \equiv \frac{d\Xi(V,T)}{d\omega} = \frac{V\omega^2}{\pi^2 c^3} \frac{\xi}{\exp\left((\hbar\omega - \mu)/kT\right) - 1}, \quad (4.98)$$

where ξ is the value of the property considered per photon and μ is the chemical potential of the radiation. The dimension of $\mathcal{P}_\Xi$ is Ξ per unit frequency.

(ii) For isotropic radiation the spectral density of the property Ξ is

$$\frac{\mathcal{P}_\Xi(\omega,V,T)}{4\pi V} \equiv \frac{1}{4\pi V}\frac{d\Xi(V,T)}{d\omega} = \frac{1}{4\pi^3 c^3} \frac{\xi\omega^2}{\exp\left((\hbar\omega - \mu)/kT\right) - 1}. \quad (4.99)$$

The dimension of (4.99) is Ξ per unit frequency per unit volume per unit solid angle.

(iii) The spectral radiance of the property is

$$\mathcal{R}_\Xi(\omega, T) \equiv c\frac{\mathcal{P}_\Xi(\omega, V, T)}{4\pi V} = \frac{1}{4\pi^3 c^2} \frac{\xi\omega^2}{\exp\left((\hbar\omega - \mu)/kT\right) - 1} \quad (4.100)$$

and has the dimension of Ξ per unit area per unit time per unit solid angle per unit frequency.

Before zproceeding to the conversion theory it is helpful to have some additional definitions. One considers a point P on the surface of a body. Let $d\Omega_{\alpha\beta}$ be an infinitesimal solid angle in direction (α, β) with vertex at point P. Here α and β are zenith and azimuth angles in respect to the surface normal. One denotes by $\mathcal{F}_\Xi(\omega, T)$ the *spectral property density flux* [dimension: property per unit area, unit time, and unit frequency] incident at P from a solid angle Ω_{inc}. Then, $\mathcal{F}_\Xi(\omega, T)$ is given by

$$\mathcal{F}_\Xi(\omega, T) = \int_{\Omega_{inc}} \mathcal{R}_{\Xi,\alpha\beta}(\omega,T)d(\alpha, \beta)\, d\Omega_{\alpha\beta}, \quad (4.101)$$

where $d(\alpha, \beta)$ is a function which describes the standard response of the receiving body to radiation of the same radiance coming from different directions. For a plane Lambertian surface, and for a small isotropic spherical receiver, one has, respectively,

$$d_{\text{plane}}(\alpha, \beta) = \cos \alpha, \quad d_{\text{sphere}}(\alpha, \beta) = 1. \tag{4.102}$$

The *spectral* quantity $\mathbf{f}_{\Xi,\text{sphere}}(\omega, T)$ of dimension property per unit time and unit frequency and received from the solid angle Ω_{inc} by a small spherical body of radius r is given by

$$\mathbf{f}_{\Xi,\text{sphere}}(\omega, T) = \pi r^2 \mathcal{F}_{\Xi,\text{sphere}}(\omega, T) = \pi r^2 \int_{\Omega_{inc}} \mathcal{R}_{\Xi,\alpha\beta}(\omega, T) \, d\Omega_{\alpha\beta}. \tag{4.103}$$

Here (4.101) and (4.102) were used. Let us suppose the body receives isotropic radiation from the finite solid angle Ω_{inc}. Then the quantity $\mathcal{R}_{\Xi}(\omega, T) \equiv \mathcal{R}_{\Xi,\alpha\beta}(\omega, T)$ does not depend on direction, and simple integration of (4.103) leads to

$$\mathbf{f}_{\Xi,\text{sphere}}(\omega, T) = \pi r^2 \mathcal{R}_{\Xi}(\omega, T) \int_{\Omega_{\text{inc}}} d\Omega_{\alpha\beta} = \pi r^2 \mathcal{R}_{\Xi}(\omega, T) \Omega_{\text{inc}}. \tag{4.104}$$

The spectral property flux $\mathbf{f}_{\Xi,\text{plane}}(\omega, T)$ received from the solid angle Ω_{inc} by a plane surface of area S_{plane} is given by

$$\mathbf{f}_{\Xi,\text{plane}}(\omega, T) = S_{\text{plane}} \mathcal{F}_{\Xi,\text{plane}}(\omega, T) = S_{\text{plane}} \int_{\Omega_{\text{inc}}} \mathcal{R}_{\Xi,\alpha\beta}(\omega, T) \cos \alpha \, d\Omega_{\alpha\beta}. \tag{4.105}$$

In the case of isotropic radiation one obtains

$$\mathbf{f}_{\Xi,\text{plane}}(\omega, T) = S_{\text{plane}} B \mathcal{R}_{\Xi}(\omega, T), \tag{4.106}$$

where the geometric factor B is given by $B \equiv \int_{\Omega_{inc}} \cos \alpha \, d\Omega_{\alpha\beta}$.

The property flux $\mathbf{F}_{\Xi,\text{sphere(plane)}}(\omega, T)$ [dimension: property per unit time] received by a small spherical body or a plane surface is

$$F_{\Xi,\text{sphere(plane)}}(T) = \int_0^\infty \mathbf{f}_{\Xi,\text{sphere(plane)}}(\omega, T) \, d\omega. \tag{4.107}$$

When fluxes of absorbed radiation are considered, the right-hand side member of (4.101) has to be multiplied by the absorptivity $a_{\alpha\beta}(\omega, T)$.

Let us apply the above general formulas for two common radiation properties, namely energy and photon number. In the case of black-body radiation $\mu = 0$. Applying (4.98) for $\Xi = E$ and $\xi = \hbar\omega$ one obtains as a particular case the Planck energy distribution per unit frequency range $\mathcal{P}_{E_{bb}}(\omega, V, T)$ (see, for instance, [28, p. 203, Eq. (60.5)]):

$$\mathcal{P}_{E_{bb}}(\omega, V, T) \equiv \frac{dE_{bb}(V, T)}{d\omega} = \frac{V\hbar}{\pi^2 c^3} \frac{\omega^3}{\exp(\hbar\omega/kT) - 1}. \tag{4.108}$$

When the body is a semiconductor, $\mu = q(E_{Fe} - E_{Fh})$, where E_{Fe} and E_{Fh} are the quasi-Fermi levels for electrons and holes, respectively. For a semiconductor solar *cell* the usual approximation is $E_{Fe} - E_{Fh} = V$, where V is the voltage across the cell. Then, (4.98) for $\xi = \hbar\omega$ and $\mu = qV$ becomes

$$\mathcal{P}_{E_{bb}}(\omega, V, T) \equiv \frac{dE_c(V,T)}{d\omega} = \frac{V\hbar}{\pi^2 c^3} \frac{\omega^3}{\exp\left((\hbar\omega - qV)/kT\right) - 1}. \quad (4.109)$$

The spectral energy radiance $\mathcal{R}_E(\omega, T)$ is obtained from (4.100):

$$\mathcal{R}_E(\omega, T) = c\frac{\mathcal{P}_E(\omega, V, T)}{4\pi V} = \frac{\hbar}{4\pi^3 c^2} \frac{\omega^3}{\exp\left((\hbar\omega - \mu)/kT\right) - 1}. \quad (4.110)$$

Similarly, the spectral number radiance $\mathcal{R}_N(\omega, T)$ is obtained by using (4.100) in the case of $\Xi = N$ and $v = 1$:

$$\mathcal{R}_N(\omega, T) \equiv c\frac{1}{4\pi V}\frac{dN(V,T)}{d\omega} = \frac{1}{4\pi^3 c^2} \frac{\omega^2}{\exp\left((\hbar\omega - \mu)/kT\right) - 1}. \quad (4.111)$$

The spectral energy fluxes $\mathbf{f}_{E,\text{sphere}}(\omega, T)$ and $\mathbf{f}_{E,\text{plane}}(\omega, T)$ received by a small spherical body of radius r and a plane surface of area S_{plane} are given, respectively, by

$$\mathbf{f}_{E,\text{sphere}}(\omega, T) = \pi r^2 \mathcal{R}_E(\omega, T) \Omega_{\text{inc}}, \quad (4.112)$$
$$\mathbf{f}_{E,\text{plane}}(\omega, T) = S_{\text{plane}} B \mathcal{R}_E(\omega, T).$$

Integration of (4.112) over the spectrum gives the energy flux $\mathbf{F}_E$ incident on the spherical or plane surface.

4.8.2 *Theory of TPV Conversion*

We now turn to a simple model of TPV conversion. The configuration is that of [29] but the theoretical approach is different. An intermediate small absorber A of spherical shape (radius r_A and temperature T_A), placed in the focus of a lens or mirror (Figure 4.5) receives radiation from the Sun (temperature T_p) with a solid angle of incidence Ω_{inc}, and from the surroundings with the solid angle $4\pi - \Omega_{\text{inc}}$. For unfocused sunlight we have $\Omega_{\text{inc}} = \Omega_s$. For focused sunlight we have $\Omega_{\text{inc}} > \Omega_s$. One places the absorber A as in Figure 4.5 in one focal point of a rotational ellipsoid whose inner wall is a perfect mirror. Into the other focal point of the ellipsoid we place the spherical photovoltaic cell C (radius r_c and temperature T_c). The solar cell receives radiation that is emitted by the absorber A into the solid

4.8. Thermophotovoltaic Conversion

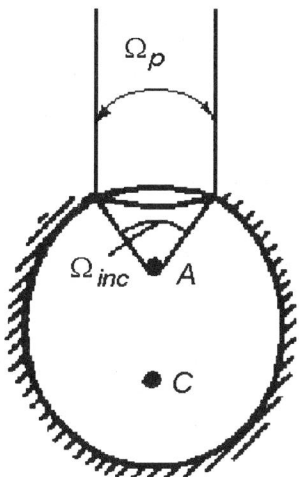

FIGURE 4.5. An ellipsoidal mirror has a spherical absorber A and a solar cell C placed at its foci for thermophotovoltaic conversion. Solar radiation is incident under the solid angle Ω_p so as to converge in A.

angle $4\pi - \Omega'_{inc}$ (after reflexion at the ellipsoid walls). Here Ω'_{inc} ($< \Omega_{inc}$) refers to the solid angle subtended at the cell.

Since both the absorber and the cell are spherical, they emit photons into a solid angle 4π.

Equation (4.107) is used now to derive the steady-state energy balance for the small spherical absorber. By neglecting the conductive and convective losses one obtains

$$\Omega_{inc} S_{TA} \int_0^\infty a\mathcal{R}_E(\omega, T_p)\, d\omega + (4\pi - \Omega_{inc}) S_{TA} \int_0^\infty a\mathcal{R}_E(\omega, T_c)\, d\omega$$
$$= 4\pi S_{TA} \int_0^\infty a\mathcal{R}_E(\omega, T_A)\, d\omega. \quad (4.113)$$

Here $S_{TA} \equiv \pi r_A^2$ and a is the absorptivity. The first and second term in the left-hand member of (4.113) refer to the energy fluxes received from Sun and cell, respectively, while the right-hand member is the energy flux emitted by the absorber.

Let us assume the absorber is black-body ($\mu_A = 0$ and $a = 1$). As the chemical potential of solar radiation is $\mu_p = 0$, then by using (4.110) one obtains (see [30, p. 230] for details of the calculation)

$$\int_0^\infty \mathcal{R}_E(\omega, T_p)\, d\omega = \frac{\sigma}{\pi} T_p^4, \quad \int_0^\infty \mathcal{R}_E(\omega, T_A)\, d\omega = \frac{\sigma}{\pi} T_A^4, \quad (4.114)$$

where σ ($= \pi^2 k^4/60\hbar^3 c^2$) is the Stefan–Boltzmann constant. The chemical potential of the radiation emitted by the cell is $\mu_c = qV$ and the second

98   4. Some Methods of Analyzing Solar Cell Efficiencies

integral from (4.113) becomes

$$\int_0^\infty \mathcal{R}_E(\omega, T_c)\, d\omega = \frac{1}{4\pi^3 c^2} \int_{E_g/\hbar}^\infty \frac{\hbar\omega^3\, d\omega}{\exp\left((\hbar\omega - qV)/kT_c\right) - 1}. \quad (4.115)$$

Here we have taken into account that the photons emitted by the cell have frequencies $\omega \geq E_g/\hbar$. Use of (4.114) enables (4.113) to become

$$\Omega_{\text{inc}}\sigma T_p^4 + \frac{(4\pi - \Omega_{\text{inc}})}{4\pi^2 c^2} \int_{E_g/\hbar}^\infty \frac{\hbar\omega^3\, d\omega}{\exp\left((\hbar\omega - qV)/kT_c\right) - 1} = 4\pi\sigma T_A^4. \quad (4.116)$$

The absorber temperature T_A can be obtained from (4.116) as a function of T_c and V. Apart from multiplying constants, relation (4.116) puts the energy flux from the pump plus that from the cell equal to the energy flux emitted by the absorber.

The spectral number flux $\mathbf{f}_N(\omega, T_A)$ of photons emitted by the absorber is given by (4.104) if one bears in mind that the absorber emits into the solid angle 4π. Then

$$\mathbf{f}_N(\omega, T_A) = 4\pi S_{T_A} \mathcal{R}_N(\omega, T_A). \quad (4.117)$$

Of the photon number flux emitted by the intermediate absorber into the solid angle 4π (i.e., $\int_{E_g/\omega}^\infty \mathbf{f}_N(\omega, T_A)\, d\omega$), the fraction $\Omega'_{\text{inc}}/4\pi$ is emitted into the same solid angle from which the solar radiation is received. This fraction cannot, therefore, be directed onto a solar cell without obscuring the incident solar radiation. Only the fraction $1 - \Omega'_{\text{inc}}/4\pi$ of the photons emitted by the absorber can therefore be maximally focused onto the solar cell.

In the simplest case the solar cell can be assumed to have the following ideal properties:

(a) its absorptivity a is 0 for $\hbar\omega < E_g$;
(b) all photons with $\hbar\omega \geq E_g$ are absorbed, i.e., absorptivity is unity for $\hbar\omega \geq E_g$;
(c) each photon absorbed generates a pair electron–hole;
(d) recombination of electrons and holes is entirely radiative;
(e) photons with $\hbar\omega < E_g$ are not reflected onto the intermediate absorber.

Consequently, the number flux $\mathbf{F}_{N,\text{abs}}$ of photons absorbed by the cell is given by (4.107) and it is

$$\mathbf{F}_{N,\text{abs}} = \left(1 - \frac{\Omega'_{\text{inc}}}{4\pi}\right) \int_{E_g/\hbar}^\infty \mathbf{f}_N(\omega, T_A)\, d\omega. \quad (4.118)$$

4.8. Thermophotovoltaic Conversion

By using assumption (e) the electrical short circuit (sc) current I_{sc} provided by the cell is given by

$$I_{sc} = q\mathbf{F}_{N,\text{abs}}. \tag{4.119}$$

The electrical short circuit current density J_{sc} of the solar cell can be defined as follows:

$$J_{sc} \equiv \frac{I_{sc}}{S_c} = \frac{q}{4S_{Tc}}\left(1 - \frac{\Omega'_{\text{inc}}}{4\pi}\right)\int_{E_g/\hbar}^{\infty} \mathbf{f}_N(\omega, T_A)\, d\omega, \tag{4.120}$$

where $S_c \equiv 4S_{Tc} \equiv 4\pi r_c^2$. By taking into account (4.117) and (4.111) one obtains

$$J_{sc} = q\,\frac{S_{TA}}{4S_{Tc}}\,\frac{4\pi - \Omega'_{inc}}{4\pi^3 c^2}\int_{E_g/\hbar}^{\infty}\frac{\omega^2\, d\omega}{\exp(\hbar\omega/kT_A)-1}. \tag{4.121}$$

In open-circuit (oc) the solar cell is in a steady state. This means that the number flux of photons absorbed by the cell $\mathbf{F}_{N,\text{abs}}$ is equal to the number flux of photons emitted by the cell $\mathbf{F}_{N,\text{em}}$. The last quantity is given by

$$\mathbf{F}_{N,\text{em}} = \int_{E_g/\hbar}^{\infty} \mathbf{f}_N(\omega, T_c)\, d\omega. \tag{4.122}$$

The steady-state solar cell equation $\mathbf{F}_{N,\text{abs}} = \mathbf{F}_{N,\text{em}}$ is written finally by

$$(1-\frac{\Omega'_{\text{inc}}}{4\pi})\frac{S_{TA}}{S_{Tc}}\int_{E_g/\hbar}^{\infty}\frac{\omega^2}{\exp(\hbar\omega/kT_A)-1}d\omega = \int_{E_g/\hbar}^{\infty}\frac{\omega^2}{\exp\big((\hbar\omega-qV_{\text{oc}})/kT_c\big)-1}d\omega. \tag{4.123}$$

The open-circuit voltage V_{oc} can be obtained by solving (4.123) numerically.

The thermophotovoltaic conversion efficiency η_{TPV} can be defined as the ratio between the output power density of the solar cell and the solar energy flux incident on the intermediate absorber. The first quantity is given by $\eta_3 J_{sc} V_{\text{oc}}$ where the fill factor η_3 can be evaluated in first approximation as a function of V_{oc} from (4.42). The last quantity can be evaluated by using (4.107), (4.113), and (4.111) and is $\Omega_{\text{inc}} S_{TA}(\sigma/\pi)T_p^4$. Finally, the efficiency is

$$\eta_{\text{TPV}} = \frac{\eta_3 J_{sc} V_{\text{oc}}}{\Omega_{\text{inc}} S_{TA}(\sigma/\pi)T_p^4}. \tag{4.124}$$

Previous studies of TPV devices using fully concentrated radiation show that [29]:

- an intermediate black-body absorber increases the efficiency of small-bandgap solar cells considerably;

- for solar cells with a bandgap of $E_g = 1$ eV and greater, an intermediate black-body absorber is disadvantageous (as an example, $\eta_{\max}(E_g = 0.3 \text{ eV}) \approx 0.30$, $\eta_{\max}(E_g = 0.5 \text{ eV}) \approx 0.33$, and $\eta_{\max}(E_g = 1.0 \text{ eV}) \approx 0.27$);
- in order to increase the efficiency, a selective absorber is to be used; the maximum efficiency in this case is 54% and corresponds to $E_g \approx 0.80$ eV, a temperature of the intermediate absorber of 1800 K and a solid angle of the incident radiation of 0.03 sr;
- The efficiency increases further if a selective mirror is used; in this case the maximum efficiency is 65% (which corresponds to $E_g \approx 0.80$ eV and a solid angle of the incident radiation of 0.03 sr). The above qualitative results were derived in [29] under neglect of the term qV in (4.116), so that the present theory may give marginally different results.

4.9 Recent Results

The potential of an ideal photovoltaic cell with charge carrier multiplication by impact ionization is discussed in [31]. Such a cell is described as a hot-carrier cell where interband equilibrium is inhibited and where thermalization is restricted to the band edge. The cell is shown to be equivalent to a thermal absorber coupled to a Carnot engine. In conjunction with a spectral filter, a finite bandgap cell with carrier multiplication can even surpass the work obtainable from a selective thermal absorber. Figures 4 and 5 of [31] show examples of the influence of impact ionization on the solar cell efficiency, for various bandgaps. The efficiency shows a maximum that is somewhat higher than without carrier multiplication and slightly shifted to a lower-gap energy. Most solar photons have not enough energy for significant carrier multiplication. A local maximum at low-gap energy appears, that is completely absent when there is no carrier multiplication. In the best case studied the efficiency is 36%, which is significantly lower than for the selective thermal absorber, 52.28% (Figure 4 of [31]). It must be stressed again that impact ionization is in any case part of the internal processes in a cell and cannot be "introduced" artificially. It can merely be "uncouraged" by amending band structures or other parameters.

New refinements in the detailed balance theory of solar cells are introduced in [32]. The key achievement here is the observation that the introduction of angle and energy restrictions for the emitted photons can yield the theoretically maximal efficiency of about 86% as was found by other workers. But in this case the maximum is independent of concentration. In the case of total absorption the maximum theoretical efficiency is shown to vary between 30.5% (for a bandgap = 1.26 eV) in the case of a 2π emission solid angle and 40.7% (for $E_g = 1.06$ eV) for a restricted emission solid

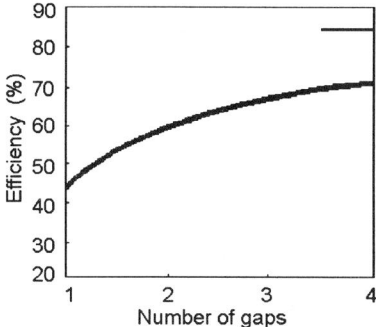

FIGURE 4.6. Maximum percentage efficiencies of a multigap solar cell whose solid angle of emission of radiation is arranged to be equal to the solid angle of incidence of the solar radiation. AM 1.5 normal direct irradiance is assumed [33]. The limit for an infinite number of gaps is indicated by a line.

angle (Figure 3 of [32]). A curve of maximum efficiencies as a function of the number of gaps in the case of a multigap system is given in Figure 4.6.

The difference between the results predicted by the standard (diode-type) solar cell equation and the detailed balance equation (Shockley–Queisser type) is studied in [34]. The authors state that the differences are mainly due to neglect of photon recycling in the diode-type equation. Without photon recycling effects, the limiting efficiency (one sun) of a GaAs solar cell is 26.8% while the true figure is found to be 30.7%, 38.7% as long as the solid angle of emission of photons from the cell is fully restricted (Table III and Figure 2 of [34]).

A suggestion, based on modeling, was made to increase the efficiency of solar cells by introducing an impurity level in the semiconductor bandgap [35]. It was suggested that the efficiency exceeds in this case not only the Shockley–Queisser efficiency for ideal solar cells, but also for that ideal two-terminal tandem cells which use two semiconductors. It is claimed that a cell with an intermediate band can reach an efficiency of 63.1% (Figure 2 of [35]).

The main result of [36] is that impact ionization in combination with carrier–carrier scattering in the absence of photon scattering in an illuminated semiconductor leads to an energy distribution of electrons in the conduction band, and of holes in the valence band which is best described by a single Fermi-distribution without splitting of quasi-Fermi-energies, but with a temperature different from the lattice temperature. To make proper use of this distribution in a solar cell, electrons and holes must be withdrawn through membranes, which are composed of narrow band, large bandgap semiconductors and which allow only electrons or holes in a narrow energy range to be transmitted. In the case of a semiconductor with bandgap $E_g \to 0$ older results [37] are restated (see also Section 2.3): for

102   4. Some Methods of Analyzing Solar Cell Efficiencies

unconcentrated sunlight, the maximum efficiency is 0.053 (Figure 2 of [36] and Eq. (22) of [37]) while at maximum concentration an efficiency 0.85 is obtained (Figure 2 of [36]).

4.10 Conclusions

In this chapter we have explained some of the theoretical methods available for discussing solar cell efficiencies. One can proceed from photon fluxes to current densities, using well-known properties of radiation-often justifiably approximated as black-body radiation. However, the presence of electron and hole systems, frequently associated with distinct quasi-Fermi levels, injects a slight departure from conventional treatments of this radiation (Sections 4.1–4.3). As also emphasized in Chapter 3, tandem cells, graded bandgaps and the utilization of impact ionization can improve the theoretical efficiencies. We approached cells with graded bandgap in this chapter by first treating the theoretical efficiencies of heterojunctions allowing at the same time for impact ionization (Sections 4.4–4.6). A theory of the efficiency of graded bandgap cells (Section 4.7) and a theory of thermophotovoltaic conversion efficiencies (Section 4.8) are also provided.

Typical theoretical effects are shown as a result of varying the energy gap of the two materials which form the heterojunction (Figure 4.3), and of increasing the probability of impact ionization in a homojunction (Figure 4.2). Figure 4.3 suggests 27% theoretical efficiency as a typical value and Figure 4.2 up to 48% for a GaAs cell. Actual multigap systems have reached about 30% under unconcentrated solar radiation as will be discussed in greater detail in our forthcoming partner publication [1].

The Fraunhofer Institute notes in the ninth edition of June 1997 [2b] top experimental, confirmed efficiencies as follows (AM1.5 global, 25 °C, irradiance of 1 kW/m^2):

- monocrystalline silicon cells 23.5%;
- silicon films and thin silicon cells 20.6%;
- multijunction 25%;
- III–IV cells 23.9%;
- monocrystalline silicon bifacial cells 20.6%; and
- photoelectrochemical cells 11.0%.

We have not discussed the effect of photon recycling. This effect lengthens the theoretical estimate of the radiation lifetime in solar cells and hence acts in a direction of greater theoretical efficiencies. However, this effect is always already part of the mechanisms in any solar cell, and its main significance resides in its effect on merely *calculated* solar cell parameters. Thus, it affects the distribution of a minority carrier. The typical effect of taking it into account is an increase by about 2% in the theoretically determined efficiencies [38].

4.11 References

[1] P. T. Landsberg and V. Bădescu: Solar energy conversion: List of efficiencies and some theoretical considerations, *Progr. Quantum Electron.* **22**, 211, 1998; **22**, 231, 1998.

[2] (a) M. A. Green, K. Emery, K. Bucher, D. L. King, and S. Igari: Solar cellq efficiency tables (version 10), *Prog. Photovoltaics Research Appl.* **5**, 265, 1997.
(b) Fraunhofer Institute for Solar Energy Systems, PV Charts, edition 9, update 1, 1997.

[3] P. T. Landsberg: An introduction to the theory of photovoltaic cells, *Solid-State Electron.* **18**, 1043, 1975.

[4] P. T. Landsberg: Non-equilibrium concepts in solar energy conversion, in: *Energy Transfer Processes in Condensed Matter* (ed. B. di Bartolo), NATO Advanced Study Institute, Plenum, New York, 1985.

[5] P. T. Landsberg and P. Baruch: The thermodynamics of the conversion of radiation energy for photovoltaics, *J. Phys. A: Math. Gen.* **22**, 1911, 1989.

[6] V. Bădescu: On the thermodynamics of the conversion of partially polarized black-body radiation, *J. Phys. III France* **2**, 1925, 1992.

[7] V. Bădescu and P. T. Landsberg: Statistical thermodynamic foundation for photovoltaic and photothermal conversion. II. Application to photovoltaic conversion, *J. Appl. Phys.* **78**, 2793, 1995.

[8] V. Bădescu, P. T. Landsberg, and A. De Vos: Statistical thermodynamic foundation for photovoltaic and photothermal conversion. III. Application to hybrid solar converters, *J. Appl. Phys.* **81**, 2692, 1997.

[9] W. Shockley and H. Queisser: Detailed balance limit of efficiency of p–n junction solar cells, *J. Appl. Phys.* **32**, 510, 1961.

[10] D. Trivich and P. Flinn: Maximum efficiency of solar energy conversion by quantum processes, in: *Solar Energy Research* (eds. F. Daniels and J. A. Duffie), Thames and Hudson, London, 1955, p. 143.

[11] P. T. Landsberg, A. De Vos, and P. Baruch: Comparison of some efficiency factors in photovoltaics, *J. Phys. Condens. Matter* **3**, 6415, 1991.

[12] S. R. Dhariwal and R. C. Sharma: Field-assisted recombination in silicon solar cells with heavily doped base: A loss mechanism for the open-circuit voltage, *Semicond. Sci. Technol.* **7**, 315, 1992.

[13] H. Pauwels and A. De Vos: Determination of the maximum efficiency solar cell structure, *Solid-State Electron.* **24**, 835, 1981.

[14] A. De Vos: Calculation of the maximum attainable efficiency of a single hetrojunction solar cell, *Energy Conversion* **16**, 67, 1976.

[15] P. T. Landsberg: *Recombination in Semiconductors*, Cambridge University Press, Cambridge, 1991, p. 251, Case 4.

[16] P. T. Landsberg, H. Nussbaumer and G. Willeke: Band–band impact ionization and solar cell efficiency, *J. Appl. Phys.* **74**, 1993.

[17] S. M. Sze: *Physics of Semiconductor Devices*, 2nd ed., Wiley, 1981, p. 86.

[18] W. Shockley: *Electrons and Holes in Semiconductors*, Van Nostrand, New York, 1950, p. 314.

[19] M. A. Green: *Solar Cells*, Prentice Hall, Englewood Cliffs, 1982, p. 80.

[20] P. T. Landsberg, J. K. Liakos, and A. De Vos: Effect of band–band impact ionization on the efficiency of heterojunction photovoltaic cells, *12th EC Photovoltaic Solar Energy Conference*, Amsterdam, 1994, p. 1343.

[21] J. K. Liakos and P. T. Landsberg: Auger recombination and impact ionization in heterojunction photovoltaic cells, *Semicond. Sci. Technol.* **11**, 1895, 1996.

[22] M. J. Adams and P. T. Landsberg: The theory of the injection laser, in: *Gallium Arsenide Lasers*, (ed. C. H. Gooch), Wiley Interscience, New York, 1969.

[23] H. C. Casey Jr. and M. B. Panish: *Heterostructure Lasers*, Academic Press, New York, 1978, Chap. 3.

[24] P. Kireev: *La Physiques des Semiconducteurs*, Editions MIR, Moscow, 1975, Chapter 76, pp. 581–591.

[25] E. O. Kane: The $k.p$ method, in: *Semiconductors and Semimetals* **1**, *Physics of III–V Compounds,* Academic Press, New York, 1966.

[26] J. E. Parrott: The limiting efficiency of an edge-illuminated multigap solar cell, *J. Phys. D* **12**, 441, 1979.

[27] J. E. Parrott: Analysis of an edge-illuminated graded-gap solar cell, *Solid State Electron. Devices* **12**, Special Issue S 79, 1978.

[28] L. Landau and E. Lifchitz: *Physique Statistique*, Editions MIR, Moscow, 1967.

[29] P. Würfel and W. Ruppel: Upper limit of thermophotovoltaic solar-energy conversion, *IEEE Trans. Electron. Devices* **ED-27**, 745, 1980.

[30] P. T. Landsberg: *Thermodynamics and Statistical Mechanics*, Dover, New York, 1990.

[31] W. Spirkl and H. Ries: Luminescence and efficiency of an ideal photovoltaic cell with charge carrier multiplication, *Phys. Rev. B* **152**, 11319–11325, 1995.

[32] G. L. Araujo and A. Marti: Absolute limiting efficiencies for photovoltaic energy conversion, *Solar Energy Mater. Solar Cells* **133**, 213–240, 1994.

[33] A. Marti and G. L. Araujo: Limiting efficiencies for photovoltaic energy conversion in multigap systems, *Solar Energy Mater. Solar Cells* **143**, 203–222, 1996.

[34] A. Marti, J. L. Balenzategui and R. F. Reyna: Photon recycling and Shockley's diode equation, *J. Appl. Phys.* **182**, 4067–4075, 1997.

[35] A. Luque and A. Marti: Increasing the efficiency of ideal solar cells by photon induced transitions at intermediate levels, *Phys. Rev. Lett.* **178**, 5014–5017, 1997.

[36] P. Würfel: Solar energy conversion with hot electrons from impact ionization, *Solar Energy Mater. Solar Cells* **146**, 43–52, 1997.

[37] V. Bădescu: Maximum conversion efficiency for the utilization of diffuse solar radiation, *Int. J. Energy* **116**, 783–786, 1991.

[38] V. Bădescu and P. T. Landsberg: Influence of photon recycling on solar cell efficiencies, *Semicond. Sci. Technol.* **12**, 1491–1497, 1997.

5
Solar buildings

L. Sertorio
G. Tinetti

ABSTRACT. A building that has the capacity to control the thermal flow interactions with the outside, the in and out flows, and in addition is able to control the internal thermal flows between adjacent zones, formally acts as a thermodynamic automaton. The automaton rule minimizes the distance of the temperature of the core (living space) from a predetermined desired temperature. The automaton rule must manage the chaotic variations of the external inputs and also its own capacity to learn whether the external inputs belong to summer or winter, etc., because according to season the automaton strategy changes. If backup injection of heating (or cooling) is necessary, how can this be operated? In fact, the building is "alive," and the injection of energy automatically perturbs the automaton logic; it may fool the automaton into making it believe that it is summer when instead it is winter (or vice versa). A problem of similar nature is known to pilots of highly automated airplanes, for instance in the process of automatic landing, where the intervention of the human action has caused disasters. It is also known in medical science where the introduction of a drug always carries basically unknown side-effects. We are studying the strategy of the "back up" for a class of thermodynamic automata.

5.1 Finalistic Systems. Introduction

The problem of stabilizing the temperature of the inside of a house near a preassigned value, is perhaps the first stationary control problem faced by man. The words stationary control need to be explained with a brief introduction.

The understanding of the natural phenomena is divided into two general branches: deterministic and finalistic. The distinction among the two is not trivial at all. We may say that deterministic is what is described by the theories of the fundamental interactions, gravity, strong and electroweak forces. A large part of the macroscopic systems are described with the language of thermodynamics which interfaces macroscopic and microscopic. From the formal viewpoint all the equations that govern the fundamental interactions can be derived from a general variational principle, where what is minimized, the action, is the integral of the Lagrangian on a fixed time interval. What is important is that the action refers to a closed system.

We may say that finalistic are those systems that appear to satisfy a purpose. Which purpose? What is important is that the system is open to an "ambient" and responds to it. Historically, the above distinction was spelled using the terms nonliving and living, as if the finalistic behavior was a signal of life or a definition of it.

From the formal viewpoint we may say that a macroscopic finalistic system obeys deterministic laws (the laws of classical mechanics, or of thermodynamics, for instance) plus a variational request which is not the minimization of an action, but of something different. The object integrated over the time is not the Lagrangian but "a cost functional." The variation is not made on the Lagrange variables but on the new redundant variables, called control variables. After this semiphilosophical start, let us quickly come down to earth with an example.

Consider a simple mechanical control system. Take a steam engine operating a shaft that works against an external irregular torque. The Watt governor regulates the pressurized vapor escape valve, using the centrifugal forces acting on two weights, which rotate with the same angular velocity of the active shaft: if the shaft velocity is higher than a preassigned value, the valve opening is decreased; if it is lower, it is increased.

If appropriately designed, the Watt governor keeps the velocity constant, despite the changes in the applied torque:

- The system is open: there is the external torque.
- The system obeys the laws of Newtonian dynamics.
- There is a redundant variable: the valve opening ξ.
- The governor is designed in such a way as to give the best solution to the additional variational requirement:

$$\delta_\xi \int_{t_1}^{t_2} |\omega(t) - \omega_0| \, dt = 0. \tag{5.1}$$

The variation is taken with respect to the opening of the vapor escape valve, ξ. This opening is the control variable of the deterministic mechanical system. The deterministic mechanical system contains the engine and the mechanical components of the governor. For ξ fixed, we have n degrees of freedom. Classical mechanics governs such problem with n equations. If the equations are linear with constant coefficients, the solution is well known. If the opening of the valve is variable, $\xi = \xi(t)$, we have $n+1$ degrees of freedom and n equations. The additional equation that closes the problem is (5.1).

Maxwell was the first to study mathematically the Watt governor, and he indicated the importance of considering the mechanical equations in the linear regime. In this regime the equations of motion are of the kind

$$\dot{\vec{x}} = \mathbf{A} \cdot \vec{x} + \vec{u}(t), \tag{5.2}$$

where the vector $\vec{x} = (x_1(t) \ldots x_n(t))$ is the phase space, $\vec{u}(t)$ is the input vector (which in the case of the Watt governor has only one component), and $\mathbf{A}$ is the $n \times n$ coefficients matrix which contains the parameter ξ. The solution of (5.2) implies the knowledge of the roots of the eigenvalues equation

$$\det|\mathbf{A} - \lambda \mathbf{1}| = 0 \tag{5.3}$$

and Maxwell pointed out that the real part of the complex roots must be negative, if the governor must satisfy (5.1), and also realized the importance of understanding two typical times, namely the decay time (the real part of the root) and the oscillatory time (the imaginary part of the root).

This is, in a nutshell, the formal problem of finalistic systems. One can easily understand that when the dynamics is nonlinear, both because the deterministic part is itself nonlinear and because the control variables appear multiplied by the deterministic variables, a finalistic problem can be very difficult. No surprise that control theory has developed significantly with the development of numerical calculation.

We end this section pointing out an important distinction among finalistic problems:

1. *Target problems*
   They can be:

   (a) *Final time*
       A problem of this kind 1(a) is typically the soft landing of a spacecraft (crashing is deterministic). In this problem the object to minimize is a target that must be reached at $t \to t_{\text{final}}$.

   (b) *Instant time*
       Problems of this kind 1(b) are the Watt governor itself, and modern sophisticated control systems for aircrafts and cars. The object to minimize is the instantaneous offset from the target.

2. *Quasi-periodic problems*
   These problems imply a logic of the finalistic system which adapts itself to the ambient according to a long-term strategy. The ambient in general is not static but rather quasi-periodic. An example is given by the seasons of an ecosystem: they are quasi-periodic. The variational requirement is

$$\delta \int_0^\tau |\vec{x}(t; \vec{u}(t), \vec{\xi}(t)) - \vec{x}_b| \, d\tau = 0. \tag{5.4}$$

In formula (5.4) $\vec{x}(t; \vec{u}(t))$ is the system configuration which depends on the input $u(t)$ and the controls $\vec{\xi}(t)$; $\vec{x}_b$ is the periodic best configuration (including a constant); and τ is the period of the input $u(t)$. The solar house we are considering is a particular case of (5.4).

5.2 The Geophysical Inputs

5.2.1 *The Incoming Solar Flux*

As we have seen a control problem consists of three parts:

- the input;
- the deterministic system; and
- the governor.

In this section we discuss the input. What, in Section 5.1, we called in $u(t) = u_1(t) \ldots u_n(t)$ is now something rather complicated. We have two main inputs at a position θ, φ (we use polar coordinates so that θ is the colatitude):

The incoming solar radiation $g_{\text{in}}(t) = g_{\text{in}}(\theta, \varphi, t)$.

The ambient temperature $T_E(t) = T_E(\theta, \varphi, t)$.

(We neglect other important climatic inputs, like wind.) These two functions are not independent of each other. In fact, in the general theory of the Earth"s surface, everything is put into motion by the incoming solar flux.

We know that this dynamics is immensely complicated. It involves several fields: $T(\vec{r}, t)$, $\vec{v}(\vec{r}, t)$, temperature and velocity of the fluid, the atmosphere, $\rho(\vec{r}, t)$ density, $P(\vec{r}, t)$ pressure, $c_i(\vec{r}, t)$ concentration of the ith component of the atmosphere, and many other fields, like the distributed optical properties (local absorption and reflection). It is not even known how many fields should be taken into consideration. The geography, land, sea, vegetation, adds to the general complexity.

Despite this complexity of events at each $\vec{r}$, t, the global energy balance is nearly stationary. This means that the total incoming radiation flux equals the total outgoing radiation flux. These sums, obviously, must be taken at the top of the atmosphere. There, the outgoing flux has two components: one short wave, the albedo, and one long wave, the emission from all places where thermalization has taken place. Inside the thin shell comprising liquid–solid surface plus the atmosphere, the motions are chaotic because the equations of motion (only in part known) are nonlinear. (In fact, the global circulation models of the atmospheric dynamics have a very short forward time validity just because of the nonlinearity.) This is the rather hopeless fundamental scenario concerning the theoretical prediction of each local $T_E(t)$ and $g_{\text{in}}(t)$. Conversely, there is a network of geological stations that can produce the recordings of $T_E(t)$ and $g_{\text{in}}(t)$ day by day.

Now both inspection of the data and theoretical considerations indicate that the fields $T_E(\theta, \varphi, t)$ and $g_{\text{in}}(\theta, \varphi, t)$ contain a regular part and a chaotic part. How the two contributions can be identified is an open problem. Nevertheless this separation is necessary. In fact, the experimental recordings are taken every few seconds, hour by hour, day after day, year after year. The year is a sensible period for the Earth. Now, which is a typical year? And within this typical year made of 365 days, which is the typical nth day? We need a typical year made of typical days for the learning process of the house, see Section 5.3. The learning process means creating a rule of behavior for a certain control, that we may call inner map control.

After the consideration of the typical year, namely $T_E^{\text{reg}}(\theta, \varphi, t)$, $g_{\text{in}}^{\text{reg}}(\theta, \varphi, t)$, and the relative map, we consider the real phenomenological year made of real phenomenological days that contain the irregular, or chaotic, geophysical fields $T_E(\theta, \varphi, t)$ and $g_{\text{in}}(\theta, \varphi, t)$. The house will need to respond to the real fields with "adaptive controls." The concept of regular and chaotic parts is fairly easy to understand considering $g_{\text{in}}(\theta, \varphi, t)$. In fact, we know the incoming flux hitting a unit surface element tangent to the Earth surface in the absence of atmosphere. This flux is

$$g_0(\theta, \varphi, t) = \sigma \frac{R_\odot^2}{d^2} T_\odot^4 \, f(t), \tag{5.5}$$

where $T_\odot$ is the temperature of the surface of the Sun

$$T_\odot \approx 5800 \text{ K};$$

σ is the radiation Stefan–Boltzmann constant

$$\sigma = 5.67 \times 10^{-8} \text{ W m}^{-2} \text{ K}^{-4};$$

$R_\odot$ is the radius of the Sun

$$R_\odot = 7 \times 10^8 \text{ m},$$
$$f(t) = \vec{d} \cdot \vec{n} \; \Theta(\vec{d} \cdot \vec{n});$$

d is the distance Sun–Earth:

$$d = 1.5 \times 10^{11} \text{ m};$$

$\vec{d}$ is the unit vector of the straight line Earth–Sun; and $\vec{n}$ is the unit vector orthogonal to such surface.

The factor $\vec{d} \cdot \vec{n} \; \Theta(\vec{d} \cdot \vec{n})$ takes into account the Earth's spin plus orbit motion, and is known [1] (see Figure 5.1):

$$\vec{d} \cdot \vec{n} = \cos\theta \sin(\omega_y t) \sin\zeta + \sin\theta \cos(\omega_y t) \cos\varphi + \sin\theta \sin\varphi \sin(\omega_y t) \cos\zeta, \tag{5.6}$$

where $\varphi = \varphi_0 + \omega_d t$ and φ_0 is an arbitrary initial position. $\Theta(x)$ is the Heaviside function (we have used the capital letter Θ instead of the lower

5.2. The Geophysical Inputs

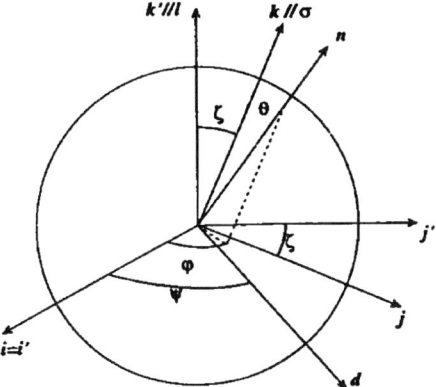

FIGURE 5.1. In this figure the vector $\vec{d}$ rotates in the orbit plane, the unit vectors $\vec{i'}$ and $\vec{j'}$ lie in the orbit plane, and the vector $\vec{k}$ is parallel to the direction of the spin σ.

case θ to distinguish the function from the angle):

$$\Theta(x) = \begin{cases} 1, & x > 0, \\ 0, & x < 0, \end{cases}$$

$$\omega_d = \frac{2\pi}{\tau_d}, \quad \text{where} \quad \tau_d = 24 \text{ h},$$

$$\omega_y = \frac{2\pi}{\tau_y}, \quad \text{where} \quad \tau_y = 365 \text{ days } 5 \text{ h } 48 \text{ min } 46 \text{ s}.$$

In (5.6) $t = 0$ is the local noon at the spring equinox. Notice that the input (5.6) is not a rational torus.

We also examine the formula given in [2] for $\vec{d} \cdot \vec{n'}$, where $\vec{n'}$ is the normal to a unit surface not tangent to the Earth. The notations in [2] are different. See Figure 5.2. The expression of $\vec{d} \cdot \vec{n'}$ is

$$\vec{d} \cdot \vec{n'} = \sin\delta \sin\phi \cos\beta - \sin\delta \cos\phi \sin\beta \cos\gamma \cos\delta \cos\phi \cos\beta \cos\omega$$
$$+ \cos\delta \sin\phi \sin\beta \cos\gamma \cos\omega + \cos\delta \sin\beta \sin\gamma \sin\omega, \quad (5.7)$$

where

ϕ Latitude; it is related to the colatitude θ by $\phi = \pi/2 - \theta$.

δ Declination, this is the angular position of the Sun at solar noon with respect to the plane of the equator, north positive. δ can be calculated with the equation

$$\delta = 0.409 \sin\left(2\pi \frac{284 + n}{365}\right).$$

112   5. Solar buildings

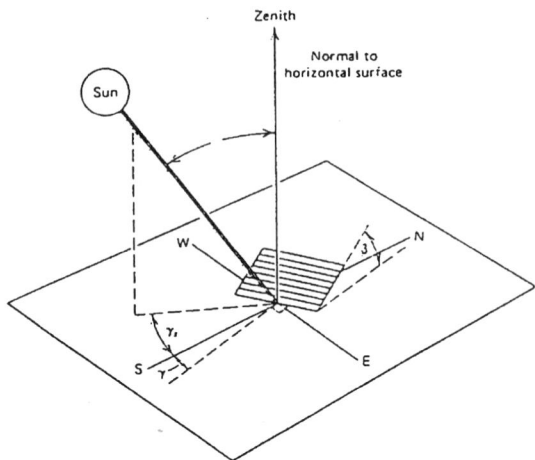

FIGURE 5.2. Zenith angle, slope, surface azimuth angle, and solar azimuth angle for a tilted surface.

The declination δ mentioned above never appears again in this chapter. In Section 5.2.2, the symbol δ is used for the penetration length.

β Slope, this is the angle between the plane surface in question and the horizontal ($0 \leq \beta \leq \pi$).

γ Surface azimuth angle, this is the deviation of the projection on a horizontal plane of the normal to the surface from the local meridian, with zero due south, east negative, west positive ($-\pi \leq \gamma \leq \pi$).

ω Hour angle, this is the angular displacement of the Sun east or west of the local meridian due to rotation of the Earth on its axis at $\pi/12$ per hour, morning negative, afternoon positive.

n Day of the year; n can easily be obtained with the help of the following table:

Jan.	$n = i$	Feb.	$n = i + 31$	Mar.	$n = i + 59$
Apr.	$n = i + 90$	May	$n = i + 120$	Jun.	$n = i + 151$
Jul.	$n = i + 181$	Aug.	$n = i + 212$	Sep.	$n = i + 243$
Oct.	$n = i + 273$	Nov.	$n = i + 304$	Dec.	$n = i + 334$

In order to compare (5.6) and (5.7) we must put $\beta = 0$, $\gamma = 0$, so that $\vec{n} = \vec{n'}$. The advantage of formula (5.7) is that it shows explicitly the relationship with the calendar.

The presence of the atmosphere produces a multitude of scattering events for the incoming photons that were produced by the Sun according to a Planck distribution (except the absorption lines due to the solar atmosphere itself. This can be neglected.).

The inelastic scattering of the incoming photons produces at ground level, where the house sits, the incoming flux that is in part direct, in part diffused, namely filtered. The effect of the filter is particularly strong where the atmosphere contains locally water molecules assembled in the form of droplets, the clouds. The clouds are formed and dissolved according to local thermodynamic events, which again contain both a regularity and a certain unpredictability.

We define the regular part of $g_{\text{in}}^{\text{exp}}$ as the flux (5.5) multiplied by $1 - \bar{\alpha}$, where $\bar{\alpha}$ is the average albedo of the Earth

$$g_{\text{in}}^{\text{reg}} = (1 - \bar{\alpha}) \cdot g_0 = (1 - \bar{\alpha}) \, \sigma \, \frac{R_\odot^2}{d^2} \, T_\odot^4 \, f(t), \tag{5.8}$$

$\bar{\alpha}$ is a number known to climatologists [3]. They evaluate

$$\bar{\alpha} = 0.3.$$

In conclusion we may put

$$g_{\text{in}}(\theta, \varphi, t) = g_{\text{in}}^{\text{reg}}(\theta, \varphi, t) + g_{\text{in}}^{\text{irr}}. \tag{5.9}$$

Notice, however, that the cloudiness over a given region may not be appropriately called chaotic. There are regions which are mainly cloudy and regions that are mainly clear. Therefore the addend $g_{\text{in}}^{\text{irr}}$ can be further divided into a regional component and a local component.

To illustrate the above discussion we present, in the following figures the graphs of the theoretical $g_{\text{in}}^{\text{reg}}$, formula (5.8), and the measured $g_{\text{in}}^{\text{exp}}$ corresponding to several different days. The measurements were taken by the GEOFIT, at Torino, Italy [5].

Note: It is convenient to use this notation for any function of t, like $T_E(t)$, $g_{\text{in}}(t)$, etc. We consider

$$\frac{1}{\tau_d} \int_{t_n}^{t_n + \tau_d} F(t') \, dt',$$

where t_n is the time corresponding to the beginning of the nth day, namely midnight of the $n - 1$ day. With this definition of t_n we adopt the self-explanatory notation

$$\frac{1}{\tau_d} \int_{t_n}^{t_n + \tau_d} F(t') \, dt' = \langle F^{(n)}(t) \rangle = F(n). \tag{5.10}$$

114   5. Solar buildings

Concerning the temperature $T_E(\theta, \varphi, t)$ (the latitude of Torino is $\theta = \pi/4$ in the language of (5.6) or $\phi = \pi/4$ in the language of (5.7)), let us show the graph of the experimental measurements taken by GEOFIT (Figure 5.6). This figure refers to the year 1992.

The actual behavior of $T_E(t)$ is visibly irregular and we need to define a regular behavior. We have two different approaches:

- *Statistical analysis.* This defines the typical year. The actual data, by definition, deviate from this statistical handicraft [8].

- *Study of a theoretical dynamical model.* This is our choice in the following. Obviously the dynamical model is interesting if it contains few parameters and if its physical content is in agreement with well-established experimental facts.

The model is constructed with the following steps:

1. We study the dry Earth. In this step we introduce two global parameters: uniform albedo $\bar{\alpha}$, and uniform long-wave opacity $\bar{\gamma}$.

2. Next we study the wet Earth. Wet means that we consider the cycle of evaporation and condensation of water, namely the cycle of the latent heat. In this step we introduce two new parameters, β and $\mathcal{N}_h$, related to the evaporation rate of water and to the local (namely at

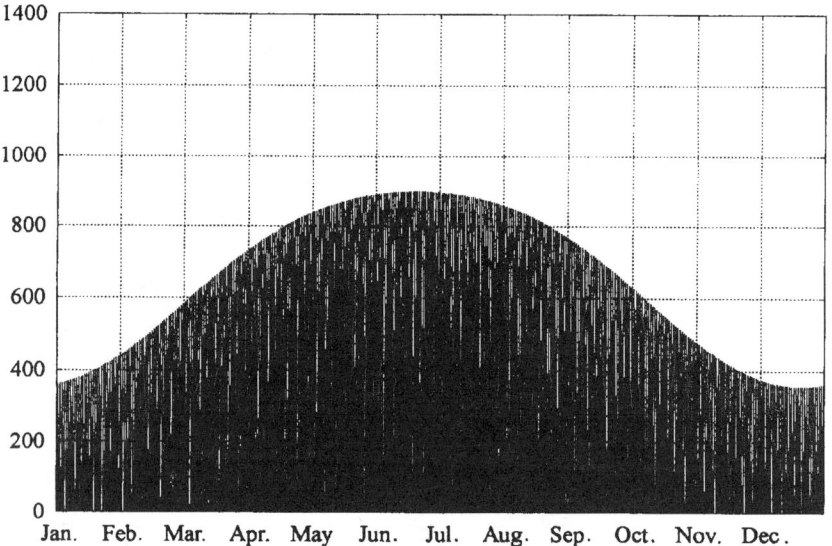

FIGURE 5.3. Graph of the theoretical $g_{\text{in}}^{\text{reg}}(\phi = \pi/4, t)$ given by formula (5.8), corresponding to $1 \leq n \leq 365$. $g_{\text{in}}^{\text{reg}}$ is given in W/m².

5.2. The Geophysical Inputs   115

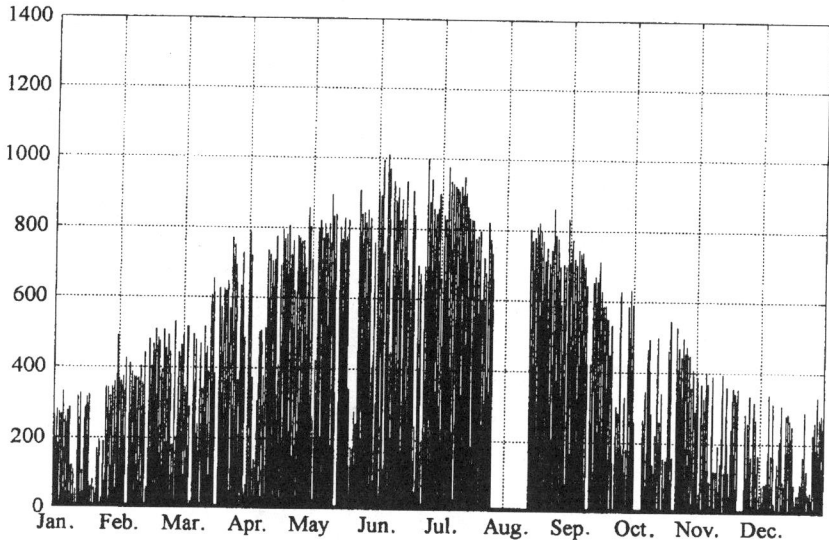

FIGURE 5.4. Measured $g_{in}^{exp}(t)$ corresponding to $1 \leq n \leq 365$. These measurements were taken by the GEOFIT, at Torino, Italy, in 1992. The empty interval is due to instrumental failure. g_{in}^{exp} is given in W/m².

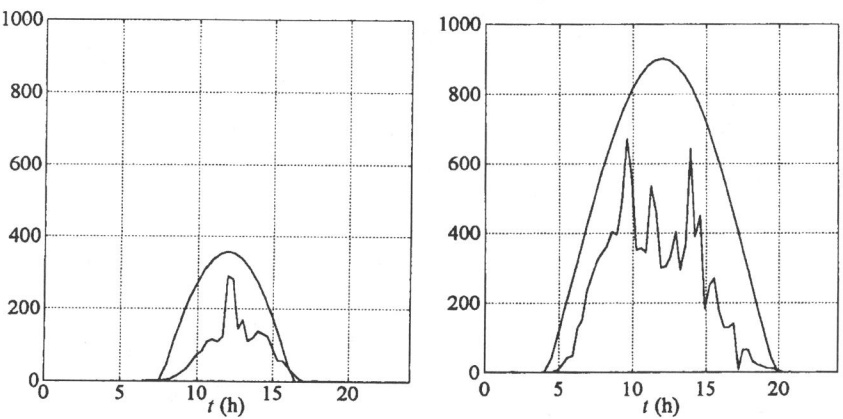

FIGURE 5.5. $g_{in}^{reg}(\phi = \pi/4, t)$ and $g_{in}^{exp}(t)$ corresponding to $n = 355$, winter solstice (the small tail on the right-hand side of the kinematical zero of $f(t)$ can be explained as diffuse radiation), and to $n = 172$, summer solstice. g_{in}^{reg} and g_{in}^{exp} are given in W/m².

116   5. Solar buildings

a given position on the Earth) content of specific humidity relative to the global average specific humidity of the whole atmosphere.

3. Finally, we take into account the transport of sensible heat. We will show that at the particular location of Torino one single parameter with dimension of temperature is able to describe the contribution of the sensible heat.

Notice that the parameters γ, β mentioned above have nothing to do with the angles γ, β appearing in formula (5.7). We will always use (5.7) with $\gamma = 0$, $\beta = 0$, so that there will never be any risk of confusion.

The above three combined steps constitute the local theoretical model. The actual day, $n = 1\ldots 365$, is compared to this theoretical day. Deviations of the actual day are consequently interpreted not as deviations from a statistical handicraft, but from a dynamical model which contains few parameters.

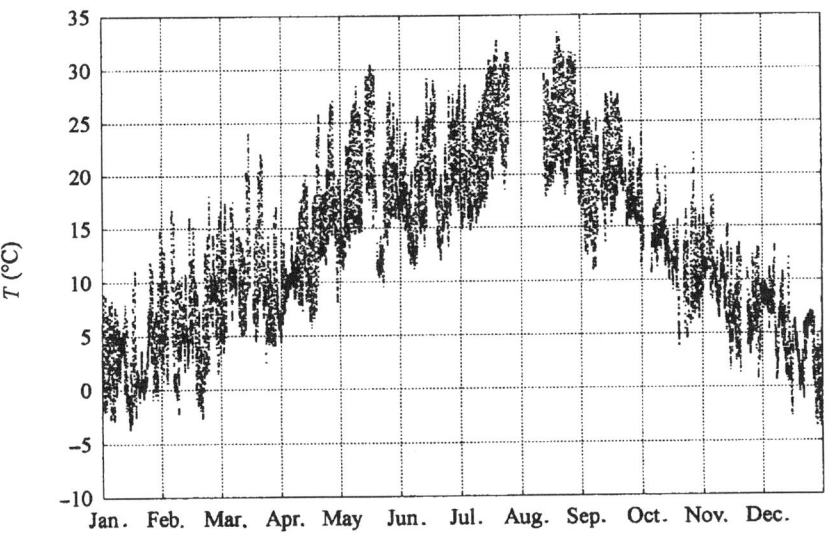

FIGURE 5.6. Measured $T_E^{\text{exp}}(t)$ corresponding to $1 \leq n \leq 365$. These measurements were taken by the GEOFIT, at Torino, Italy, in 1992. $T_E^{\text{exp}}(t)$ is given in °C. The empty interval is due to instrumental failure.

5.2.2 The Equation for T_E^{dry}

Is the Fourier equation in spherical symmetry with boundary conditions specified by the incoming flux (5.8) and an outgoing flux $(1 - \bar{\gamma})\,\sigma\,T^4$?

$$c_v \rho \frac{\partial T}{\partial t} = \kappa \left[\frac{\partial^2 T}{\partial r^2} + \frac{2}{r}\frac{\partial T}{\partial r} + \frac{1}{r^2}\left(\frac{\partial^2 T}{\partial \theta^2} + \frac{1}{\tan\theta}\frac{\partial T}{\partial \theta} + \frac{1}{\sin^2\theta}\frac{\partial^2 T}{\partial \varphi^2}\right) \right]$$
$$+ \left[g_{\text{in}}^{\text{reg}}(\theta,\varphi,t) - (1-\bar{\gamma})\sigma T^4\right]\hat{\delta}(r-R). \tag{5.11}$$

In (5.11):

$r,\ \theta,\ \varphi$	polar coordinates	
c_v	specific heat	1680 J kg^{-1} K^{-1},
κ	Fourier constant	0.42 W m^{-1} K^{-1},
ρ	density	2000 kg m^{-3},
σ	Stefan–Boltzmann constant	5.670×10^{-8} W m^{-2} K^{-4},
$\bar{\gamma}$	long-wave opacity	dimensionless,
$\bar{\alpha}$	reflectivity (albedo)	dimensionless,
$\hat{\delta}(x)$	Dirac function	$[\hat{\delta}] = 1/x$,
R	Earth radius	6.38×10^6 m.

Equation (5.11) describes the rigid circulation of heat on the Earth. The importance of this equation is manifold:

1. It is a simplified reference for the understanding of the real heat circulation, which is known to be in large part carried by the complicated fluid motions.

2. It provides a theoretical treatment of the concept of temperature penetration along the variable r, for $r < R$.

3. It provides a very good agreement with experiment of the time delay of the local surface temperature with respect to the incoming local solar radiation.

In references [1] and [4] it has been found that ∇T, the gradient of T along r, decreases exponentially from the value at $r = R$. We can define the penetration length

$$\delta \approx 2\sqrt{\frac{2\kappa}{\omega_d\,\rho\,c_v}}, \tag{5.12}$$

for $\omega_d = 2\pi/\tau_d$ we have $\delta = 20$ cm. The same number is found experimentally [3]. Given δ we can work with the tangential equation associated to (5.11), namely the Fourier equation relative to a shell of thickness δ:

$$c_v \rho \delta \frac{\partial T}{\partial t} = \kappa \delta \left[\frac{1}{R^2}\left(\frac{\partial^2 T}{\partial \theta^2} + \frac{1}{\tan\theta}\frac{\partial T}{\partial \theta} + \frac{1}{\sin^2\theta}\frac{\partial^2 T}{\partial \varphi^2}\right) \right]$$
$$+ g_{\text{in}}^{\text{reg}} - (1-\bar{\gamma})\sigma T^4. \tag{5.13}$$

118   5. Solar buildings

We can moreover simplify (5.13) by neglecting the Laplacian term, since it is multiplied by the factor $1/R^2$. We obtain the graph of Figure 5.7. Comparison of Figure 5.7 with the experimental data, Figure 5.6, shows

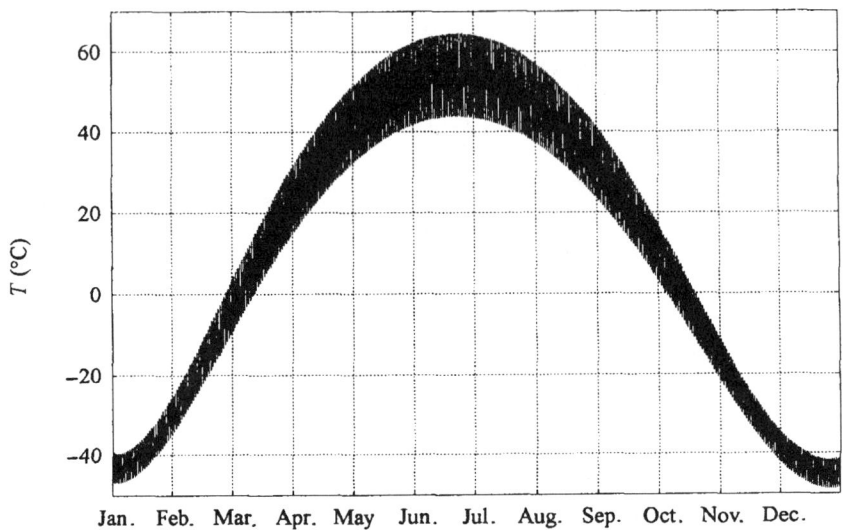

FIGURE 5.7. $T_E^{\mathrm{dry}}(t)$, numerical solution of (5.13) having neglected the Laplacian term.

that the dry model equation gives the wrong daily oscillations and the wrong annual oscillation. On the other hand, the dry model gives the right time delay between the input $g_{\mathrm{in}}(t)$ and the solution $T_E(t)$. Figure 5.8 shows the situation for the day $n = 194$.

Moreover the thermal wave penetration along $r > R$ has a time delay. This delay has been accurately studied in [1]:

$$\Delta \tau_r = \sqrt{\frac{c_v \, \rho}{2\omega_y \, \kappa}} \cdot (R - r). \tag{5.14}$$

The concept of thermal penetration with time delay can be found in the mathematical literature, see, for instance, reference [6], which however refers to the simpler unidimensional case.

5.2.3 *The Equation for T_E^{wet}*

Let us consider the global latent heat cycle. We have two experimental informations [3] the total mass of H_2O in the atmosphere is constant in the

5.2. The Geophysical Inputs

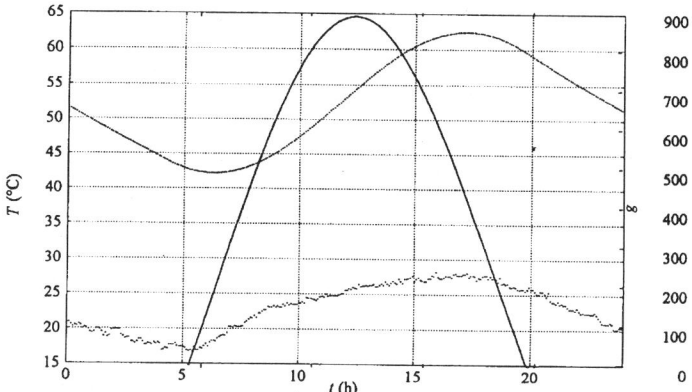

FIGURE 5.8. $g_{in}^{reg}(t)$, $T_E^{exp}(t)$ and $T_E^{dry}(t)$ (having neglected the Laplacian term) corresponding to $n = 194$.

present global equilibrium state and is

$$M_{H_2O} = 13.1 \times 10^{15} \text{ kg}, \tag{5.15}$$

which is equivalent to a uniform layer of ≈ 2.5 cm of water covering the globe.

Water in the atmosphere does not sit motionless, but instead changes continuously from liquid to a gaseous phase. The second experimental information in fact concerns the global (integral over the terrestrial surface) average (integral over one year) precipitation P_r:

$$P_r = 1 \text{ m}.$$

It follows that M_{H_2O} has a residence time of ≈ 9 days, or that M_{H_2O} evaporates ≈ 40 times per year. We indicate with $\dot{M}_{H_2O}^{\uparrow}$ the rate of evaporation, we have

$$\dot{M}_{H_2O}^{\uparrow} = \frac{M_{H_2O}}{9 \text{ days}} = \frac{13.1}{9 \cdot 2.4 \cdot 3.6} \times 10^{11} \text{ kg/s} = 0.168 \times 10^{11} \text{ kg/s}. \tag{5.16}$$

The condensation rate is $\dot{M}_{H_2O}^{\downarrow} = \dot{M}_{H_2O}^{\uparrow}$ because the global state is stationary

$$\dot{M}_{H_2O} = \dot{M}_{H_2O}^{\downarrow} - \dot{M}_{H_2O}^{\uparrow} = 0.$$

To this movement of water (water cycle) corresponds the global power $P_\mathcal{L}$ (where $\mathcal{L}$ stands for latent heat):

$$P_\mathcal{L} = \dot{M}_{H_2O}^{\uparrow} \mathcal{L}. \tag{5.17}$$

Using the fact that
$$\mathcal{L} = 2.5 \times 10^6 \text{ J/kg},$$
we have
$$P_\mathcal{L} = 0.42 \times 10^{17} \text{ W}, \tag{5.18}$$

where the power (5.17) comes from. The external power is the global power of the incoming flux
$$P_\gamma = \pi R^2 (1-\alpha) \cdot C, \tag{5.19}$$
where C is the solar constant
$$C = 1350 \text{ W/m}^2.$$

Putting in the numbers we have
$$P_\gamma = 1.21 \times 10^{17} \text{ W} \tag{5.20}$$

and finally
$$\eta = \frac{P_\mathcal{L}}{P_\gamma} = \frac{0.42}{1.21} = 0.347. \tag{5.21}$$

This indicates that almost 35% of solar flux goes into the water cycle, latent heat branch (there is also a sensible heat branch). This information is the starting point for the following detailed study of the thermodynamic effects of the water cycle. Consider in fact the following wet tangential equation

$$c_v \rho \delta \frac{\partial T}{\partial t} = g_{\text{in}}^{\text{reg}} - (1-\bar{\gamma})\sigma T^4 + \dot{q}. \tag{5.22}$$

Equation (5.22) is nothing other than (5.13) without the Laplacian term and with the additional source term $\dot{q} = \dot{q}(\theta, t)$. We can write, generalizing (5.17),
$$\dot{q}(\theta, t) = \dot{\mu} \cdot \mathcal{L}, \tag{5.23}$$

where $\dot{\mu}(\theta, t)$ is the mass flow of water and $\mathcal{L}$ is the latent heat. We make the hypothesis that
$$\dot{\mu} = \beta \cdot \frac{T_s - T}{T_0 - T}. \tag{5.24}$$

The physical meaning of (5.24) is the following: $\dot{\mu}$ is the evaporation rate per square meter ($[\dot{\mu}] = \text{kg/m}^2 \text{ s}$) due to the local disequilibrium between the surface temperature $T(\theta, t)$ and the temperature of saturated vapor $T_s(\theta, t)$ at the actual partial pressure of H_2O, p_{H_2O}; T_0 is the boiling temperature of H_2O at p_{atm} (atmospheric pressure, $p_{\text{atm}} = 1.015 \times 10^5$ Pa), namely $T_0 = 373$ K. The denominator $(T_0 - T)$ represents the fact that the surface evaporation process becomes a volume process at the boiling point. For $T \to T_0$, $T_s \to T_0$, β represents the evaporation rate at the boiling regime. We do not have a laboratory evaluation of β; we will find β

5.2. The Geophysical Inputs

with calculations indicated in the following. The important finding is that β appears to be a constant. We anticipate

$$\beta = 1.25 \times 10^{-3} \text{ kg/m}^2 \text{ s}. \tag{5.25}$$

We show now that we can express $T_s(\theta, t)$ at a given place θ with the determination of one single parameter. We have for $\theta = \pi/4$, namely Torino, the experimental measurement of the specific humidity $h_s(t)$, Figure 5.9. The specific humidity is defined by

$$h_s = \frac{m_{H_2O}}{m_{atm}}.$$

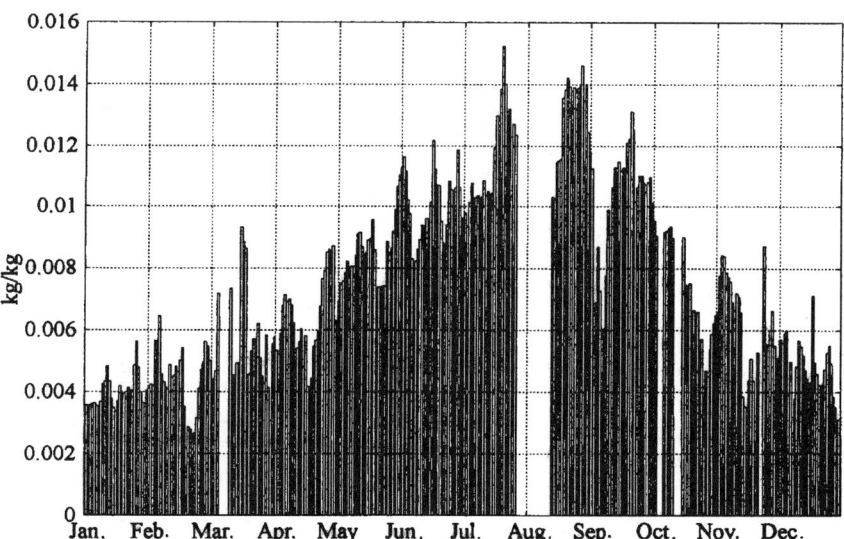

FIGURE 5.9. Experimental specific humidity in Torino. The empty interval is due to experimental failure.

In order to appreciate this curve, we take as reference value the global specific humidity

$$h_s^{\text{glob}} = \frac{M_{H_2O}}{M_{atm}},$$

where M_{H_2O} is the total mass of water in the atmosphere already seen in the preceding pages, and M_{atm} is the total mass of the atmosphere [3]:

$$M_{atm} = 5.15 \times 10^{18} \text{ kg}.$$

The global specific humidity is therefore

$$h_s^{\text{glob}} = 2.54 \times 10^{-3} \text{ kg/kg}. \tag{5.26}$$

122 5. Solar buildings

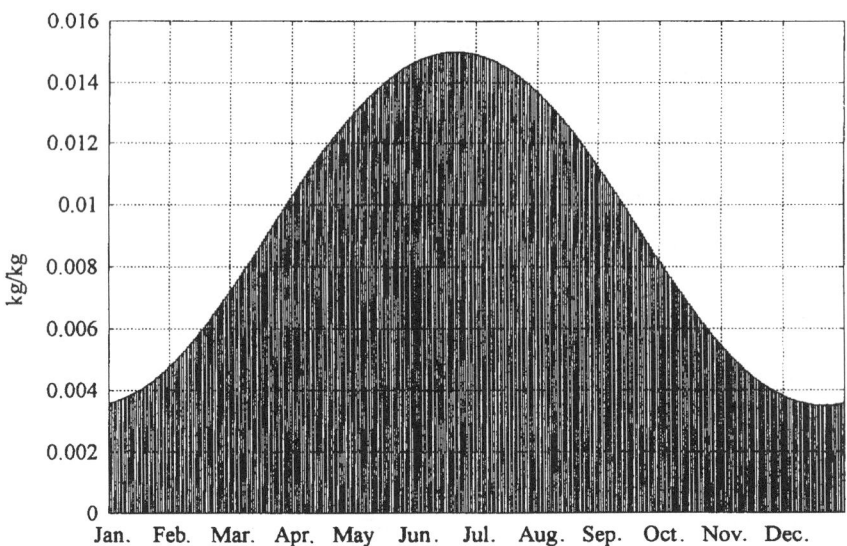

FIGURE 5.10. $h_s^{\text{reg}}(n)$ given by (5.27).

Notice that in the geophysical literature the specific humidity is expressed in g/kg. There is then a factor 10^{-3} between our natural definition and the common usage.

By inspection of the experimental curve (Figure 5.9), we make the following phenomenological guess

$$h_s^{\text{reg}}(n) = \mathcal{N}_h \cdot h_s^{\text{glob}} \cdot f(n). \tag{5.27}$$

The normalization factor $\mathcal{N}_h$ has been estimated as

$$\mathcal{N}_h = 16. \tag{5.28}$$

At this point, having the expression for $h_s^{\text{reg}}(n)$, we can calculate $T_s^{\text{reg}}(n)$. In fact from the specific humidity we have the partial pressure of H_2O which is given by the formula

$$p_{H_2O} = \frac{p_{\text{atm}} \cdot h_s}{0.622}. \tag{5.29}$$

(Notice that (5.29) comes from the equation of state $pV = \mathcal{N}kT$, plus the relationships $M_{H_2O} = \mathcal{N}_{H_2O} \cdot \mu_{H_2O}$; $M_{\text{atm}} = \mathcal{N}_{\text{atm}} \cdot \mu_{\text{atm}}$; $\mu_{H_2O} = 18$ amu; $\mu_{\text{atm}} = 28.9$ amu). Second, given the partial pressure of water, we use the

5.2. The Geophysical Inputs

equations relating p_{H_2O} to T_s:

$$T_s^{gl} = -0.622 \cdot \frac{\mathcal{L}^{gl}}{R_d \cdot \ln(p_{H_2O}/C^{gl})} \quad \text{if} \quad p_{H_2O} > 611.29 \text{ Pa}, \quad \text{gas-liquid},$$

$$T_s^{gs} = -0.622 \cdot \frac{\mathcal{L}^{gs}}{R_d \cdot \ln(p_{H_2O}/C^{gs})} \quad \text{if} \quad p_{H_2O} < 611.29 \text{ Pa}, \quad \text{gas-solid}, \tag{5.30}$$

where [3]:

$$R_d = 287 \text{ J/kg K}.$$

The constants C^{gs} and C^{gl} can be determined normalizing (5.30) to the tables of the *Handbook* [7]:

$$C^{gl} \approx 2.546 \times 10^{11} \text{ Pa}, \quad C^{gs} \approx 3.635 \times 10^{12} \text{ Pa}.$$

Inserting (5.30) into (5.24) we have the explicit expression of $\dot{q}$. The resulting equation (5.22) is solved numerically. From this solution we can evaluate $\dot{q}$, which is shown in Figure 5.11. We call $\dot{q}^{(+)}$ the positive part of $\dot{q}$ (condensation) and $\dot{q}^{(-)}$ the negative part of $\dot{q}$ (evaporation). Integrating then $\dot{q}^{(\pm)}$ over the year we find the values

$$\dot{q}^{(+)} = 1.26 \times 10^7 \text{ J/m}^2, \quad \dot{q}^{(-)} = 1.55 \times 10^7 \text{ J/m}^2, \tag{5.31}$$

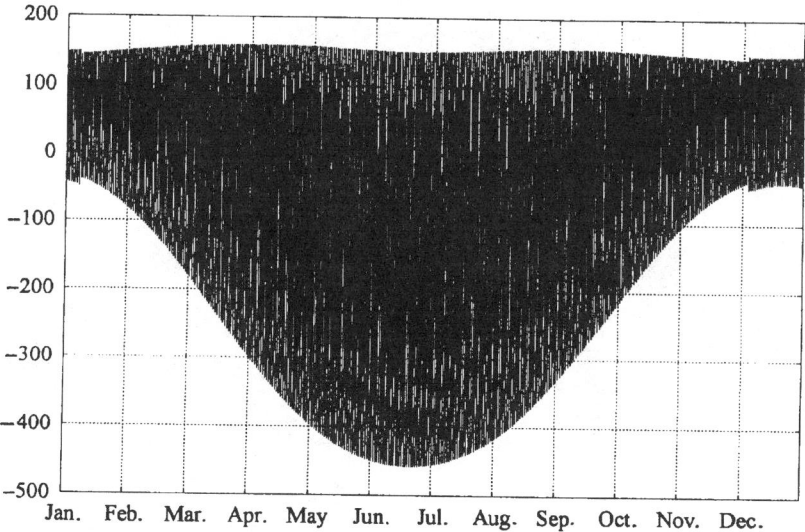

FIGURE 5.11. $\dot{q}(t)$ evaluated for Torino corresponding to $1 \leq n \leq 365$. $\dot{q}$ is given in W/m^2

which, if compared with the integral over the year of the radiation power at $\theta = \pi/4$, give

$$\eta^{(+)} = \frac{\dot{q}^{(+)}}{P_\gamma(\theta = \pi/4)} = \frac{1.26 \times 10^7 \text{ J/m}^2}{4.62 \times 10^7 \text{ J/m}^2} = 0.27, \quad \eta^{(-)} = \frac{\dot{q}^{(-)}}{P_\gamma(\theta = \pi/4)} = 0.33. \tag{5.32}$$

The above numbers are in qualitative agreement with the global number (5.21). The asymmetry between evaporation and condensation is not surprising. In fact, the global balance of the water cycle does not imply the local balance.

The examination of $T_E^{\text{wet}}(t)$ shows that $T_E^{\text{wet}}(t)$ gives a good regular representation of the experimental T provided it is augmented to a constant value of 2 °C. We define in conclusion

$$T_E^{\text{reg}}(t) = T_E^{\text{wet}}(t) + 2 \text{ °C} \tag{5.33}$$

We attribute this constant contribution to the transport of sensible heat. $T_E^{\text{reg}}(t)$ is shown in Figure 5.12.

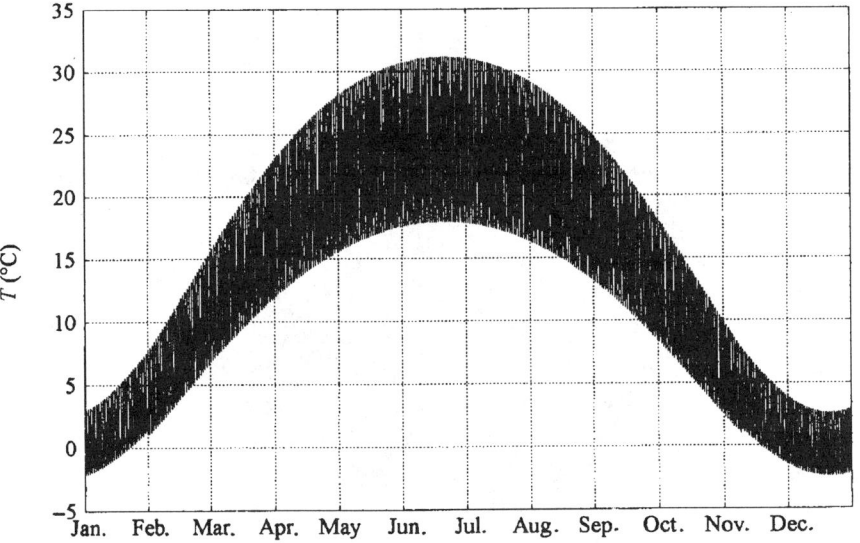

FIGURE 5.12. $T_E^{\text{reg}}(t)$ corresponding to $1 \leq n \leq 365$. $T_E^{\text{reg}}(t)$ is given in °C.

We conclude this section with the evaluation of T_g, namely the temperature of the ground below the surface. According to Figure 5.14 of the next section, we study a model house which rests on the ground simply at a depth of 0.5 m. The thermodynamic behavior, due to a particular way of digging, is not important at this point. For simplicity, we evaluate $T_g(t)$ using (5.33) with the value of the penetration length $\delta = 0.5$ m. We show $T_g(t)$ in Figure 5.13.

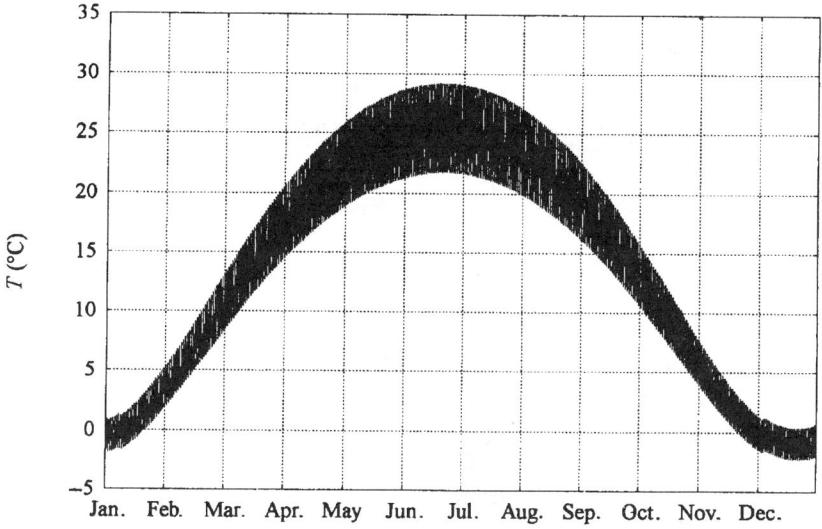

FIGURE 5.13. $T_g(t)$ corresponding to $1 \leq n \leq 365$. $T_g(t)$ is given in °C.

5.3 The Model of the Solar House

5.3.1 General Remarks on the Model with Fixed Controls

We consider here two houses:

(a) house with peripheral wall; external insulation (Figure 5.14(a)); and

(b) house with internal wall; external insulation (Figure 5.14(b)).

We do not consider the common house, with external wall and internal insulation (Figure 14(c)), because it is not interesting thermodynamically. The house in both topology (a) and (b) can be described by three thermodynamic elements:

1. *Interior space*
   With volume V_1, specific heat c_1, density ρ_1, thermal inertia $\mu_1 = V_1 c_1 \rho_1$, and temperature $T_1(t)$.

126   5. Solar buildings

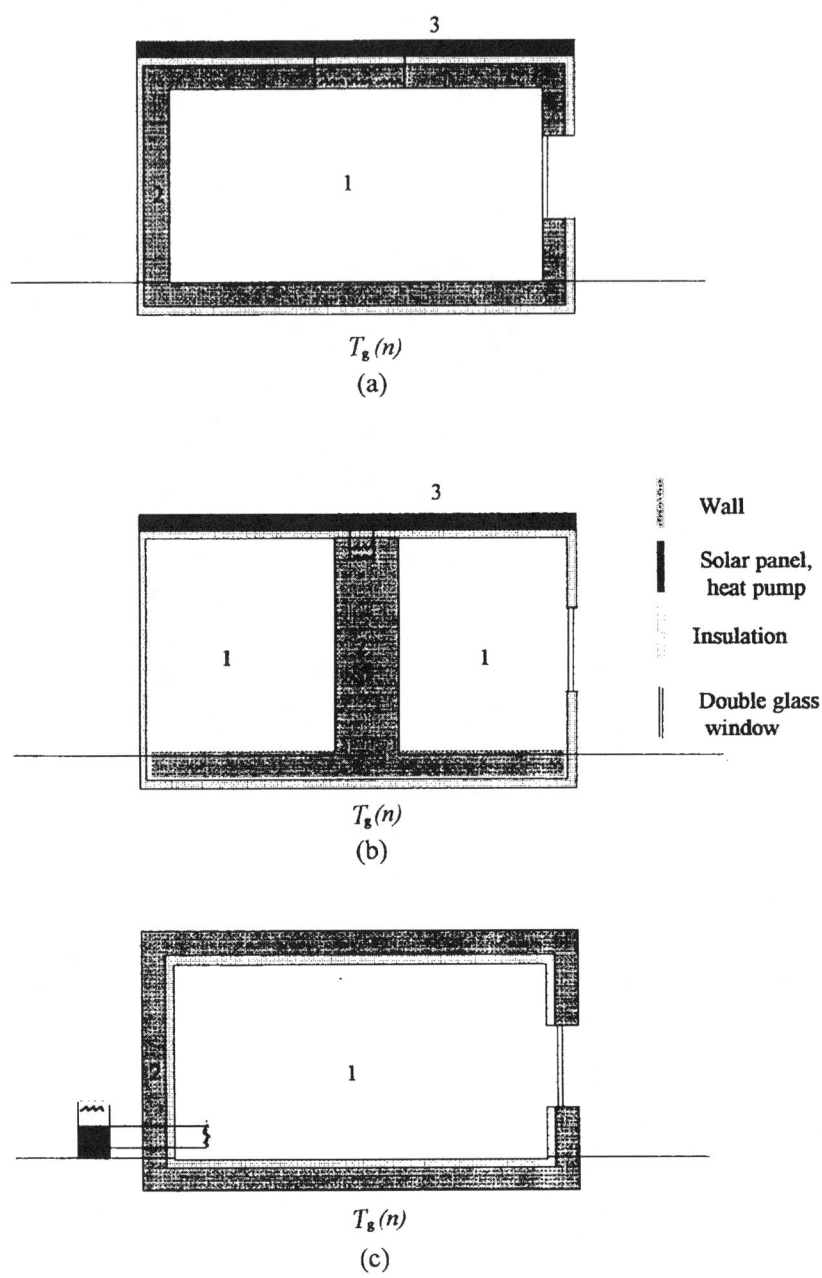

FIGURE 5.14. (a) House with peripheral wall; external insulation. (b) House with internal wall; external insulation. (c) House with peripheral wall; internal insulation, the wrong house; the little box on the left-hand side represents the heat pump.

2. *Wall*
   With the word wall we mean the massive load-bearing structure, with volume V_2, specific heat c_2, density ρ_2, thermal inertia $\mu_2 = V_2 c_2 \rho_2$, and temperature $T_2(t)$.

3. *Solar heat exchanger (solar panel)*
   With volume V_3, specific heat c_3, density ρ_3, thermal inertia $\mu_3 = V_3 c_3 \rho_3$, and temperature $T_3(t)$.

To start our analysis we take for both (a) and (b) the same triplet of μ_1, μ_2, and μ_3. Notice that c_1, c_2, c_3, ρ_1, ρ_2, ρ_3, are restricted to rather narrow ranges by the physical properties of air, concrete, and metal (for instance copper). On the contrary V_1, V_2, V_3 can vary, but not in any particular way. There are relationships between interior volume and the mass of the wall, between the external surface and the maximum surface of the solar panel. Concerning the heat flow across windows and insulating layers, there is considerable variability. It is indeed very interesting to study the nonequilibrium thermodynamics of a house with a given ratio between window surface and peripheral wall; with a given kind of window glasses (single, double, etc.), and with a given thickness of external insulation. To start our analysis we need to make a reasonable choice.

We show in Figure 5.15 the flow diagrams of the thermal interactions. Direct comparison of Figure 5.15 with Figure 5.14(a) and (b), makes the flow diagrams self-evident. From Figure 5.15 (a) we obtain the differential system (5.34) for case (a):

$$\mu_1 \dot{T}_1 = -\Phi_{12} - \Phi_{1E} - S_{1E}\epsilon\sigma(T_1^4(t) - T_{\text{sky}}^4(t)),$$

$$\mu_2 \dot{T}_2 = -\Phi_{23}u(t) + \Phi_{12} - \Phi_{2E} - \Phi_{2g}, \tag{5.34}$$

$$\mu_3 \dot{T}_3 = \Phi_{23}u(t) - \Phi_{3E} + S_{3E}g_{\text{in}}(t) - S_{3E}\epsilon\sigma(T_3^4(t) - T_{\text{sky}}^4(t)).$$

From Figure 5.15(b) we obtain the differential system (5.35) for case (b):

$$\mu_1 \dot{T}_1 = -\Phi_{12} - \Phi_{1E} - \Phi'_{1E} - S_{1E}\epsilon\sigma(T_1^4(t) - T_{\text{sky}}^4(t)),$$

$$\mu_2 \dot{T}_2 = -\Phi_{23}u(t) + \Phi_{12} - \Phi_{2g}, \tag{5.35}$$

$$\mu_3 \dot{T}_3 = \Phi_{23}u(t) - \Phi_{3E} + S_{3E}g_{\text{in}}(t) - S_{3E}\epsilon\sigma(T_3^4(t) - T_{\text{sky}}^4(t)),$$

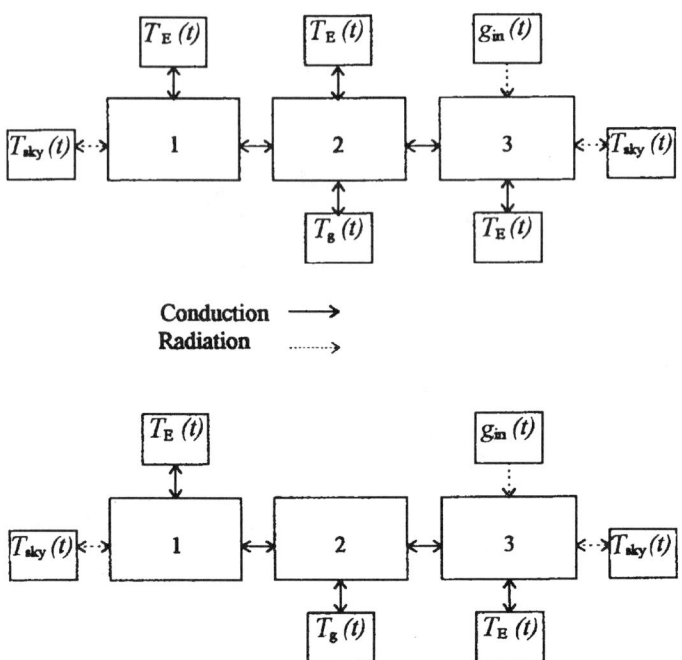

FIGURE 5.15. (a) Flow diagrams of the thermal interactions for the house with external wall, and (b) for the house with internal wall.

where
$$\Phi_{12} = \alpha_{12} S_{12}(T_1(t) - T_2(t)),$$
$$\Phi_{1E} = \alpha_{1E} S_{1E}(T_1(t) - T_E(t)),$$
$$\Phi'_{1E} = \alpha_{2E} S_{2E}(T_1(t) - T_E(t)),$$
$$\Phi_{23} = \alpha_{23} S_{23}(T_2(t) - T_3(t)),$$
$$\Phi_{2E} = \alpha_{2E} S_{2E}(T_2(t) - T_E(t)),$$
$$\Phi_{2g} = \alpha_{2g} S_{2g}(T_2(t) - T_g(t)),$$
$$\Phi_{3E} = \alpha_{3E} S_{3E}(T_3(t) - T_E(t)).$$

<u>Note:</u> Until now we have taken S_{3E} (surface of the panel) and S_{23} (surface coupling T_3 and T_2) different to each other. In the following, for simplicity, we will put
$$S_{23} = S_{3E} = S.$$

5.3. The Model of the Solar House

In systems (5.34) and (5.35) we have input variables, variables, and parameters. See Table 5.1.

In the next section, S will be a control variable: $S \to S(t)$. In Section 5.3.2 we put $u \to u(t) = \begin{cases} 0 \\ 1 \end{cases}$ according to a control rule that will be discussed there. In this section we put $u = 1$.

Systems (5.34) and (5.35) are nonlinear, due to the terms T_1^4 and T_3^4. Nevertheless, these nonlinearities are benign and an analytic approach can be attempted by linearizing systems (5.34) and (5.35), assuming that

$$T_1 \to \langle T_1 \rangle + \tau_1; \qquad T_3 \to \langle T_3 \rangle + \tau_3 \tag{5.36}$$

and considering τ_i small such that we accept the approximation:

$$T_1^4 \approx \langle T_1 \rangle^4 + 4\langle T_1 \rangle^3 \tau_1, \qquad T_3^4 \approx \langle T_3 \rangle^4 + 4\langle T_3 \rangle^3 \tau_3. \tag{5.37}$$

Now systems (5.34) and (5.35) can be put in the matrix form:

$$\dot{\mathbf{T}} = \mathbf{A} \cdot \mathbf{T} + \mathbf{B}, \tag{5.38}$$

where

$$\mathbf{T} = \begin{pmatrix} \tau_1 \\ \tau_2 \\ \tau_3 \end{pmatrix}, \qquad \mathbf{A} = \begin{pmatrix} a_{11} & a_{12} & a_{13} \\ a_{21} & a_{22} & a_{23} \\ a_{31} & a_{32} & a_{33} \end{pmatrix}, \qquad \mathbf{B} = \begin{pmatrix} b_1 \\ b_2 \\ b_3 \end{pmatrix},$$

and where, for case (a),

$$a_{11} = -\frac{\{\alpha_{12} S_{12} + \alpha_{1E} S_{1E} + 4 S_{1E} \epsilon \sigma \langle T_1^3 \rangle\}}{\mu_1},$$

$$a_{12} = \frac{\alpha_{12} S_{12}}{\mu_1},$$

$$a_{13} = 0,$$

$$a_{21} = \frac{\alpha_{12} S_{12}}{\mu_2},$$

$$a_{22} = -\frac{\{\alpha_{12} S_{12} + \alpha_{2E} S_{2E} + \alpha_{2g} S_{2g} + \alpha_{23} S\}}{\mu_2}, \tag{5.39}$$

$$a_{23} = \frac{\alpha_{23} S}{\mu_2},$$

$$a_{31} = 0,$$

$$a_{32} = \frac{\alpha_{23} S}{\mu_3},$$

$$a_{33} = -\frac{\{\alpha_{23} S + \alpha_{3E} S + 4 S \epsilon \sigma \langle T_3^3 \rangle\}}{\mu_3},$$

TABLE 5.1. Input variables, variables, parameters, and controls.

	Input variables	
$T_g(t)$	temperature of the ground	K
$T_E(t)$	outdoor temperature	K
$g_{\text{in}}(t)$	solar radiation	W m^{-2}
$T_{\text{sky}}(t)$	$= 0.0552\, T_E^{1.5}$	K
(An exhaustive explanation of $T_{\text{sky}}(t)$ is given in [2].)		

	Variables	
$T_1(t)$	temperature, indoor	K
$T_2(t)$	temperature, wall	K
$T_3(t)$	temperature, panel	K

	Parameters	
μ_i	$= \rho_i V_i c_i$	J K^{-1}
c_1	specific heat, air	1007 J kg^{-1} K^{-1}
ρ_1	density, air	1.161 kg m^{-3}
V_1	volume, air	600 m^3
c_2	specific heat, wall	2000 J kg^{-1} K^{-1}
ρ_2	density, wall	2000 kg m^{-3}
V_2	volume of the wall	141.58 m^3
c_3	specific heat, copper	385 J kg^{-1} K^{-1}
ρ_3	density, copper	8960 kg m^{-3}
V_3	volume, panel	$S \cdot \Delta x;\ \ \Delta x = 10^{-3}$ m
α_{12}	Newton constant, indoor–wall	10 W m^{-2} K^{-1}
α_{1E}	Newton constant, indoor–outdoor	2.6 W m^{-2} K^{-1}
α_{3E}	Newton constant, outdoor–panel	2.6 W m^{-2} K^{-1}
α_{2E}	Newton constant, wall–outdoor	0.66 W m^{-2} K^{-1}
α_{23}	Newton constant, wall–panel	60 W m^{-2} K^{-1}
α_{2g}	Newton constant, pavement–ground	0.66 W m^{-2} K^{-1}
S_{12}	surface indoor air–wall	410 m^2
S_{1E}	window surface	30 m^2
S_{2g}	surface pavement–ground	125.08 m^2
S_{2E}	surface wall–outdoor	349.48 m^2
σ	Stefan–Boltzmann constant	5.670×10^{-8} W m^{-2} K^{-4}
ϵ	glass–window emissivity	0.9, dimensionless

	Controls	
S	panel surface	m^2
u	control rule	

5.3. The Model of the Solar House

and for case (b):

$$a_{11} = -\frac{\{\alpha_{12}S_{12} + \alpha_{1E}S_{1E} + \alpha_{2E}S_{2E} + S_{1E}\epsilon\sigma\langle T_1^3\rangle\}}{\mu_1},$$

$$a_{12} = \frac{\alpha_{12}S_{12}}{\mu_1},$$

$$a_{13} = 0,$$

$$a_{21} = \frac{\alpha_{12}S_{12}}{\mu_2},$$

$$a_{22} = -\frac{\{\alpha_{12}S_{12} + \alpha_{2g}S_{2g} + \alpha_{23}S\}}{\mu_2}, \tag{5.40}$$

$$a_{23} = \frac{\alpha_{23}S}{\mu_2},$$

$$a_{31} = 0,$$

$$a_{32} = \frac{\alpha_{23}S}{\mu_3},$$

$$a_{33} = -\frac{\{\alpha_{23}S + \alpha_{3E}S + S\epsilon\sigma\langle T_3^3\rangle\}}{\mu_3}.$$

Notice that **B** contains the input terms.

We consider now the homogeneous equation associated to (5.38):

$$\dot{\mathbf{T}} = \mathbf{A} \cdot \mathbf{T} \tag{5.41}$$

and the secular equation

$$\det(\mathbf{A} - \lambda \mathbf{1}) = 0. \tag{5.42}$$

We find that the roots of (5.42) are negative; therefore the solutions of the original systems (5.34) and (5.35) in both cases are attracted by the inputs in a regular way.

In view of the foregoing discussion, it is useful to consider the daily average of the equations of motion. We assume that

$$T_i^{(n)}(t) = T_i^{(n)}(t + \tau_d), \quad i = 1, 2, 3, \quad n = 1, \ldots, 365,$$

and we integrate systems (5.34) and (5.35) over the daily period for each given n and we sum. We get in case (a):

$$\begin{aligned}S(n)g_{\text{in}}(n) = &\ \alpha_{1E}S_{1E}(T_1(n) - T_E(n)) + \alpha_{2E}S_{2E}(T_2(n) - T_E(n)) \\&+ \alpha_{2g}S_{2g}(T_2(n) - T_g(n)) + \alpha_{3E}S(n)(T_3(n) - T_E(n)) \\&+ S_{1E}\epsilon\sigma(T_1^4(n) - T_{\text{sky}}^4(n)) + S(n)\epsilon\sigma(T_3^4(n) - T_{\text{sky}}^4(n)),\end{aligned}$$

and we get in case (b):

$$\begin{aligned}S(n)g_{in}(n) &= \alpha_{1E}S_{1E}(T_1(n)-T_E(n))+\alpha_{2E}S_{2E}(T_1(n)-T_E(n))\\&+\alpha_{2g}S_{2g}(T_2(n)-T_g(n))+\alpha_{3E}S(n)(T_3(n)-T_E(n))\\&+S_{1E}\epsilon\sigma(T_1^4(n)-T_{sky}^4(n))+S(n)\epsilon\sigma(T_3^4(n)-T_{sky}^4(n)).\end{aligned}$$

Notice that according to the notation used so far

$$T_i(n) = \langle T_i^{(n)}(t)\rangle, \text{ etc. } 1 \le n \le 365. \tag{5.43}$$

These daily balance equations will be useful in the discussion of the annual control.

5.3.2 *The Annual Control*

After these preliminaries we consider the control problem. The variational requirement is

$$\delta \int_{t_n}^{t_n+\tau_{day}} |T_i(t') - T_b|\, dt' = 0, \quad 1 \le n \le 365, \tag{5.44}$$

T_b is the constant desired temperature for the inside of the house (to fix the idea $T_b \approx 22°C$). The variation is considered with respect to the control variables. Let us consider first as control variable the surface of the solar panel: $S(n), 1 \le n \le 365$.

The solar panel with variable surface may act as a heater or as a refrigerator. We distinguish the regimes:

- **Heating**
  The solar panel of surface $S(n)$ is activated only when it may work as a heater, namely when $T_3 > T_2$.

- **Refrigerating**
  The solar panel of surface $S(n)$ is activated only when it may work as a refrigerator, namely when $T_3 < T_2$. These two regimes are described by two rules for $u(t)$

$$u(t) = \begin{cases} \Theta(T_3-T_2) = u_h & \text{if } \dfrac{1}{\tau_{day}}\int_{t_n}^{t_n+\tau_d} T_1(t)\, dt < T_b, \\[2ex] \Theta(T_2-T_3) = u_c & \text{if } \dfrac{1}{\tau_{day}}\int_{t_n}^{t_n+\tau_d} T_1(t)\, dt > T_b. \end{cases} \tag{5.45}$$

According to (5.45) the equations of motion change.

5.3. The Model of the Solar House

We analyze first the problem with regular inputs $g_{\text{in}}^{\text{reg}}$, T_E^{reg}. We solve numerically the equations of motion (5.34) and (5.35) with $u(t)$ given by (5.45), for a given n and for values of S changing by steps of 1 m^2. We stop the process when (5.44) is satisfied and we find in this way

$$S = S^{\text{reg}}(n). \tag{5.46}$$

These calculations show that:

- The daily oscillation of $T_1(t)$ remains within a small range if μ_2/μ_1 is appropriate.

- In the domain $1 \leq n \leq 365$ we have several subdomains indicated in Figure 5.16; the meaning of these subdomains will be analyzed in Section 5.4. We have adopted the convention of negative values of $S^{\text{reg}}(n)$ to indicate the refrigerating regime. Figure 5.17 is the annual behavior of $S^{\text{reg}}(n)$ for house (b).

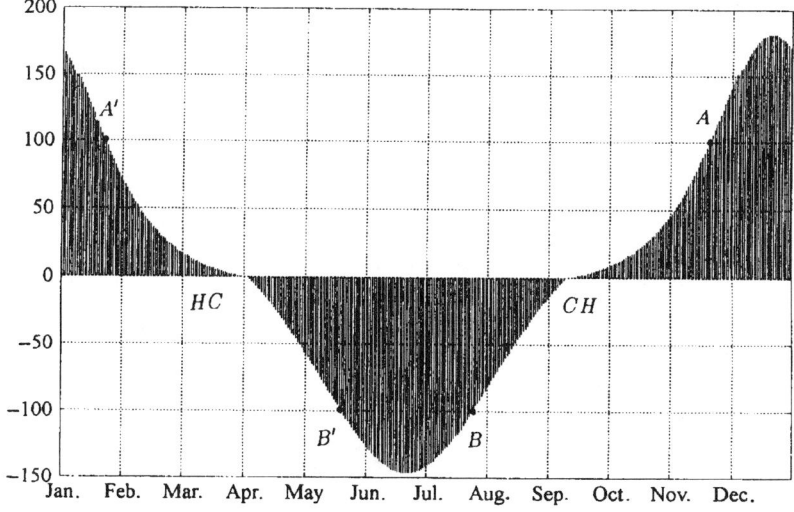

FIGURE 5.16. Annual behavior of $S^{\text{reg}}(n)$ for house (a). We have adopted the convention of negative values of $S^{\text{reg}}(n)$ to indicate the refrigerating regime.

Concerning the daily oscillations, we show the day $n = 323$ (day A, Figure 5.18, and the day $n = 140$ (day B'), Figure 5.19. These two figures refer to the solutions of system (5.34) (house (a)); in the scale of these figures the differences between house (a) and house (b) are hardly visible. Figure 5.20 (day A) and Figure 5.21 (day B') show the behavior of $T_1(t)$ and $T_2(t)$ on fine scale for house (a) and for house (b). We see that house (a) has a better control efficiency than house (b).

134   5. Solar buildings

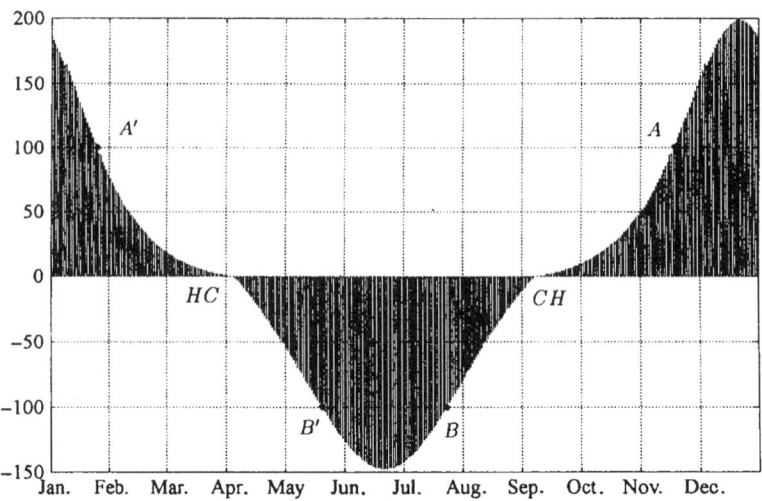

FIGURE 5.17. Annual behavior of $S^{\text{reg}}(n)$ for house (b). We have adopted the convention of negative values of $S^{\text{reg}}(n)$ to indicate the refrigerating regime.

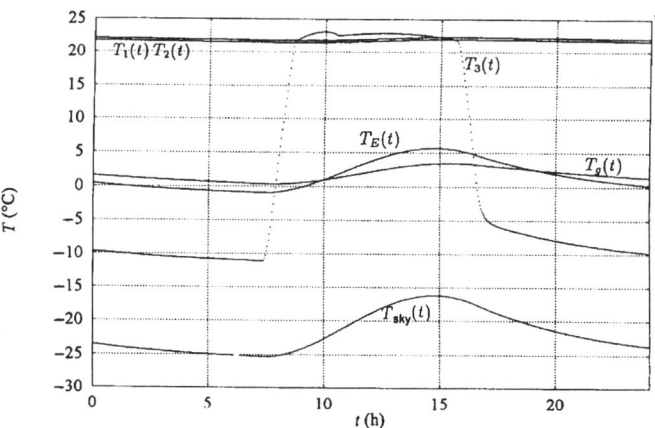

FIGURE 5.18. $T_1, T_2, T_3, T_E, T_{\text{sky}}$, and T_g calculated for house (a), $S(n = 323) = 100\text{m}^2$.

5.4 Backup and Adaptive Controls

With the study of the preceding section, the house has received a regular education and is by now endowed with an inner map of the external events,

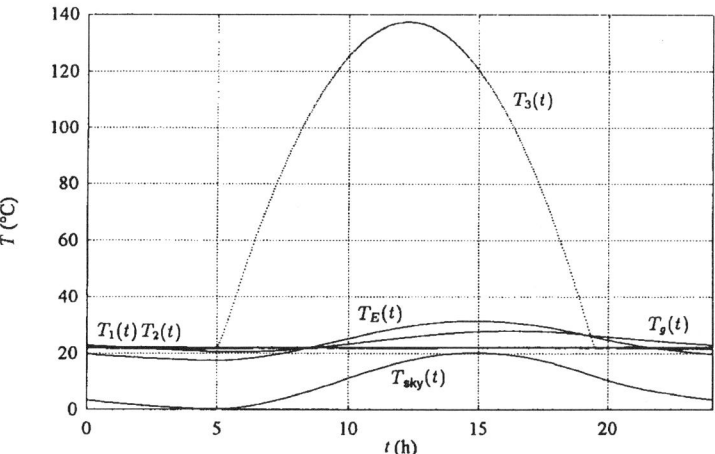

FIGURE 5.19. $T_1, T_2, T_3, T_E, T_{\text{sky}}$, and T_g calculated for house (a), $S(n = 140) = 100\text{m}^2$.

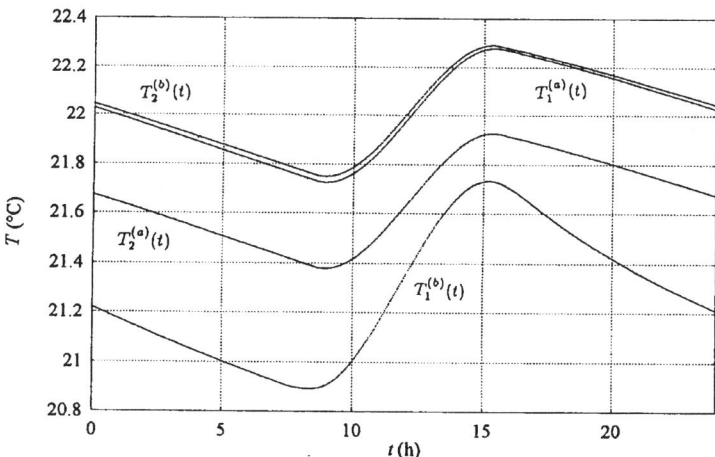

FIGURE 5.20. Behavior of T_1 and T_2 for house (a) and for house (b), $S(n = 323) = 100 \text{ m}^2$.

namely the function $S(n)$. But the good education is not sufficient to cope with the chaotic part of the regular external events.

The phenomenological inputs differ from the regular inputs both during short time intervals, for instance, several fluctuations during τ_d, and long time intervals, for instance, fluctuations involving several days. We know from the study of the preceding sections that if the ratio μ_2/μ_1 is large (in

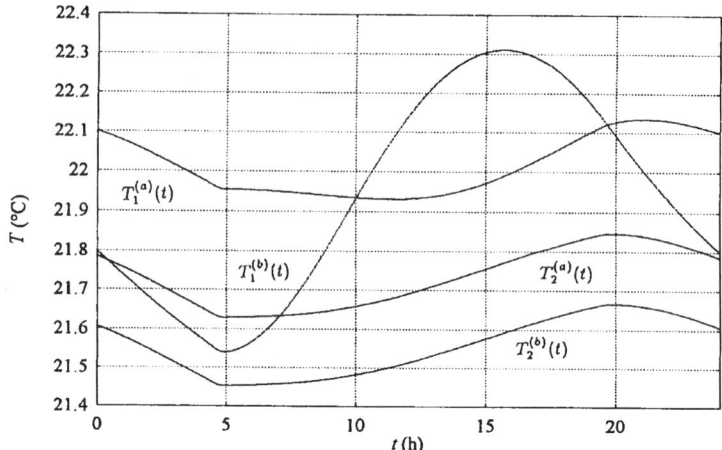

FIGURE 5.21. Behavior of T_1 and T_2 for house (a) and for house (b), $S(n=140) = 100$ m^2.

our calculations ≈ 800) the daily variations of the inputs produce variations of T_2 and T_1 of the order of 1 K. We are consequently concerned with fluctuations on the longer time scale. In this section we limit ourselves to general remarks. In fact, the strategy of adaptive controls is a difficult problem that cannot be included in the extent of this chapter. Let us refer to the circle of the days 22. We have two regions that have quite different thermodynamical behaviors.

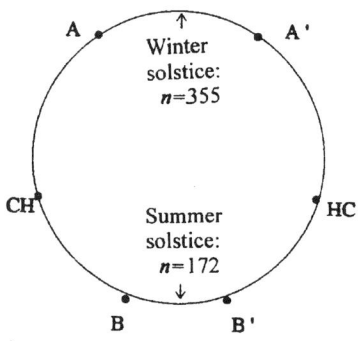

FIGURE 5.22.

1. The domain $I = AA' + B'B$.

   In the domain AA' the solar panel is unable to satisfy the control requirement (5.44) and therefore additional heating back-up is necessary. In the domain BB' the refrigerating back-up is necessary. In Figure 5.23 we show the directions of the heat pump flows. The back-up rule is

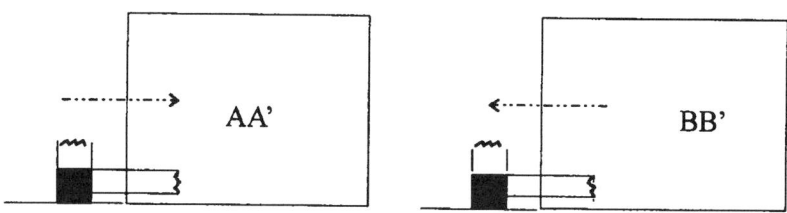

FIGURE 5.23. Domain AA' and domain BB'.

$$q_0^{(+)} = \alpha S(T_+ - T_1)\Theta(T_b - T_1),$$
$$q_0^{(-)} = \alpha S(T_- - T_1)\Theta(T_1 - T_b), \quad (5.47)$$
$$[q_0^{(\pm)}] = W,$$

T_+ is a fixed value, $T_+ > T_b$; T_- is a fixed value, $T_- < T_b$.

The energy necessary to power $q_0^{(\pm)}$ comes from an external source. A source which is not the solar source is by definition a fossil source. Equation (5.47) is the well-known rule of the thermostat. The thermostat is the automaton of zeroth-order complexity. It responds uniquely to the input $(T_1 - T_b)$ and ignores the dynamics of the house.

2. The domain $II = A'B' + BA$.
   In this domain the solar panel is able to satisfy the control requirement (5.44). In $A'HC + CHA$ the panel is in the heating mode while in $HCB' + BCH$ the panel is in the refrigerating mode.
   If the inputs are regular the control problem has been solved in Section 5.3.2. If the inputs are not regular, we have two options:
   - Solving systems (5.34) and (5.35) with the data evaluated at each value of the time t. This can be done numerically as in Section 5.3.1, finding $S(n)$ by direct calculation. Alternatively, one may use the formalism of Pontryagin [9], [10]. This approach is cumbersome. Notice that each house must be equipped with a dedicated computer.
   - Finding an automaton rule which governs the solar panel with a "strategy," rather than with a brute-force solution of the equations. Take as an example the best control machine that we

138   5. Solar buildings

know: man. Man solves complex problems of behavior by not solving equations but inventing "strategies."

After these comments let us emphasize what is the difficulty of the house problem. We have a distance in the input space

$$\delta \vec{u} = \vec{u} - \vec{u}^{\text{reg}}, \tag{5.48}$$

where

$$\vec{u} = (T_E(t), g_{\text{in}}(t))$$

and

$$\vec{u}^{\text{reg}} = (T_E^{\text{reg}}(t), g_{\text{in}}^{\text{reg}}(t)).$$

Corresponding to $\delta \vec{u}$ the solar panel changes its surface by an amount

$$\delta S = S(n) - S^{\text{reg}}(n). \tag{5.49}$$

(Notice that according to the rules adopted in Section 5.3, the surface of the panel changes by steps of 1 m². The step could be different, depending on the engineering of the house.)

The interesting observation is that the correspondence (5.48) to (5.49) is of the kind *many to one*; in fact, $\delta = \vec{u}$ is a complex configuration, δS is simply a number. To appreciate the complexity of $\delta \vec{u}$, we list four kinds of configurations:

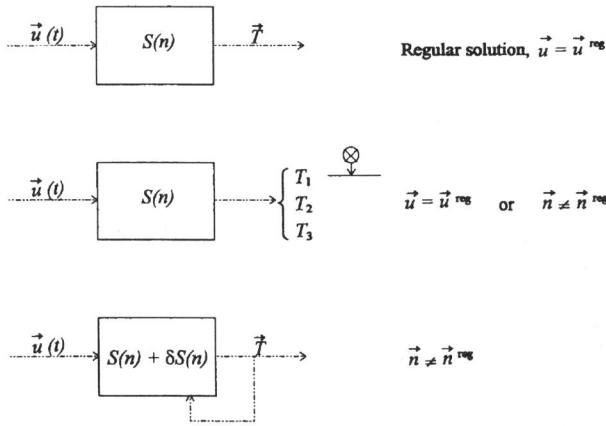

FIGURE 5.24. Map of the regular input, back-up and adaptive control.

$$T_E > T_E^{\text{reg}}, \quad g_{\text{in}} > g_{\text{in}}^{\text{reg}}, \qquad \text{warm and clear,}$$

$$T_E > T_E^{\text{reg}}, \quad g_{\text{in}} < g_{\text{in}}^{\text{reg}}, \qquad \text{warm and cloudy,}$$

$$T_E < T_E^{\text{reg}}, \quad g_{\text{in}} > g_{\text{in}}^{\text{reg}}, \qquad \text{cold and clear,}$$

$$T_E < T_E^{\text{reg}}, \quad g_{\text{in}} < g_{\text{in}}^{\text{reg}}, \qquad \text{cold and cloudy.}$$

(5.50)

To a given $\delta \vec{u}$ the dynamics of the house produces in its own phase space

$$\delta \vec{T} = \delta T_1, \delta T_2, \delta T_3,$$

$\delta \vec{T}$ exists and is sufficiently bounded because the governing equations are dissipative of benign nature (the free system is a sink).

We end this section presenting the flow diagram of the "dynamical system house" for the various regimes. The study of the adaptive control goes beyond the limits of this chapter and will be the object of further work.

5.5 References

[1] L. Sertorio and G. Tinetti: Entropy production for closed, open and third kind systems, *La Rivista del Nuovo Cimento* **22**, 1999.

[2] J. A. Duffie and W. A. Beckman: *Solar Engineering of Thermal Processes*, Wiley, 1980.

[3] A. H. Oort and J. P. Peixoto: *Physics of Climate*, American Institute of Physics, New York, 1992.

[4] F. D'Isep and L. Sertorio: Maximum mechanical work from the steady energy circulation on a finite body, *Il Nuovo Cimento* **67**, 1982.

[5] We thank Professor Longhetto, University of Torino, for having provided these data.

[6] A. N. Tichonov and A. A. Samarskij: *Uravnenija Matematičheckeskoi Fiziki*, Nauka, 1977.

[7] *Handbook of Chemistry and Physics*, CRC Press, Boca Raton, edition 1997, 1998.

[8] There is a rich bibliography on this subject; we thank Professor Bădescu, coauthor in this book, for information on this branch of research.

[9] L. Pontryagin, V. Boltyanskii, R. Gamkrelidzee and E. Mishchenko: *The Mathematical Theory of Optimal Processes*, New York, 1962.

[10] W. L. Brogan: Optimal control theory applied to systems described by partial differential equations, in: *Advances in Control Systems,* (ed. C. T. Leondes), New York, 1968.

Part II

Conversion of Thermal and Chemical Energy

6
Discrete Hamiltonian Analysis of Endoreversible Thermal Cascades

S. Sieniutycz
R. S. Berry

ABSTRACT. Endoreversible multistage processes which yield mechanical work are optimized by a relatively little-known discrete maximum principle of Pontryagin's type. A discrete optimization approach extends the classical method, well known for continuous systems in which a Hamiltonian is maximized with respect to controls. Equations of dynamics which follow from energy balance and transfer equations are difference constraints for optimizing work. Irreversibilities caused by the energy transport are essential. Variation of efficiency is analyzed in terms of the heat flux. Enhanced bounds for the work released from an engine system or added to a heat-pump system are evaluated. Lagrangians of work functionals, canonical equations, and structure of the Hamiltonian function are all discrete characteristics which reach their continuous conterparts in the limit of an infinite number of stages. For a finite-time passage of a resource fluid between two given temperatures, optimality of an irreversible process manifests itself as a connection between the process duration and an optimal intensity expressed in terms of the Hamiltonian. Extremal performance functions that describe extremal work are found in terms of final states, process duration, and number of stages. A *discrete* extension of classical thermal exergy to systems with a finite number of stages and a finite holdup time of a resource fluid is one of the main results. This extended exergy, that has an irreversible component, simplifies to the classical thermal exergy in the limit of infinite duration and an infinite number of stages. The extended exergy exhibits a hysteretic property as a decrease of maximum work received from a multistage engine system and an increase of minimum work added to a heat-pump system, two properties which are particularly important in high-rate regimes.

6.1 Introduction: Multistage Novikov–Curzon–Ahlborn Process

We are dealing with a finite-time extension of the classical problem of maximizing the work produced by a nonequilibrium engine system operating with two flowing fluids or a fluid and a bath, i.e., an infinite reservoir. The fluids have finite thermal conductivity, and hence nonvanishing thermal resistances in their boundary layers and the irreversible properties of any such heat-transfer system. The process operates at a steady state, as represented in the multistage system, shown schematically in Figure 6.1, for the case of a fluid and a bath. For the purpose of optimization, a standard representation of this multistage process as a discrete optimal control problem is given in terms of a set of difference equations. The multistage process is, in fact, a sequence of Novikov–Curzon–Ahlborn processes (NCA processes; [1], [2]). These repeat in all the stages of the cascade; the only differences among them are the temperatures of each stage, which must be consistent with energy balance. The multistage system contains: the driving fluid, which flows with gradually decreasing temperatures $T^1 \ldots T^N$ through the stages; the environment at the constant temperature T^e; the boundary layers which act as the thermal conductances; and the set of the Carnot engines, $C^1 \ldots C^N$, which generate the mechanical work at each stage n.

At each stage the bulk of the fluid is well mixed, and a mixing scale is assigned to the fluid flow. The fluxes of the work produced per cycle are summed up, so that a cumulative average power output $\mathbf{W}$ is obtained at the last stage. The power output per unit flux of the driving fluid, $W = \mathbf{W}/G$, has units of work per unit mass, and describes the specific work associated with the steady-flow process of power production. In classical thermodynamics such processes involve fluids and Carnot engines only; no boundary layers are considered. Their continuous quasistatic limit leads to the classical thermal exergy for a reversible process; it has no rate term. It is the irreversible multistage process with sequence of NCA machines that leads to a finite-time counterpart of the available energy (exergy) of the driving fluid.

By definition, the work released in the engine mode is positive. An optimization problem for the cascade can be stated for the extremum work $W = \mathbf{W}/G$, consistent with the extremum of power $\mathbf{W}$. For the cascade of engines, where the work is released, the work W has to be maximized. For the cascade of heat-pumps, where the work is supplied to the system, the minimization of the negative of the work, $(-W)$, is required. When one of the boundary states of the resource fluid is the state of equilibrium with the environment (e.g., $T^N = T^e$), the maximal specific work represents an extended or finite-time exergy of the flowing fluid. We shall see that this extended exergy simplifies to the classical thermal exergy in the reversible limit when the overall-stage conductances, g^n, tend to infinity.

An active, work-producing, heat exchange process between two fluids is characterized by destruction of the initial nonequilibrium structure and an approach of the resource fluid to equilibrium with the reservoir fluid. For this process the problem of the maximum work delivered in a finite time is a relevant optimization problem. This (first) version of the process is called

6.1. Introduction: Multistage Novikov–Curzon–Ahlborn Process 145

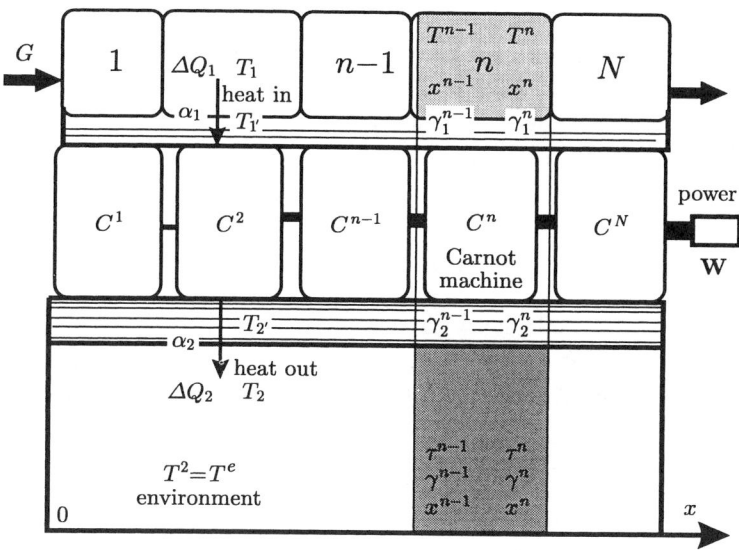

FIGURE 6.1. Scheme of power production in a sequence of Novikov–Curzon–Ahlborn (NCA) engines and the basic terminology.

the engine mode. In the second version, called the heat-pump mode, the system leaves thermodynamic equilibrium and work is added. The system works then as a heat pump, and the associated optimization problem is that of minimizing the work supplied in a finite time. In classical, reversible thermodynamics, the associated magnitudes of the mechanical work are the same for the two operations. Finite-time thermodynamics, a theory for irreversible processes, is a schema showing that (and why) the magnitude of the heat-pump-mode work is always larger than that of the engine-mode work.

We assume a steady-state process in the cascade. Each stage n, which is described by its difference equations (Section 6.3) as a distinct "black-box" unit, has nonetheless its own internal structure shown in Figure 6.1 characteristic of that stage. This structure is composed of the following elements: a well-mixed cell of bulk fluid (the shadowed upper part of the stage n; with the inlet temperature T^{n-1} and outlet temperature T^n); the thermal resistances of the two boundary layers; a Carnot engine C^n; and a contact zone with an isothermal bath (the shadowed lower part of the stage). The bath is assumed to be thermally homogeneous at the outset and retains its homogeneity for all time because it is infinite. The mixing is assumed ideal within each nth fluid cell, consistent with the thermal homogeneity assumed for the pipeline fluid within each stage and the equality of the temperatures of the fluid inside the stage and at the outlet from the stage.

146   6. Discrete Hamiltonian Analysis of Endoreversible Thermal Cascades

Thermal homogeneities of upper and lower parts of each Carnot engine C^n (at each stage n) are assured by the requirement that the circulating fluid is well mixed, to achieve the spatial isothermicity of its high-T and low-T runs within each Carnot loop. The driving fluid and the circulating fluid have only finite residence times at each process stage.

6.2 A Single Stage with the Driving Heat Flux as a Control Variable

The analysis of a single stage must precede the cascade optimization. The left side of Figure 6.1, the region surrounding the second Carnot engine, shows the nomenclature used for describing a single stage. Two finite thermal resistances are represented by the two boundary layers, each characterized by its own heat transfer coefficient, α_1 and α_2. They are contained between the resource and bath fluids and the working fluid circulating in the Carnot part of the stage. The resistances are, of course, reciprocals of the corresponding conductances, $g_1 = \alpha_1 a_1$ and $g_2 = \alpha_2 a_2$, where a_1 and a_2 are the exchange surface areas at the stage. Clearly, one can express the analysis equally well in terms of the conductances g_i or the resistances $1/g_i$. When the system is in the engine mode, the hot fluid releases heat at a high temperature T_1, which reaches the engine part at $T_{1'}$. The low-grade heat is released by the engine fluid at the low temperature $T_{2'}$ to the bath fluid, and reaches the bath at a low temperature $T_2 = T^e$. In a more general case of two fluids, each flowing with a finite mass flux, both fluid temperatures T_1 and T_2 vary along the path, however we restrict ourselves here to the case $T_2 = T^e$. Clearly, for the engine mode, the efficiency of the power production at a stage, $\eta = w/q_1$, is smaller than the efficiency η_C of the Carnot cycle operating between T_1 and T_2.

The entropy balance of the Carnot machine in a single NCA unit, that is, for any stage of the cascade,

$$\frac{q_1}{T_{1'}} = \frac{q_2}{T_{2'}} \tag{6.1}$$

and the energy balance combined with the definition of the first-law efficiency

$$q_2 = q_1 - w = q_1(1 - \eta) \tag{6.2}$$

yield the stage efficiency given by the Carnot formula

$$\eta = 1 - \frac{T_{2'}}{T_{1'}}. \tag{6.3}$$

This efficiency is of course lower than that of a Carnot engine working between the extremal temperatures T_1 and T_2 because the engine itself operates on the reduced temperature difference $T_{1'} - T_{2'}$. The temperatures

6.2. A Single Stage with the Driving Heat Flux as a Control Variable

$T_{1'}$ and $T_{2'}$ are unknown, but they may be expressed in terms of the temperatures T_1 and T_2 and a control variable at the stage under consideration. The choice of the control variable is in principle arbitrary; for example, the control may be the power output, w, the heat flux, q_1, the efficiency, η, or others. Below we demonstrate an analysis of a one-stage NCA process in terms of the heat flux, q_1, initiated in [3]. The results of the analysis will next be adopted to the cascade process consisting of N stages, with arbitrary N.

If Newton's law, heat flow is proportional to the temperature difference, describes the heat exchange, as we assume here, the temperature $T_{1'}$ in terms of the driving heat flux q_1 is

$$T_{1'} = T_1 - \frac{q_1}{g_1}. \tag{6.4}$$

To obtain the second intermediate temperature one substitutes (6.4) into the reversible entropy balance following from (6.1)

$$\frac{g_1(T_1 - T_{1'})}{T_{1'}} = \frac{g_2(T_{2'} - T_2)}{T_{2'}} \tag{6.5}$$

and solves the above result with respect to the temperature $T_{2'}$. This operation yields

$$T_{2'} = \frac{T_2}{1 - q_1/[g_2(T_1 - g_1^{-1}q_1)]}. \tag{6.6}$$

Hence the heat flux q_2 in terms of q_1 equals

$$q_2 = g_2(T_{2'} - T_2) = g_2 T_2 \frac{q_1}{g_2(T_1 - g_1^{-1}q_1) - q_1} \tag{6.7}$$

and the corresponding equation for the mechanical power is

$$w = q_1 - g_2 T_2 \frac{q_1}{g_2(T_1 - g_1^{-1}q_1) - q_1}. \tag{6.8}$$

This expression can also be transformed into the form

$$w = q_1\left(1 - \frac{g_2 T_2}{g_2(T_1 - g_1^{-1}q_1) - q_1}\right) = q_1\left(1 - \frac{T_2}{T_1 - q_1(g_1^{-1} + g_2^{-1})}\right), \tag{6.9}$$

where the last expression contains the reciprocal of the overall conductance of heat transfer. This conductance is defined in the traditional way as the harmonic mean

$$g = \left(\frac{1}{g_1} + \frac{1}{g_2}\right)^{-1} = \frac{g_1 g_2}{g_1 + g_2}. \tag{6.10}$$

Consequently, we obtain a simple equation linking the power w with the heat flux q_1,

$$w = q_1\left(1 - \frac{T_2}{T_1 - g^{-1}q_1}\right) \tag{6.11}$$

and the corresponding first-law efficiency of the stage is

$$\eta = 1 - \frac{T_2}{T_1 - g^{-1}q_1} \equiv 1 - \frac{T_2}{T_1 + u}, \quad (6.12)$$

where $u \equiv -g^{-1}q_1$ is a measure of the heat added to the fluid 1. It has units of temperature and is positive for fluid heating (heat pump) and negative for fluid cooling (engine). It will be used throughout this chapter as a suitable control variable, equivalent to $(-q_1)$ whenever g is constant. The efficiency of an engine stage deviates monotonically from the Carnot efficiency with the finite heat flux q_1. For a quasistatic heat transfer, i.e., for very low q_1, the efficiency is that of the Carnot engine. Yet the efficiency vanishes for a sufficiently large flux q_1, equal to $g(T_1 - T_2)$, which corresponds to pure heat conduction and no power production at all. The power expression, (6.11), deviates from the Carnot model due to the nonvanishing q_1. The Carnot efficiency, $\eta_C = 1 - T_2/T_1$, is achieved when the effect of the overall resistance g^{-1} is negligible or the flux q_1 is very low. The extremum power at the stage of interest corresponds to

$$\frac{\partial w}{\partial q_1} = -\frac{\partial w}{\partial (gu)} = \frac{(T_1 - g^{-1}q_1)^2 - T_1 T_2}{(T_1 - g^{-1}q_1)^2} = 0. \quad (6.13)$$

Hence the driving heat flux at extremum power conditions (designated by superscript 0) is

$$(q_1)^0 = -(gu)^0 = g(T_1 - \sqrt{T_1 T_2}). \quad (6.14)$$

The second derivative of w is negative at the extremum point, and the extremum is the maximum. Note that the extremum value of q_1 is lower than the purely conductive heat $q_1 = g(T_1 - T_2)$ which corresponds to zero efficiency η. When (6.14) is substituted into the efficiency formula, (6.12), the wellknown NCA efficiency for maximum power,

$$\eta^0 = 1 - \sqrt{\frac{T_2}{T_1}} \quad (6.15)$$

is obtained. Substituting (6.14) into (6.11) yields the extremum power

$$w^0 = g(\sqrt{T_1} - \sqrt{T_2})^2 \quad (6.16)$$

which is a familiar expression known from a standard treatment of NCA process [3], [4]. Change of control variable yields an equivalent description. Thus it is permissible to choose a convenient control variable, typically one preferred for technical reasons. Use of the driving heat as a control has some virtues for making physical interpretation simple: with this choice, the reversible point is the point at which the control variable q_1 is zero, and the irreversibility description in terms of q_1 leads to formulas similar to

6.3. Applying Single-Stage Formulas to a Multistage Process

those used frequently for describing "inactive" heat transfer. On the other hand, transition from one control to another is possible. For example, the inverted form of (6.12) can be used to present nonoptimal results (e.g., power w, temperatures $T_{1'}$ and $T_{2'}$, etc.) in terms of the efficiency control η instead of the heat flux control q_1,

$$q_1 = g\left(T_1 - \frac{1}{(1-\eta)}T_2\right). \tag{6.17}$$

(See [4].) In the efficiency representation, the power produced, $w = \eta q_1$, is

$$w = g\eta\left(T_1 - \frac{T_2}{(1-\eta)}\right). \tag{6.18}$$

The heat flux q_1 is also the most convenient variable to characterize properties of the stage. These are defined by the two unique points, the so-called *short-circuit point* (with $\eta = 0$), at which $q_1 = q_2 = g\Delta T$, and the so-called *open-circuit point* ($\eta = \eta_C$), at which $q_1 = q_2 = 0$. At each of these two points the power $w = 0$; the power may be produced only in the range of efficiencies between 0 and η_C. These points divide each characteristics into three ranges: the range of a normal work of the engine, the range of a refrigerator, and the range of a heat pump. Some works [3]–[6] discuss consequences of changing or extending the heat exchange model, an issue which we will not consider here.

6.3 Applying Single-Stage Formulas to a Multistage Process

We add the stage number superscript, n, to all symbols in the above formulas when we apply them to a cascade. Yet, since in the infinite reservoir case T_2 is only a parameter and not a state variable, we can also simplify our designation for the state variable T_1^n and heat q_1^n by rejecting the unnecessary subscript 1; thus we will use the symbols T^n and q^n for the resource fluid temperature and driving heat flux at the stage n. We shall also designate the constant heat-sink temperature as T^e rather than T_2^n. With these changes, (6.17), which links the heat q^n and the efficiency η^n, becomes

$$u^n = \frac{T^e}{1-\eta^n} - T^n, \tag{6.19}$$

where $u^n \equiv -(g^{-1})^n q^n$. Equation (6.12) with $T_2 = T^e$ and superscripted by n is the inverse of (6.19). The efficiency takes the simple form

$$\eta^n = 1 - \frac{T_e}{T^n - (g^n)^{-1}q^n} = 1 - \frac{T^e}{T^n + u^n}. \tag{6.20}$$

Equations (6.19) and (6.20) show that the efficiency of power production at an engine stage decreases with the intensity variables q^n or $(-u^n)$. While (6.20) is inverse of (6.19), they are not equally suitable for describing cascades in which there are some extensive quantities specific to the cascade as a whole, and not to a single stage. These are cumulative quantities which may play the roles of state coordinates in the cascade optimization when there are limits on the consumption of certain inputs and on the corresponding fluxes. Additivity of an underlying physical quantity is necessary to make the definition of a cumulative variable self-consistent and useful.

Since the optimization problem is that of maximizing W^n at the end of the process, i.e., for $n = N$, it is convenient to associate the variable W with a zeroth state coordinate. Equation (6.22) below is a discrete rate formula describing this coordinate. An equation for the state variable describing the cumulative work is obtained from an expression for the power delivered at the stage n. We define the cumulative power per unit mass flow, $W^n \equiv \Sigma w^n / G$, a quantity whose units are those of specific work. When (6.18) and (6.20) are applied at the stage n with the identifications $T_2 = T^e$ and $T_1 = T^n$ and we introduce a nondimensional conductance θ^n at the stage n, such that

$$\theta^n \equiv \frac{g^n}{Gc}, \qquad (6.21)$$

then we obtain a state equation for the specific work in terms of the control u^n:

$$W^n - W^{n-1} = -c\left(1 - \frac{T^e}{(T^n + u^n)}\right) u^n \theta^n$$

$$= -c\left(1 - \frac{T^e}{T^n}\right) u^n \theta^n - cT^e \frac{(u^n)^2}{T^n(T^n + u^n)} \theta^n. \qquad (6.22)$$

The second line of this expression splits the work production rate into a classical term and a dissipative term. The total work is the sum of these expressions over the stages. Taking into account that the conductance g^n may be expressed as the product of an overall coefficient of heat transfer and a corresponding overall transfer area, we identify the nondimensional conductance θ^n as the so-called number of transfer units at the stage n, a well-known engineering quantity. The ratio of the stage length ℓ^n to θ^n is then a quantity with units of length; it is identified with the so-called height of the transfer units at the stage n, H_{TU}^n. Again, this is a wellknown quantity in heat transfer theory. From its definition and (6.21), H_{TU}^n equals $Gc\ell^n / g^n$. The smallness of this quantity is a good indicator of the quality of transfer at the stage; the quantities H_{TU}^n exhibit very small values for large conductances g^n. However, it will be sufficient to use the dimensionless variable θ^n only. One may note that the variable, θ^n, appears linearly in (6.22), a fact we will use in optimizing the model. To do this, we first need the second discrete equation of state, which describes the driving heat.

To find the equation of state for the temperature T^n, we define the cumulative driving heat Q^n for the first n stages of the cascade: $Q^n = \Sigma q^k$, where $k = 1, 2, \ldots, n$. This quantity satisfies the equality $Q^n - Q^{n-1} = -Gc(T^n - T^{n-1}) = q^n$, where, because of (6.21) and the definition $u^n \equiv -(g^{-1})^n q^n$, the heat q^n is the product of $-g^n$ ($= -Gc\theta^n$) and u^n. The comparison of these two expressions yields the second equation of state

$$T^n - T^{n-1} = u^n \theta^n. \qquad (6.23)$$

This equation tells us that $u^n = -q^n/g^n$, i.e., the negative of the heat received by the nth Carnot engine (in units of temperature), at the same time plays the role discrete rate of the temperature change, $\Delta T^n / \Delta \tau^n$.

To close the model, we will regard θ^n as the increment of an independent variable. This variable, which we designate by τ^n, is nondimensional, of definite (positive) sign, and measures the cumulative number of heat transfer units. The variable τ^n satisfies the definition $\tau^n = \Sigma \theta^k$ for $k = 1, 2, \ldots, n$. Thus, the third equation of state is simply

$$\tau^n - \tau^{n-1} = \theta^n. \qquad (6.24)$$

Equation (6.24) is essential in all problems with the constrained sum of θ^n. This corresponds to the constrained total transfer area, because the single-stage transfer areas are contained in g^n and hence in θ^n. This also corresponds to a constraint on the total length.

In accordance with the terminology used in optimization, (6.22)–(6.24) are the discrete equations of state for the cascade. They contain on their right-hand sides the state variables and controls. An optimizing procedure for a discrete problem requires specification of an optimization criterion, a complete set of the state equations (as above), and possibly some local constraints at the stage in question. A general scheme of optimization of cascade processes is considered in the next section, and the theory developed is then applied to CAN cascades.

6.4 Pontryagin's Structure of Optimal Control

For the NCA process considered here, and its continuous limit, we now show the application of a discrete optimization theory based on the so-called discrete maximum principle with a constant Hamiltonian [7]–[12]. The existence of this principle has been shown for discrete models which are linear and homogeneous with respect to a particular unconstrained control variable. This control may be an unconstrained interval of one of the state variables, such as our interval of the dimensionless time, θ^n, or an unconstrained interval of a parameter. The model's linearity with respect to θ^n is crucial for the formal similarity of the necessary conditions for optimality in discrete and continuous processes because, broadly speaking, this linearity causes vanishing of the second order and higher terms in Taylor expansions of characteristic functions for discrete processes. Consequently, discrete formulas acquire the structures already known for continuous processes.

Our task here is a brief analysis of these properties for discrete (multistage) processes with a finite number of stages, although some results at the continuous limit (for an infinite number of stages) will also be discussed. For discrete processes, the standard discrete theory of optimal control [13], [14] does not predict any special similarity of discrete and continuous descriptions, e.g., such characteristic features of the continuous theory as constancy of an autonomous Hamiltonian or the Hamilton–Jacobi equation [15], [16]. This is because discrete descriptions are generally not reducible to continuous ones in the limit of an infinite number of discrete units. To assure us that a discrete model converges to the continuous limit and symplectic properties, one must restrict oneself to a special class of discrete processes in which finite intervals of state variables are not constrained explicitly, and the allowed constraints can affect only ratios of differences of the state variables at any stage. To satisfy these requirements, a discrete set of the state equations and an optimization algorithm must be linear with respect to a particular decision, θ^n. Whenever a discrete model has a structure linear in θ^n, a remarkable similarity emerges between necessary optimality conditions in both continuous and discrete cases. In particular, optimal values of a performance coordinate satisfy Hamilton and Hamilton–Jacobi formalisms in the discrete case and a discrete maximum principle emerges in a form analogous to that known for continuous systems. In the case of NCA cascades, it is an optimal work function that satisfies these relationships. The outline of an abstract analytical description with occasional references to the NCA cascade is given below.

Applying the usual description based on ordinary difference equations, the optimization theory for (6.22)–(6.24) of the NCA cascade can be represented by a general set of equations

$$x_i^n - x_i^{n-1} = f_i^n(\mathbf{x}^n, \mathbf{u}^n)\theta^n. \tag{6.25}$$

Here $i = 0, 1, \ldots, s, s+1$ and $f_{s+1}^n = 1$ for the $(s+1)$th coordinate of the state vector x^n, consistent with (6.24) for $\mathbf{x}^n$ which is the unconstrained interval of the variable $\tau \equiv x_{s+1}^n$. In the case of our NCA cascade, for which $s = 1$, the state vector $\mathbf{x}^n$ is three-dimensional: $x_0^n \equiv W^n$, $x_1^n \equiv T^n$, and $x_2^n \equiv \tau^n$, for each stage n. The set (6.25) with arbitrary f_{s+1}^n may be considered as a proper generalization of the original model which still preserves the linearity of the model with respect to a free control θ^n. For this set the optimization problem is stated as maximizing the zeroth state coordinate for $n = N$ when the initial states of $\mathbf{x}^n$ are fixed. To derive the necessary optimality conditions for discrete processes, (6.22)–(6.24) or the general equation (6.25), Bellman's method of dynamic programming (DP) can be applied [17], [18]. A recent treatment of dynamic programming of

6.4. Pontryagin's Structure of Optimal Control

NCA cascades is helpful [19] as it allows one to pass from DP results to the discrete maximum principle which we will apply in the next part of this chapter.

Let us define the optimal performance function for a general n-stage process

$$I^n(\mathbf{x}^n) \equiv \max \sum_1^n f_0^n(\mathbf{x}^n, \mathbf{u}^n)\theta^n = -\min \sum_1^n -f_0^n(\mathbf{x}^n, \mathbf{u}^n)\theta^n, \quad (6.26)$$

where f_0^n is a profit generation function that follows from (6.22). In this formulation one assumes that $\mathbf{x} = (x_1, x_2, \ldots, x_{s+1} = t)$, i.e., that the performance coordinate has been excluded from the coordinates of the state vector. In the case of our NCA cascade, the profit intensity is just the intensity of work produced, and the definition of the optimal work production function has a special form of (6.26):

$$I^N[T^0, \tau^0, T^N, \tau^N] = \max \sum_{n=1}^N f_0^n \theta^n = -\min \sum_{n=1}^N (-f_0^n \theta^n), \quad (6.27)$$

where we have written it to the whole cascade and pointed out the boundary condition making the coordinates of the final state (T^N, τ^N) fixed. The work production intensity follows from (6.22) as

$$\begin{aligned} f_0^n &\equiv -c\left(1 - \frac{T^e}{T^n + u^n}\right) u^n \\ &= -c\left(1 - \frac{T^e}{T^n}\right) u^n - cT^e \frac{(u^n)^2}{T^n(T^n + u^n)}. \end{aligned} \quad (6.28)$$

This shows that the function I^N incorporates the minimum entropy production with a negative sign. The definition of I pays attention to whether the terminal thermodynamic states are initial or final in a process, so it differs from some other possibilities that might, for example, have identical end temperatures.

For a general n-stage process, in the stages $1, 2, \ldots, n$ described by (6.25) and (6.26), Bellman's recurrence equation is

$$I^n(\mathbf{x}^n) = \max_{\mathbf{u}^n, \theta^n} \{f_0^n(\mathbf{x}^n, \mathbf{u}^n)\theta^n + I^{n-1}(\mathbf{x}^n - \mathbf{f}^n(\mathbf{x}^n, \mathbf{u}^n)\theta^n)\}. \quad (6.29)$$

This represents the most popular form of recurrence equations for multistage optimization. Starting with $I^0 = 0$, the sequence of the optimal functions $I^1, I^2, \ldots, I^n, \ldots, I^N$ can be obtained by a well-established recurrence procedure [17] in which extremizations on the right sides are carried out with respect to controls at constant coordinate values of the final state x^n. Analytical solutions are seldom possible. However, a transformation to the maximum principle algorithm, described below, increases the likelihood of obtaining the solution in an analytical form.

154   6. Discrete Hamiltonian Analysis of Endoreversible Thermal Cascades

To pass to a maximum principle, one writes (6.29) in the form

$$\max_{u^n,\theta^n}\{f_0^n(\mathbf{x}^n,\mathbf{u}^n)\theta^n - (I^n(\mathbf{x}^n) - I^{n-1}(\mathbf{x}^n - \mathbf{f}^n(\mathbf{x}^n,\mathbf{u}^n)\theta^n))\} = 0, \quad (6.30)$$

which enables us to include variations of the final coordinates of state [8], [20]. For an unconstrained θ^n and possibly constrained u^n, the set equivalent to (6.30) has the form of the three equations

$$f_0^n(\mathbf{x}^n,\mathbf{u}^n)\theta^n - (I^n(\mathbf{x}^n) - I^{n-1}(\mathbf{x}^n - \mathbf{f}^n(\mathbf{x}^n,\mathbf{u}^n)\theta)) = 0, \quad (6.31)$$

$$f_0^n(\mathbf{x}^n,\mathbf{u}^n) - \frac{\partial I^{n-1}}{\partial \mathbf{x}^{n-1}} \cdot \mathbf{f}^n(\mathbf{x}^n,\mathbf{u}^n) = 0, \quad (6.32)$$

and

$$\max_{u^n}\{f_0^n(\mathbf{x}^n,\mathbf{u}^n)\theta^n - (I^n(\mathbf{x}^n) - I^{n-1}(\mathbf{x}^n - \mathbf{f}^n(\mathbf{x}^n,\mathbf{u}^n)\theta))\} = 0. \quad (6.33)$$

In (6.32) the stationarity condition for the extremal intervals θ^n was applied. Whenever θ^n is finite and positive, then, after using (6.31) and (6.32) in (6.33), the last equation can be transformed to

$$\max_{u^n}\left\{f_0^n(\mathbf{x}^n,\mathbf{u}^n) - \frac{\partial I^{n-1}}{\partial \mathbf{x}^{n-1}} \cdot \mathbf{f}^n(\mathbf{x}^n,\mathbf{u}^n)\right\} = 0, \quad (6.34)$$

which represents a maximum principle for the so-called enlarged Hamiltonian

$$\mathbf{H}^{n-1} \equiv f_0^n(\mathbf{x}^n,\mathbf{u}^n) - \frac{\partial I^{n-1}}{\partial \mathbf{x}^{n-1}} \cdot \mathbf{f}^n(\mathbf{x}^n,\mathbf{u}^n)$$

$$= f_0^n - \sum_{i=1}^{s} \frac{\partial I^{n-1}}{\partial x_i^{n-1}} f_i^n - \sum_{i=1}^{s} \frac{\partial I^{n-1}}{\partial \tau^{n-1}} \quad (6.35)$$

with respect to the controls $\mathbf{u}^n$. As $f_{s+1}(=1)$ is $\mathbf{u}$-independent, this optimality condition can also be expressed in terms of the energylike Hamiltonian which does not contain the partial derivative of I^{n-1} with respect to the time,

$$H^{n-1} \equiv f_0^n - \sum_{i=1}^{s} \frac{\partial I^{n-1}}{\partial x_i^{n-1}} f_i^n. \quad (6.36)$$

Equation (6.36) represents the Hamiltonian (6.35) without the $\mathbf{u}$-independent term $\partial I^{n-1}/\partial \tau^{n-1}$.

Next the so-called adjoint variables are defined, which are the partial derivatives of I^n with respect to the state coordinates x_i^n,

$$z_i^{n-1} \equiv -\frac{\partial I^{n-1}(\mathbf{x}^{n-1})}{\partial x_i^{n-1}} \quad (6.37)$$

6.4. Pontryagin's Structure of Optimal Control

($i = 1, \ldots, s, s+1$). In the extended space including the performance coordinate the Hamiltonian (6.35) may be written as the scalar product of all rates and their adjoint variables

$$\mathbf{H}^{n-1}(\mathbf{x}^n, z_0^{n-1}, \mathbf{z}^{n-1}, \mathbf{u}^n) \equiv \sum_{v=0}^{s+1} z_v^{n-1} f_v^n$$

$$= z_0^{n-1} f_0^n + \sum_{i=1}^{s} z_i^{n-1} f_i^n + z_\tau^{n-1}, \quad (6.38)$$

where $z_0^{n-1} = 1$, $z_i^{n-1} = -\partial I^{n-1}/\partial x_i^{n-1}$, for $i = 1, 2, \ldots, s$ and $n = 1, 2, \ldots, N$. Moreover, $z_s^{n-1} \equiv z_\tau^{n-1} = -\partial I^{n-1}/\partial \tau^{n-1}$ and $f_{s+1} = 1$. This Hamiltonian must be a maximum with respect to the controls $\mathbf{u}^n$ which maximize a performance index of the profit type, such as the work production criterion, $W^N = \Sigma f_0^n \theta^n$, whose optimal function I^N is defined by (6.27). In an equivalent formulation, one minimizes a performance index of the cost type, such as the work consumption criterion, $(-W^N) = \Sigma(-f_0^n \theta^n) = \Sigma L^n \theta^n$, whose optimal function may be defined as $R^N \equiv -I^N$. In that formulation, the unchanged Hamiltonian (6.38), which may operate with $-L^n$ in place of f_0^n, still has to attain its maximum with respect to the controls $\mathbf{u}^n$, or a modified Hamiltonian $\mathbf{H}'^{n-1} \equiv -\mathbf{H}^{n-1}$ has to attain the minimum with respect to $\mathbf{u}^n$. Thus the original maximization problem for $W^N = \Sigma f_0^n \theta^n$ requires maximizing the Hamiltonian $\mathbf{H}^{n-1}$, and the equivalent problem for minimization of $(-W^N) = \Sigma L^n \theta^n$ requires minimizing the modified Hamiltonian $\mathbf{H}'^{n-1}$ ($\equiv -\mathbf{H}^{n-1}$), which contains L^n or $(-f_0^n)$ in place of f_0^n in (6.38).

Differentiating the bracketed expression in (6.30) to determine its stationarity conditions with respect to the final space and time coordinates, we obtain an optimal difference set which is canonical with respect to two sort of equations, one defining the changes of state and one the corresponding changes of the adjoint variables. Using the most popular energylike Hamiltonian, (6.36), expressed in terms of the adjoint variables,

$$H^{n-1}(\mathbf{x}^n, \mathbf{z}^{n-1}, \mathbf{u}^n, \tau^n) \equiv f_0^n(\mathbf{x}^n, \mathbf{u}^n, \tau^n) + \sum_{i=1}^{s} z_i^{n-1} f_i^n(\mathbf{x}^n, \mathbf{u}^n, \tau^n), \quad (6.39)$$

the maximum principle is contained in the equations

$$\frac{x_i^n - x_i^{n-1}}{\theta^n} = \frac{\partial H^{n-1}}{\partial z_i^{n-1}}, \quad (6.40)$$

$$\frac{z_i^n - z_i^{n-1}}{\theta^n} = -\frac{\partial H^{n-1}}{\partial x_i^{n-1}}, \quad (6.41)$$

$$\frac{\max H^n - \max H^{n-1}}{\theta^n} = \frac{\partial H^{n-1}}{\partial \tau^n}, \quad (6.42)$$

$$z_\tau^{n-1} + \max_{\mathbf{u}^n} H^{n-1}(\mathbf{x}^n, \mathbf{z}^{n-1}, \mathbf{u}^n, \tau^n) = 0 \qquad (6.43)$$

($n = 1, \ldots, N$, $i = 1, \ldots, s$, and $l = 1, \ldots, r$). Equation (6.40) constitutes the Hamiltonian form of the state equations, and (6.41) is its adjoint equation. Equation (6.42) describes the Hamiltonian interval at stage n, whereas (6.41) states that the enlarged Hamiltonian $\mathbf{H}^{n-1} = H^{n-1} + z_\tau^{n-1}$ of the extremal process is always constant and equal to zero. Equation (6.41) includes the necessary condition for the stationary optimality of the decision vector $\mathbf{u}^n$, if its optimal value falls in the interior of the allowable range $\mathbf{U}$.

The boundary conditions are determined by the vanishing stationarity conditions for the extremum of $x_0^N \equiv W^N$ with respect to the end state coordinates and end times. Setting to zero the respective partial derivatives of the optimal function I^n for one end, either $n = 0$ or for $n = N$, and invoking the definition of adjoint variables (6.37) yields, for the free-end state variables,

$$z_i^N = 0, \quad i \neq \beta, \qquad z_i^0 = 0, \quad i \neq \alpha, \qquad (6.44)$$

and for free-end time τ,

$$H^N = 0, \qquad H^0 = 0. \qquad (6.45)$$

Equation (6.44) prescribes the boundary conditions for the adjoint vector z^n at the end and at the beginning of the process, and (6.45) prescribes boundary conditions for the Hamiltonian H^n at the end and at the beginning of the process. As expressed in (6.44), if the βth component of the final state vector, x_i^N, or the αth component of the initial state vector, x_i^0, is fixed, the respective component of the adjoint vector, z_β^N or z_α^0, is undetermined. Analogously, if the boundary time, τ^N or τ^0, is fixed, the respective Hamiltonian, H^N or H^0, is undetermined. When the number of stages tends to infinity the discrete algorithm becomes the Pontryagin maximum principle [8], [9]. Below we consider the application of the above algorithm to the NCA cascade.

6.5 Work Maximizing in NCA Cascades by Discrete Maximum Principle

We return to the multistage NCA process in a cascade system shown in Figure 6.1. Its optimal performance function I^N results from (6.27) and (6.28) as

$$I^N[T^0, T^N, \tau^N, \tau^0] \equiv \max W^N = -\min(-W^N)$$

$$= \max_{\{u^n, \theta^n\}} \sum_{n=1}^{N} \left\{ -c\left(1 - \frac{T^e}{u^n + T^n}\right) u^n \theta^n \right\}. \qquad (6.46)$$

6.5. Work Maximizing in NCA Cascades by Discrete Maximum Principle

The sum in (6.46) may be regarded as a discrete functional which must be maximized subject to the difference constraints, (6.23) and (6.24),

$$T^n - T^{n-1} = u^n \theta^n, \qquad (6.23)$$

$$\tau^n - \tau^{n-1} = \theta^n. \qquad (6.24)$$

Equation (6.46) reveals the boundary condition that fixes the coordinate of the final state (T^N, τ^N). The sign of I^N defines the working mode for the cascade as a whole. In agreement with our conventions, I^N is positive in work production modes; this refers to the range of the fluid changes in which $u < 0$ whenever $T > T^e$; in this case the engine process assures the fluid's cooling, whereas the heat-pump process and inequality $u > 0$ assures the fluid's heating. Thus in engine modes, $W > 0$ and $I^N > 0$. In heat-pump modes, $W < 0$, so working with a function $R^N = -I^N$ may be more convenient. However, the direction of any process in the cascade is determined by the positivity of the entropy production rather than by the sign of any work function. Of special attention are two processes: the one which starts with $T^0 = T^e$ and terminates at an arbitrary $T^N = T$ and the one which starts at an arbitrary $T^0 = T$ and terminates at T^e. In these cases the functions I^N are generalizations for the heat-pump mode and engine mode of the classical thermal exergy in discrete processes with finite durations.

The Hamiltonian (6.39) for the NCA cascade contains one nontrivial adjoint variable, the temperature adjoint z^{n-1} and one control $u^n \equiv -(g^{-1})^n q^n = \Delta T^n / \Delta \tau^n$. The Hamiltonian has the form

$$H^{n-1} = z^{n-1} u^n - c\left(1 - \frac{T^e}{u^n + T^n}\right) u^n. \qquad (6.47)$$

According to (6.43), H^{n-1} is a maximum in the optimal process (Figure 6.2). The figure shows possible realizations of the optimal process. The case for $\eta < \eta_C$ refers to fluid cooling and the case for $\eta < \eta_C$ to the fluid heating. As in the limiting continuous case, a suitable parameter to characterize these realizations for finite rates is $p = z/c - \eta_C$ rather than the temperature adjoint itself; the analogy is complete, so this issue is omitted here, see [21]. Moreover, (6.42) proves that the Hamiltonian is constant along any optimal discrete path. This is the consequence of the autonomous nature of the model, which does not contain explicitly the time τ^n. Thus $H^{n-1} = H$, where H is a constant, and we investigate a subclass of all solutions for the same $H^{n-1} = H$. This is described by the equations

$$\frac{T^n - T^{n-1}}{\theta^n} = \frac{\partial H^{n-1}}{\partial z^{n-1}} = u^n, \qquad (6.48)$$

$$\frac{z^n - z^{n-1}}{\theta^n} = -\frac{\partial H^{n-1}}{\partial T^n} = -\frac{cT^e}{(u^n + T^n)^2} u^n, \qquad (6.49)$$

$$z^{n-1} u^n - c\left(1 - \frac{T^e}{u^n + T^n}\right) u^n = H, \qquad (6.50)$$

158   6. Discrete Hamiltonian Analysis of Endoreversible Thermal Cascades

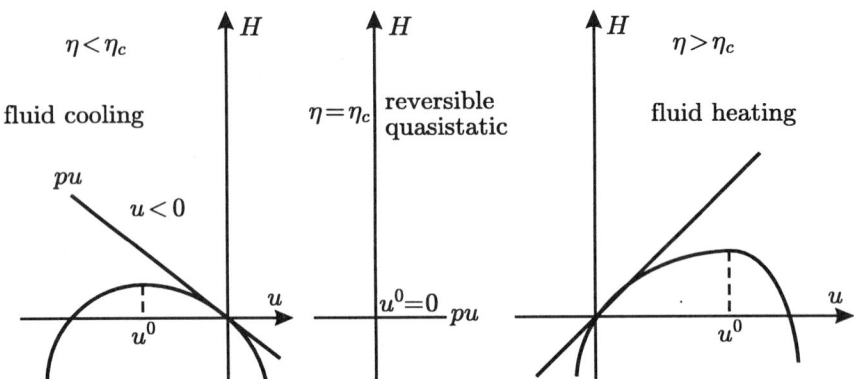

FIGURE 6.2. The discrete Hamiltonian H^{n-1}, (6.47), for an arbitrary stage n (stage index neglected) as a function of the discrete rate u, exhibits all the properties of the continuous Pontryagin Hamiltonian, for three cases of $p = z/c - \eta_C$. Optimal rates are those maximizing H.

and
$$u^n = \arg\left\{\max_{u^n} H^{n-1}(T^n, z^{n-1}, u^n)\right\}, \qquad (6.51)$$

where the last equation states that H^{n-1} satisfies the maximum condition of (6.43). Since the discrete rate u^n is unconstrained, we evaluate the stationary extremum condition for the Hamiltonian (6.47)

$$\frac{\partial H^{n-1}}{\partial u^n} = z^{n-1} - c\left(1 - \frac{T^e T^n}{(u^n + T^n)^2}\right) = 0, \qquad (6.52)$$

which should be the maximum condition. Thus the optimal control u^n should satisfy the inequality

$$\frac{\partial H}{\partial (u^n)^2} = \frac{\partial^2 f_0}{\partial u^2} = -2cT^e \frac{T^n}{(T^n + u^n)^3} < 0. \qquad (6.53)$$

This means that a temperature-like quantity $T' \equiv T + u$ must be positive at each stage of the process. While this requirement does not affect the temperature adjoints z^n, it imposes a constraint on the cooling rates u^n, preventing them from becoming too high, namely, they should satisfy the inequality $u^n + T^n > 0$ or $u^n > -T^n$. This means that the dimensionless cooling rates $\xi^n \equiv u^n/T^n$ faster than -1 cannot be realized in the system. This would seem to provide a physical limitation on the instantaneous power production intensities in stages with decreasing temperature (e.g., those working as engines at $T > T^e$), that could not release the mechanical energy faster than allowed by the limiting cooling rate $u^n/T^n = -1$.

6.5. Work Maximizing in NCA Cascades by Discrete Maximum Principle 159

However, since there are other conditions which are more restrictive than this one, the practical limitations resulting from the present condition are irrelevant. These other conditions will be discussed later.

Now we will find a solution of the equations for the optimal set, (6.48)–(6.52). In a general case of an arbitrary optimization model the condition $H^{n-1} = H^n = H$ means restriction to a subclass of the total family of the optimal trajectories. Such a subclass is characterized by a common intensity H and various indefinite (free) durations. In our case one deals with the process with the state variable T^n and a constant number of stages N. The addition of the condition $H^n = H$ restricts one to a definite trajectory with optimally adjusted intervals of time, θ^n.

On eliminating z^{n-1} from the two last equations, the integral of discrete motion is obtained in a form

$$\frac{(u^n)}{(u^n + T^n)^2} = \frac{H}{cT^e}. \tag{6.54}$$

This is, in fact, the structure of the Hamiltonian function of continuous processes [21], which is not surprising in view of the Pontryagin structure of the discrete optimization algorithm. With these constraints one obtains two realizations of the optimal control which correspond, respectively, to increasing and descreasing temperatures T^n with time τ^n, or, equivalently, to the heating and cooling strategies

$$\frac{u^n}{u^n + T^n} = \pm \left(\frac{H}{cT^e}\right)^{1/2}, \tag{6.55}$$

where the upper sign refers to heating and the lower, to cooling. Note that the constancy of H implies here the constancy of the ratio $\xi^n \equiv u^n/T^n$ which is the ratio of our discrete rate of the temperature change u^n and the temperature T^n. The intensity constant ξ in terms of the dimensionless Hamiltonian $H' \equiv H(cT^e)^{-1}$ is

$$\xi = \pm \frac{(H')^{1/2}}{1 - \pm(H')^{1/2}}. \tag{6.56}$$

The solution $u^n = \xi T^n$ represents the feedback control for this optimal control problem, where the controlling quantity u^n is adjusted by specifying its link with the system's state. Incorporating the first canonical equation (6.48), we obtain

$$\frac{T^n - T^{n-1}}{\theta^n} = \xi T^n. \tag{6.57}$$

This equation means that the discrete rate in the extremal process changes proportional to the temperature, the result analogous to the logarithmic formula obtained for the optimal intensity of the temperature change in continuous systems [21]–[23]. The above difference equation should be solved simultaneously with the equation

$$\frac{T^e T^n}{(u^n + T^n)^2} - \frac{T^e T^{n+1}}{(u^{n+1} + T^{n+1})^2} = \frac{T^e}{(u^n + T^n)^2} u^n \theta^n, \quad (6.58)$$

which was obtained from the second canonical equation, (6.49), and the extremum condition of H, (6.52). Since $u^n \theta^n = T^n - T^{n-1}$, the right side of this equation can be written in terms of the discrete change of the temperature and then simplified. This leads to the following relationship linking temperatures and rates for two arbitrary stages n and $n+1$:

$$\frac{T^{n-1}}{(u^n + T^n)^2} = \frac{T^{n+1}}{(u^{n+1} + T^{n+1})^2}. \quad (6.59)$$

After using in this $u^n = \xi T^n$ for the stages n and $n+1$ one gets

$$\frac{T^{n-1}}{(T^n)^2(\xi+1)^2} = \frac{T^{n+1}}{(T^{n+1})^2(\xi+1)^2} \quad (6.60)$$

which implies a simple result

$$(T^n)^2 = T^{n-1} T^{n+1}. \quad (6.61)$$

This equation means that the interstage temperatures T^n between the stages n and $n+1$ are geometric means of the boundary temperatures of any two-stage subprocess. Use of this formula with boundary conditions for T^0 and T^N yields the interstage temperatures in terms of the boundary temperatures

$$T^1 = (T^N)^{1/N}(T^0)^{(N-1)/N-1}, \quad (6.62)$$
$$T^2 = (T^N)^{2/N}(T^0)^{2(N-1)/N-1}, \quad (6.63)$$
$$T^3 = (T^N)^{3/N}(T^0)^{3(N-1)/N-2}, \quad (6.64)$$

$$\cdots \cdots$$

$$T^{N-1} = (T^N)^{(N-1)/N}(T^0)^{[(N-1)^2/N-(N-2)]}$$
$$= (T^N)^{(N-1)/N}(T^0)^{1/N}. \quad (6.65)$$

The optimal trajectory, (6.57), satisfies the constant intensity condition in the form

$$\xi(H) = \frac{T^n - T^{n-1}}{\theta^n T^n} = \frac{T^{n+1} - T^n}{\theta^{n+1} T^{n+1}} \cdots. \quad (6.66)$$

Writing (6.66) as

$$\xi = (\theta^n)^{-1}(1 - T^{n-1}/T^n) = (\theta^{n+1})^{-1}(1 - T^n/T^{n+1}) \cdots \quad (6.67)$$

and substituting into it (6.61) yields

$$\xi = (\theta^n)^{-1}\left(1 - \sqrt{\frac{T^{n-1}}{T^{n+1}}}\right) = (\theta^{n+1})^{-1}\left(1 - \sqrt{\frac{T^{n-1}}{T^{n+1}}}\right) \cdots, \quad (6.68)$$

6.5. Work Maximizing in NCA Cascades by Discrete Maximum Principle

thus leading to the conclusion that the time intervals are equal along an extremal

$$\theta^n = \theta^{n+1} = \cdots = \theta = \tau^N/N. \tag{6.69}$$

With (6.65) and (6.69) and the first equality in (6.67), the intensity constant ξ can be linked with the Nth stage data as

$$\xi = \frac{N}{\tau^N} \frac{T^N - T^{N-1}}{T^N} = \frac{N}{\tau^N}\left[1 - \left(\frac{T^0}{T^N}\right)^{1/N}\right]. \tag{6.70}$$

The last expression of this equation expresses the optimal intensity in terms of the boundary temperatures, total duration, and total number of stages. Introducing the function $\xi(H)$, (6.56), into (6.70) links the total duration τ^N with the Hamiltonian H

$$\tau^N = N\left[1 - \left(\frac{T^0}{T^N}\right)^{1/N}\right]\left(\pm\left(\frac{H}{cT^e}\right)^{-1/2} - 1\right). \tag{6.71}$$

The solution of this equation with respect to H, which is

$$H = cT^e\left(\frac{\tau^N}{N[1 - (T^0/T^N)^{1/N}]} + 1\right)^{-2}, \tag{6.72}$$

defines H as a cost of the process time or the numerical value of the Lagrange multiplier associated with the time constraint. This gives the numerical value of H which has to be used when the set of equations (6.47)–(6.51) is solved for a given duration τ^N. The applicability of H as the Lagrange multiplier of the constraint imposed on the total holdup time is described in the next section.

For the N-stage process the optimal work function is obtained in the form

$$\begin{aligned}I^N &= c(T^0 - T^N) + cT^e N\left[1 - \left(\frac{T^0}{T^N}\right)^{1/N}\right] \\ &\quad - cT^e \frac{\{N[1 - (T^0/T^N)^{1/N}]\}}{\tau^N + N[1 - (T^0/T^N)^{1/N}]}.\end{aligned} \tag{6.73}$$

This is consistent with the optimal trajectory satisfying (6.61)–(6.65). An efficient way to obtain this function is by dynamic programming because the results of analysis for $N = 3$ in hand make it possible to predict the form of solution for an arbitrary N [19], [24]. Here, however, the solution is obtained by a different method, using the discrete maximum principle and the properties of the Hamiltonian as the Lagrange multiplier, as described below.

6.6 The Hamiltonian as the Lagrange Multiplier of a Time Constraint

The power of the constancy of the discrete Hamiltonian can be seen particularly well when the constancy condition of the Hamiltonian, $H^n = H$, is applied to determine a sequence of work functions I^n or its economic analogues. This is frequently done by the standard dynamic programming [19], [24]. Here, however, we would like to present a modified approach with the time coordinate eliminated as it implies the equality $H^n = H$. Since, as shown by (6.43), for any optimal problem with free time intervals, the constant H equals the negative of the time adjoint z_τ, the former can replace the latter as the Lagrangian multiplier of the constraint imposed on the total duration $\tau^N - \tau^0$. Most frequently the multiplier is applied for autonomous systems; then $\tau^0 = 0$ is assumed and the constancy of the multiplier H is used to avoid the time as the state variable, i.e., for the purpose of reducing the dimensionality. However, there are also other benefits of the procedure, as it involves distinguishing among various sorts of costs, such as the work costs, the holdup time costs, or the size costs. These are closely related to the exploitation costs and the investment costs basic in economic evaluations.

In the case considered here, however, the Lagrange multiplier method requires that one maximizes the modified work criterion, $W_*^N = W^N - H\tau^N$, of "net profit type"

$$W_*^N = \sum_1^N \left[-c \left(1 - \frac{T^e}{u^n + T^n}\right) u^n - H \right] \theta^n$$

$$= \sum_1^N \left[-c \left(1 - \frac{T^e}{u^n + T^n}\right) u^n - H \right] \frac{(T^n - T^{n-1})}{u^n} \quad (6.74)$$

instead of the original criterion W^N contained in (6.46). The equation of state (6.23) was used in the second line of (6.74) to eliminate θ^n. The equation's form implies a particularly suitable definition of the optimal cost function, as the negative maximum net profit

$$R_*^N[T^0, T^N, \tau^N - \tau^0] \equiv -\max W_*^N = \min(-W_*^N)$$

$$= \min_{\{u^n, \theta^n\}} \sum_1^N \left(c \left(1 - \frac{T^e}{u^n + T^n}\right) u^n + H \right) \frac{(T^n - T^{n-1})}{u^n}. \quad (6.75)$$

Note that $R_*^N \neq -I^N$ of (6.73), because of the presence of H in (6.75). The maximizing in (6.75) is done with the fixed boundary temperatures T^0 and T^N, but for an undetermined total duration, τ^N. Of course, based on the results of the method used in the previous section, we expect that the optimal duration will follow as a function of T^0, T^N, and the total

6.6. The Hamiltonian as the Lagrange Multiplier of a Time Constraint 163

number of stages, N, in the form of (6.73), but we will have to show this in the context of the present method. The free time intervals, which are implicit in the second line of (6.74), are assured by unrestricted controls u^n and T^{n-1}. Formally, a dynamic programming equation searches here for the two decisions at the stage, the pair u^n and θ^n or the pair u^n and T^{n-1}. However, as shown below, the constancy of the extremal Hamiltonian expressed in the rate form, $H^{n-1}(u^n, T^n)$ of (6.54), causes a simplification of the search for an optimum.

Indeed, the constancy of the Hamiltonian written in the form of (6.55), or its equivalent form explicit with respect to u^n,

$$u^n = \frac{\pm (H/cT^e)^{1/2}}{1 - \pm (H/cT^e)^{1/2}} T^n \equiv \xi T^n, \tag{6.76}$$

defines a constraint that links the discrete rate u^n with the temperature T^n at each stage n. In effect, when dealing with the second line of (6.74), the dynamic programming procedure is not only one-dimensional but it effectively requires only one minimization with respect to T^{n-1} at each stage. This is shown below. The nonclassical work function defined by (6.75) is obtained by the optimization of the process between the two arbitrary temperatures T^0 and T^N.

The first optimal function ($N = 1$) is determined from the equation

$$R_*^1 = \min_{u^1} \left\{ \left(c \left(1 - \frac{T^e}{u^1 + T^1} \right) u^1 + H \right) \frac{(T^1 - T^0)}{u^1} \right\} \tag{6.77}$$

for u^1 satisfying (6.76) for $n = 1$. It is not difficult to verify that the result of minimizing the right-hand side of (6.77) with respect to u^1 is just (6.54) or (6.76) for $n = 1$. Indeed, the differential calculus yields

$$\left(\frac{cT^e}{(u^1 + T^1)^2} - \frac{H}{(u^1)^2} \right)(T^1 - T^0) = 0, \tag{6.78}$$

which leads to (6.54) and (6.76) in the case of $n = 1$. We will show that the further minimizations with respect to u^n can be eliminated by the knowledge of the constancy of H^{n-1} and (6.77), which defines the analytical structure of each H^{n-1} ($n = 1, 2, 3, \ldots, N$). For heating of the fluid we use the upper sign in (6.76), corresponding to an increase in time of a "driving" temperature $T^n(1 + \xi)$ and the related increase of the power supplied to the heat pump. Relations analogous to (6.78), differing only by the stage index, will be omitted at the subsequent stages.

Substituting into (6.77) the solution of (6.78), i.e., the function u^1 of (6.76), yields the optimal work function for the process between the temperatures T^0 and T^1

$$R_*^1 = c(T^1 - T^0) \left(1 - \frac{T^e}{T^1} \left[1 - \pm \sqrt{\frac{H}{cT^e}} \right]^2 \right), \tag{6.79}$$

164   6. Discrete Hamiltonian Analysis of Endoreversible Thermal Cascades

whose ξ representation is

$$R_*^1 = c(T^1 - T^0)\left(1 - \frac{T^e}{T^1(1+\xi)^2}\right). \tag{6.80}$$

This shows that for heating processes (positive ξ), the function of the modified work, R_*^1, increases with the intensity ξ. For cooling processes (the usual engine modes if $T > T^e$), the function is negative in agreement with its definition. The cooling rates (negative ξ) must satisfy $-1 < \xi < 0$ as required by the sufficient condition for maximizing the work W. For sufficiently low cooling rates, i.e., when ξ is close to zero but still negative, the function R_*^1 is negative. This is because we have defined R_*^n as a cost-type function whereas cooling the fluid at a low rate in the engine mode is the profit-producing process.

When $n = 2$ the standard dynamic approach involves minimizing the sum

$$\begin{aligned} e_*^2 = {} & \left[c\left(1 - \frac{T^e}{u^2 + T^2}\right)u^2 + H\right]\frac{(T^2 - T^1)}{u^2} \\ & + c(T^1 - T^0)\left[1 - \frac{T^e}{T^1}\left(1 - \pm\sqrt{\frac{H}{cT^e}}\right)^2\right] \end{aligned} \tag{6.81}$$

by a suitable choice of the two decisions, the rate u^2 and the interstage temperature T_1. Again, it follows that the minimizing of (6.81) with respect to u^2 can be replaced by using u^2 of (6.76). This allows one to apply at stage 2 the cost expression analogous to (6.79) obtained at stage 1, with only a difference in indices. Consequently, one finds a two-stage work function with interstage temperatures as the only controls

$$\begin{aligned} w_*^2 = {} & c(T^2 - T^1)\left[1 - \frac{T^e}{T^2}\left(1 - \pm\sqrt{\frac{H}{cT^e}}\right)^2\right] \\ & + c(T^1 - T^0)\left[1 - \frac{T^e}{T^1}\left(1 - \pm\sqrt{\frac{H}{cT^e}}\right)^2\right]. \end{aligned} \tag{6.82}$$

By induction, the work function of the n-stage subprocess follows:

$$w_*^N = \sum_1^N c(T^n - T^{n-1})\left[1 - \frac{T^e}{T^n}\left(1 - \pm\sqrt{\frac{H}{cT^e}}\right)^2\right]. \tag{6.83}$$

This criterion should be minimized by a suitable choice of the interstage temperatures, $T_1, \ldots, T^k, \ldots, T^{N-1}$. This approach to the solution is much easier than that based on an original recurrence equation which involves

6.6. The Hamiltonian as the Lagrange Multiplier of a Time Constraint

the time variable [19]. In that previous case there were the two decisions, u^n and θ^n, and the two state variables, T^n and τ^n, at each stage n. It is the knowledge of the constancy of the Hamiltonian and its structure that lead us to the problem with only one state variable T^n and one decision T^{n-1} at the nth stage.

Minimizing the sum (6.82) with respect to T^1 yields the geometric mean rule for the interstage temperature $T^1 = (T^2 T^0)^{1/2}$ as an H-independent optimality condition. This leads to the optimal work function of the two-stage subprocess

$$R_*^2 = c(T^2 - T^0) - 2cT^e \left(1 - \sqrt{\frac{T^0}{T^1}}\right) \left(1 - \pm\sqrt{\frac{H}{cT^e}}\right)^2. \quad (6.84)$$

For $T^0 = T^e$ this function describes a transformed exergy of the two-stage process, which may be shown to be the Legendre transform of the time-dependent exergy, A^2, related to the original optimal work function defined by (6.46). The word "transformed" also means that this exergy has incorporated the time constraint by virtue of the H term of (6.74).

Proceeding in this way for an arbitrary N, one obtains the temperature distribution in the form $T^{n-1} = (T^n T^{n-2})^{1/2}$ and the optimal function of transformed work

$$R_*^n = c(T^N - T^0) - 2cT^e N \left[1 - \left(\frac{T^0}{T^1}\right)^{1/N}\right] \left(1 - \pm\sqrt{\frac{H}{cT^e}}\right)^2. \quad (6.85)$$

With the identification $T^0 = T^e$ this function represents the Legendre-transformed exergy of the N-stage process which has a finite intensity corresponding to H. From this equation we also obtain the Legendre-transformed work at the limit of an infinite number of stages, N,

$$R_*^\infty = c(T^f - T^i) - cT^e \ln(T^f/T^i) \left(1 - \pm\sqrt{\frac{H}{cT^e}}\right)^2. \quad (6.86)$$

For the vanishing intensity, $H = 0$, the function (6.85) acquires the form

$$R_*^N = c(T^N - T^0) - cT^e N \left[1 - \left(\frac{T^0}{T^N}\right)^{1/N}\right] \quad (6.87)$$

which, at the limit of an infinite N, becomes the classical minimal work of the continuous process

$$R_*^\infty = c(T^f - T^i) - cT^e \ln(T^f/T^i). \quad (6.88)$$

From each work function R_*, the negative of the function I, considered earlier, can be obtained by the Legendre transformation of R_* with respect to the Hamiltonian H. Especially duration-dependent exergies can be recovered from H-based "asterisk exergies" by their Legendre transformations with respect to the Hamiltonian H. In the quasistatic case ($H = 0$) R and $-I$ coincide. Thus only in the classical quasistatic case are we left with the one kind of exergy. We will find the duration related work functions below. Equations (6.87) and (6.88) can be written as follows:

$$R_*^N = c(T^N - T^0) - cT^e N \left[1 - \left(\frac{T^0}{T^N}\right)^{1/N}\right]$$

$$+ cT^e N \left[1 - \left(\frac{T^0}{T^N}\right)^{1/N}\right]\left[1 - \left(1 - \pm\sqrt{\frac{H}{cT^e}}\right)^2\right] \quad (6.89)$$

and

$$R_*^\infty = c(T^f - T^i) - cT^e \ln(T^f/T^i)$$

$$+ cT^e \ln(T^f/T^i) \left[1 - \left(1 - \pm\sqrt{\frac{H}{cT^e}}\right)^2\right], \quad (6.90)$$

where the classical work has been singled out in the first line of each equation. The second-line terms describe a positively defined cost of the rate penalty. Both equations prove that the rate penalty cost becomes larger when the process becomes more intense.

As follows from the general theory [7] and Section 6.4, it is the Legendre transformation that governs the transformation from an H-related work function (with asterisk) to a τ^f-related work function. A consequence of these properties, which follow from the definition $W_*^N = W^N - H\tau^N$ contained in (6.74), is the optimal process time which appears as the partial derivative $\partial R^N/\partial H$. Thus, the differentiation of (6.89) recovers (6.71) for the optimal duration of the Nth stage process

$$\tau^N = N\left[1 - \left(\frac{T^0}{T^N}\right)^{1/N}\right]\left(\pm\left(\frac{H}{cT^e}\right)^{-1/2} - 1\right), \quad (6.71)$$

At the continuous limit one finds

$$\tau^f = \ln(T^f/T^i)\left[\pm\left(\frac{H}{cT^e}\right)^{-1/2} - 1\right], \quad (6.91)$$

which agrees with the expression $\tau^f = \ln(T^f/T^i)/\xi$ with ξ defined in (6.76). These results show that, in an optimal process, the vanishing of H is associated with infinitely long duration. For fixed heat transfer coefficients, this case would correspond to infinite contact areas and infinite investment costs. However, the optimal values of τ^f, which correspond to a finite minimum of the sum of the work costs and the area costs, are associated with finite intensities H or ξ. The formulas yield short durations for high intensities in H.

The duration-related functions of optimal work follow from the asterisk expressions after subtracting the term $H\tau^n$ as expressed by the pertinent Legendre transformation

$$R^N \equiv -I^N = R_*^N - H\, \partial R_*^N/\partial H = R_*^N - H\tau^N. \tag{6.92}$$

This may be verified for R_*^N satisfying (6.89), whose negative Legendre transform or the function I^N in terms of H is

$$I^N \equiv -R^N = c(T^0 - T^N) + cT^e N \left[1 - \left(\frac{T^0}{T^N}\right)^{1/N}\right]$$

$$- N\left[1 - \left(\frac{T^0}{T^N}\right)^{1/N}\right] cT^e \sqrt{\frac{H}{cT^e}}. \tag{6.93}$$

In terms of duration τ^N the same quantity is described by (6.73).

6.7 Limiting Continuous Process

For continuous process one obtains limiting expressions describing the extremal work functions. The limiting form of (6.73) for infinite N is

$$I = c(T^f - T^i) - cT^e \ln \frac{T^i}{T^f} - cT^e \frac{[\ln(T^i/T^f)]^2}{\tau^f - \ln(T^i/T^f)}, \tag{6.94}$$

where the indices i and f refer to the initial and final states. This is consistent with an earlier analysis of continuous systems [21]. Associated with this equation is a generalized exergy which is the maximal work I^∞ with $T^i = T$ and $T^f = T^e$ for the engine mode and the minimal work $-I^\infty$ with $T^i = T^e$ and $T^f = T$ for the heat-pump mode

$$A = c(T - T^e) - cT^e \ln \frac{T}{T^e} \pm cT^e \frac{[\ln(T/T^e)]^2}{\tau^f \pm \ln(T/T^e)}, \tag{6.95}$$

where the upper sign refers to the heat-pump mode. In an analogous way, an exergy form can be obtained for the discrete equation (6.73). Our results here add nontrivial examples to earlier works on finite-time exergies [24], [25], [27] and extend our earlier continuous results [20], [24], [27] to genuine discrete systems [29]. Enhanced bounds resulting from the exergy (6.95) are illustrated in Figure 6.3 and discussed in Section 6.8.

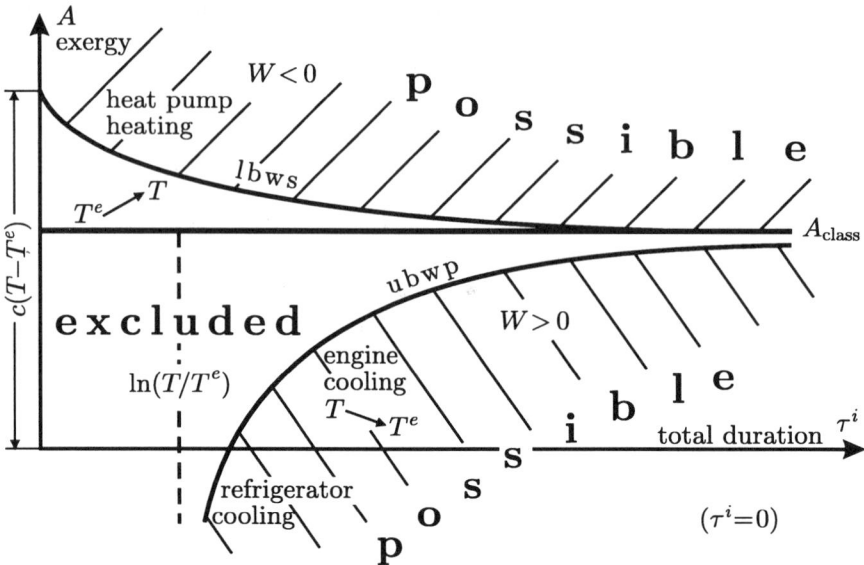

FIGURE 6.3. Finite-time exergy A of limiting continuous processes ($N \to \infty$) prohibits processes from operating below the heat-pump mode line which is the lower bound for work supplied (lbws) and/or above the engine mode line which is the upper bound for work produced (ubwp). These bounds, more realistic than those based on the classical, reversible definition of exergy, are the result of finite rates consistent with finite duration of the process.

6.8 Concluding Remarks

For finite rate processes, both discrete and continuous, it is possible (and useful) to introduce two types of optimal work functions which we call R and R_* or I and I_*. The first depends on the boundary states, the number of stages, and the duration, and the second depends on the boundary states, the number of stages, and the Hamiltonian. The same is applicable to generalized exergies. This dualism has its counterpart in classical mechanics, where two different action functions exist, one a duration-based action and the other a Hamiltonian-based action (sometimes called the abbreviated action), with the two linked by a Legendre transformation.

Enhanced thermodynamic bounds limiting the work delivered or consumed in a finite time constitute the main benefit resulting from generalized exergies. Because they include the role of dissipative factors and the dynamic nature of bounds, these are stronger, more useful, and informative than those stemming from the classical exergy.

To understand the problem of bounds and their distinction for work production and consumption, recall that the work-producing process is the inverse of the work-consuming process (the final state of the second process is the initial state of the first, and conversely), when we have fixed the durations of the two processes to be the same. In thermostatics the two bounds on the work, the bound on the work produced and that on the work consumed, coincide. However, the static limits are often too far from reality to be really useful. The generalized exergy provides bounds stronger than those predicted by the classical exergy. They do not coincide for processes of work production and work consumption, and they are "thermokinetic" rather than "thermostatic" bounds. Only for infinitely long durations or for processes with excellent transfer (an infinite number of transfer units) do the thermokinetic bounds reduce to their classical thermostatic limits. The hysteresis, or divergence of the bounds (Figure 6.3), proves that many idealized processes allowed by classical thermostatics are prohibited by the more severe and realistic constraints of thermokinetics. The hysteretic effect, caused by the dissipation and the associated increase of the exergy supplied in the pump mode (and the decrease of exergy released in the engine mode), reveals the extent to which thermokinetic bounds are stricter than thermostatic bounds. The greater the mean rate we demand, the greater is the range of performance excluded by limits predicted by the generalized exergy.

A real process, which does not apply the optimal protocol but has the same boundary states and duration as the optimal sequence, requires a real work supply that can only be larger than the finite-rate limit obtained by optimization. Similarly, the real work delivered from a nonequilibrium work-producing system (with the same boundary states and duration but with a suboptimal control) can only be lower than the corresponding finite-rate limit. Thus the two bounds for a process and its inverse, which coincide in thermostatics, diverge in thermodynamics, at a rate that grows with any of the indices of deviation from static behavior, ξ, σ, L, or H. For sufficiently high values of the rate indices, the work consumed may far exceed the classical work; the work produced can even vanish. Thus we can confront and surmount the limitations of applying classical thermodynamic bounds to real processes. This is a direction with many open opportunities, especially for separation and chemical systems.

Acknowledgment

One of the authors (S.S.) acknowledges an invitation and partial support from the University of Chicago.

6.9 References

[1] I. I. Novikov: The efficiency of atomic power stations (a review), *J. Nuclear Energy* II **7**, 125–128, (1958). English translation from *Atom Energy* **3**, 409–412, (1957).

[2] F. L. Curzon and B. Ahlborn: Efficiency of Carnot engine at maximum power output, *Amer. J. Phys.* **43**, 22–24, (1975).

[3] B. Andresen, P. Salamon, and R. S. Berry: Thermodynamics in finite time: extremals for imperfect heat engines, *J. Chem. Phys.* **66**, 1571–1577, (1977).

[4] A. De Vos: Endoreversible thermodynamics of solar energy conversion, Oxford University Press, Oxford, pp. 29–51, 1992.

[5] L. Chen and Z. Yan: The effect of heat-transfer law on performance of a two-heat-source endoreversible cycle, *J. Chem. Phys.* **90**, 3740–3743, (1989).

[6] J. M. Gordon: Maximum power point characteristics of heat engines as a general thermodynamic problem, *Amer. J. Phys.* **57**, 1136–1142, (1989).

[7] S. Sieniutycz: The constancy of Hamiltonian in a discrete optimal process, *Reports of Inst. of Chem. Engng. Warsaw Tech. Univ.* vol. **2**, pp. 399–429, 1973.

[8] S. Sieniutycz: *Optimization in process engineering*, 1st ed., Wydawnictwa Naukowo Techniczne, Warsaw, 1978.

[9] Z. Szwast: Discrete algorithms of maximum principle with constant Hamiltonian and their selected applications in chemical engineering, PhD Thesis, Institute of Chemical Engineering at the Warsaw University of Technology, Warsaw, 1979.

[10] Z. Szwast: Enhanced version of a discrete algorithm for optimization with a constant Hamiltonian, *Inż. Chem. Proc.* **3**, 529–545, (1988).

[11] S. Sieniutycz and Z. Szwast: *Practice in Optimization: Process Problems*, Wydawnictwa Naukowo Techniczne, Warsaw, 1982.

[12] S. Sieniutycz and Z. Szwast: A discrete algorithm for optimization with a constant Hamiltonian and its application to chemical engineering, *Internat. J. Chem. Engng.* **23**, 155–166, (1983).

[13] L. T. Fan and C. S. Wang: *The Discrete Maximum Principle, a Study of Multistage System Optimization*, Wiley, New York, 1964.

6.9. References

[14] V. G. Boltyanski: *Optimal Control of Discrete Systems*, Nauka, Moscow, 1973.

[15] L. T. Fan: *The Continuous Maximum Principle, a Study of Complex System Optimization*, Wiley, New York, 1966.

[16] W. Findeisen, J. Szymanowski, and A. Wierzbicki: *Theory and Computational Methods of Optimization*, Panstwowe Wydawnictwa Naukowe, Warsaw, 1980.

[17] R. E. Bellman: Adaptive Control Processes: a Guided Tour, Princeton University Press, Princeton, 1961.

[18] R. Aris: *Discrete Dynamic Programming*, Blaisdell, New York, 1964.

[19] S. Sieniutycz: Endoreversible modeling and optimization of multistage thermal machines by dynamic programming, in: *Advances in Recent Finite-Time Thermodynamics* (ed. C. Wu), Nova Science, New York, 1999.

[20] V. G. Boltyanski: *Mathematical Methods of Optimal Control*, Nauka, Moscow, 1969.

[21] S. Sieniutycz: Carnot problem of maximum work from a finite resource interacting with environment in a finite time, *Physica A* **264**, 234–263, (1999).

[22] P. Salamon, P. A. Nitzan, B. Andresen, and R. S. Berry: Thermodynamics in finite time IV: Minimum entropy production and the optimization of heat engines, *Phys. Rev.* **A 21**, 2115–2129, (1980).

[23] B. Andresen and J. Gordon: Optimal heating and cooling strategies for heat exchanger design, *J. Appl. Phys.* **71**, 76–79, (1992).

[24] R. S. Berry, V. A. Kazakov, S. Sieniutycz, Z. Szwast, and A. M. Tsirlin: *Thermodynamic Optimization of Finite Time Processes*, Wiley, Chichester, 2000.

[25] B. Andresen, M. H. Rubin, and R. S. Berry: Availability for finite time processes. General theory and model, *J. Phys. Chem.* **87**, 2704–2713, (1983).

[26] V. A. Mironova, A. M. Tsirlin, V. A. Kazakov, and R. S. Berry: Finite-time thermodynamics. Exergy and optimization of time-constrained processes, *J. Appl. Phys.* **76**, 629–636, (1994).

[27] R. S. Berry: Exergy and optimization of time-constrained processes, *Periodica Polytech. Ser. Phys. Nucl. Sci.* **2**, 5–14, (1994).

[28] S. Sieniutycz: Hamilton–Jacobi–Bellman theory of dissipative thermal availability, *Phys. Rev.* **E 56,** 5051–5064, (1997).

[29] A. De Vos, J. Chen and B. Andresen: Analysis of combined systems of two endoreversible engines, *Open Sys. Inform. Dynam.* **4,** 3–14, (1997).

7
Optimal Piston Paths for Diesel Engines

J. M. Burzler
P. Blaudeck
K. H. Hoffmann

ABSTRACT. The performance of a Diesel engine is analyzed for a model which includes losses due to mechanical friction and heat losses through the cylinder walls. Using the work output of the Diesel engine as an objective, the optimal piston trajectories for the compression and power stroke are determined simultaneously. Results for a linear approximation of the heat leakage are compared to a more realistic, empirical heat transfer law due to Annand. Optimal operating conditions are found and discussed and significant improvements in the engine's efficiency relative to conventionally designed engines are obtained.

7.1 Introduction

Improvements in the design of internal combustion engines to increase their efficiency are becoming more and more intense for economical and ecological reasons. The following approach puts its focus not primarily on the technical realization of the engine as such but on the thermodynamic process taking place inside the engine. Very much in the tradition of early thermodynamics it is thus not our idea to repeat the nearly 100 years of careful engineering and optimization, but to abstract the engines enough to make them treatable yet to include at the same time all major loss terms by also employing empirical knowledge on the heat transfer so that the results of the analysis would be useful in guiding which loss terms could be most easily reduced.

In this spirit the Otto engine was investigated by M. Mozurkewich and R. S. Berry [1] and later the Diesel engine by K. H. Hoffmann, S. Watowich, and R. S. Berry [2]. Both engines have a four-stroke cycle of successive intake, compression, power, and exhaust stroke. They operate at a constant number of revolutions per minute and with a fixed fuel consumption per cycle. Both models are endoreversible [3] in that the working fluid is in internal equilibrium. The working fluid consists of an ideal gas. In addition, both models incorporate the for real engines dominating loss mechanisms [4] friction, pressure drop, and heat leak.

For these models, and for very similar dissipative light-driven engines [5], [6], the thermodynamic processes are optimized to yield maximum work per cycle. The exhaust and intake are of minor importance for the overall performance of the engine since the changes in pressure and temperature are at least one order of magnitude smaller than during the compression and power stroke. The compression stroke, for instance, results in a strong increase of the gas temperature up to about 900 ... 1500 K. Therefore we only consider the compression and power stroke here and optimize both strokes together.

One of the most important loss terms in an internal combustion engine is the heat leakage through the cylinder walls because of large differences between the temperatures of the working fluid and of the cylinder walls. In the literature, linear Newton-type laws are commonly used to model heat transfer processes. However, such a simple description does not capture all the details of the heat leakage from the working fluid through the surfaces of the engine's cylinder.

In real combustion engines, thermal energy flows to the cylinder's walls by heat conduction and by convection of the working fluid. During the combustion of the working fluid, the luminous flame causes an additional contribution to the heat leakage by radiating energy toward the cooler cylinder walls. In effect, thermal energy leaves the cylinder and is further conducted through the cylinder wall to the cooling system of the engine. Consequently, the heat transfer rate is subject to quite large variations during the cycle and depends on various properties of the working fluid as well as on flow patterns inside the cylinder.

In principle, the governing hydrodynamic and material equations for the fluid flow in the cylinder could be solved to obtain a detailed picture of the heat transfer process, yet such an approach would result in complex and expensive computations requiring a precise knowledge of the combustion process and design features. These details are usually not well known and would restrict the analysis to a narrow class of designs. Repeated attempts have been made to characterize the complicated heat transfer processes in combustion engines in a more realistic manner without the drawbacks of first-principle methods.

Engineers have addressed the problem by adjusting relatively simple parameterized heat transfer formulas to experimental data. These formulas are based on a macroscopic view of the heat transfer process and typically contain basic gas properties and dimensionless, scalable quantities for the description of the fluid flow patterns inside the cylinder [10], [11], [12], [13], [14]. In the present study, the widely used Annand formula [15] is employed to model the heat conduction on an empirical basis. Annand reviewed various other formulas and measurements on the heat conduction in combustion engines and developed his expression, which is in fairly good agreement with a wide range of measurements.

In the following we investigate the effects of the heat leakage and other loss mechanisms on the engine's performance by comparing a model with a simple, Newton-type heat leakage rate to a model with an Annand-type description of the heat leakage. We also consider heat losses due to radiation. We demonstrate how the compression and power stroke of the engine are optimized simultaneously by using optimal control theory [7], [8] and a Monte Carlo method [9]. The optimal control theory [7], [8] leads to a set of coupled nonlinear differential equations with boundary values. These equations are solved numerically to obtain optimal piston trajectories for different descriptions of the heat leakage. The results for the path-optimized Diesel engines are discussed by comparing them to results for engines with sinusoidal piston trajectories.

7.2 Model

In principle, a Diesel engine consists of a cylinder with a movable piston. Here we assume an ideally shaped cylinder with a bore d where the volume V of the cylinder is related to the distance x between the head of the piston and the top of the cylinder as

$$V = \pi d(d/2 + x). \tag{7.1}$$

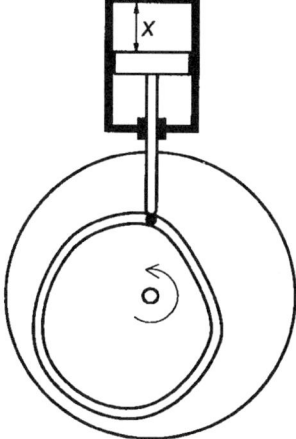

FIGURE 7.1. A disk with an engraved curve guides the piston of the engine's cylinder on a nonsinusoidal trajectory and controls the distance x between the top of the cylinder and the head of the piston.

176 7. Optimal Piston Paths for Diesel Engines

TABLE 7.1. Parameters are all in SI units for a Diesel engine model with stroke volume 375 cm³ and approximately 3600 revolutions per minute.

Cylinder	
Bore (inside diameter of cylinder) [m]	$d = 7.98 \times 10^{-2}$
Piston position [m]	x
Max. piston position (bottom dead center) [m]	$x_f = 8.0 \times 10^{-2}$
Min. piston position (top dead center) [m]	$x_0 = 0.5 \times 10^{-2}$
Volume [m³]	$V = \pi x d^2/4$
Wall area [m²]	$A = \pi d(d/2 + x)$
Average wall temperature [K]	$T_w = 600$
Friction coefficient [N s m⁻¹]	$\alpha = 12.9$
Time for compression and power stroke [s]	$t_{\text{tot}} = 17 \times 10^{-3}$

Working fluid	
Temperature [K]	T
initially [K]	$T_0 = 300$
Number of moles [mol]	N
initially [mol]	$N_i = 0.0144$
after total combustion [mol]	$N_f = 0.0157$
Constant-volume heat capacity [J K⁻¹ mol⁻¹]	C
initially	$C_i = 2.5\,R$
after total combustion	$C_f = 3.35\,R$

Heat leakage	
Linear heat-transfer coefficient [W m⁻² K⁻¹]	$k_n = 1305$
Thermal conductivity [J/K⁻¹ m⁻¹ s⁻¹]	$\kappa = 3.17 \times 10^{-4} T^l$
with exponent	$l = 0.772$
Viscosity [kg m⁻¹ s⁻¹]	$\mu = 4.5 \times 10^{-7} T^m$
with exponent	$m = 0.645$
Reynolds number [1]	$\mathcal{R} = \rho \bar{v} d/\mu$
density [kg/m³]	$\rho = 29 \times 10^{-3} N/V$
average piston speed [m/s]	$\bar{v} = 2(x_f - x_0)/t_{\text{tot}}$
Gas constant [J K⁻¹ mol⁻¹]	$R = 8.3144$
Stefan–Boltzmann constant [W m⁻² K⁻⁴]	$\sigma = 5.67 \times 10^{-8}$
Convective constant [1]	$a = 0.540$
Exponent of the Reynolds number [1]	$b = 0.7$
Radiative constant during power stroke [1]	$c = 0.1$

Combustion	
Ignition time [s]	t_z
Burn time [s]	$t_b = 1.0 \times 10^{-3}$
Heat of combustion [J mol⁻¹]	$\mathcal{Q}_c = 5.57 \times 10^4$

Often, especially in textbooks, the volume changes of cyclic operating internal combustion engines are modeled by cycles which are split into branches where certain state variables, like volume, pressure, or temperature, are constant [16], [17], [18], [19]. In real engines, however, the piston is connected to a crankshaft and the motion of the piston during the compression and power stroke is well approximated by a sinusoidal function

$$x(t) = \tfrac{1}{2}(x_f + x_0) + \tfrac{1}{2}(x_f - x_0)\cos(2\pi t/t_{\text{tot}}), \qquad (7.2)$$

if the distance between the crank shaft and the cylinder is large. Here, x_f and x_0 are the respective maximal and minimal positions of the piston and t_{tot} is the total duration of the compression and power stroke together. Note that in the following an engine with sinusoidal piston motion is denoted as a conventional engine.

In contrast to the conventional engine we consider a mode of operation where the piston path $x(t)$ is taken as a time-parameterized function to optimize the work output of the engine. One can imagine that a nonsinusoidal piston path is achieved by using a disk with an engraved slit as in Figure 7.1 to guide the piston on the desired path.

7.2.1 Combustion

In modern Diesel engines, fuel is injected into the cylinder at the end of the compression stroke and evaporates in the hot compressed air. After a short delay the fuel ignites and starts to burn rapidly causing a sharp rise in temperature and pressure. The remaining fuel burns relatively slowly as it evaporates and diffuses into oxygen-rich regions. In moderately and heavily loaded engines, the combustion process continues well until the end of the power stroke.

The finite combustion rate is one of the main features of a Diesel engine and is here approximated by a time-dependent reaction coordinate

$$\xi(t) = 1 + [(t - t_z)/t_b - 1]\exp[(t_z - t)/t_b]. \qquad (7.3)$$

with a characteristic combustion time t_b [20]. The reaction coordinate $\xi(t)$ describes the extent of the combustion such that $\xi(t \leq t_z) = 0$ before the ignition time t_z and that total combustion ($t \to \infty$) gives $\xi = 1$. This simple model was found by approximating the heat production curves of real Diesel engines investigated by Kleinschmidt [20]. The working fluid is treated as an ideal gas with time-dependent

mole number $\qquad N(t) = N_i + (N_f - N_i)\xi(t), \qquad (7.4)$
heat capacity $\qquad C(t) = C_i + (C_f - C_i)\xi(t). \qquad (7.5)$

The rate of heat produced during the combustion is described by the heating function

$$h(t) = \mathcal{Q}_c N_i \cdot \dot{\xi}(t), \qquad (7.6)$$

where Q_c is the heat per mole of the air–fuel mixture. These formulas are a continuous approximation for the combustion process in real engines [20].

7.2.2 Frictional Losses

On a well-lubricated sliding surface, the frictional forces are proportional to the piston velocity v [4]. This yields a frictional loss rate of mechanical energy which is proportional to the square of v:

$$w_f = \alpha v^2. \tag{7.7}$$

In our model, the friction is assumed to have the above simple form and a constant friction coefficient α during the compression and power stroke. We further assume that the heat generated by friction is immediately removed by the engine's cooling system and does not contribute to the engine's heat production as a low-grade heat source.

7.2.3 Conductive and Convective Heat Leak

The compression and power stroke in a Diesel engine is subject to heat losses through the walls of the cylinder. This loss is an important one and there exists a number of approaches to calculate the total heat flow through the inside surface of the cylinder.

The simplest is to assume steady-state conditions where the heat transfer rate q_w is linear and can be approximated as proportional to the exposed internal surface area of the piston cylinder

$$A = \pi d(d/2 + x) \tag{7.8}$$

and to the driving temperature difference between the fluid and surface

$$q_{c,n} = A k_n (T - T_w). \tag{7.9}$$

For simplicity, both the fluid and wall temperature are supposed to be space averaged values. In addition, the wall temperature is constant in time. A constant heat transfer coefficient k_n corresponds to the so-called "Newton-type" heat transfer law for quasistatic conditions. While such conditions are certainly not found in reprociating combustion engines, the Newton-type heat transfer law can serve as a first approximation and has been widely used in the literature primarily due to its simplicity.

A higher level of sophistication for the description of the heat leakage is achieved with an expression due to Annand [15] which in principle accounts for both, the conductive and convective character of the heat transfer, by considering basic flow patterns inside the cylinder and fluid properties. Annand's formula for the heat conduction rate can be written analogous to the linear heat transfer law

$$q_{c,a} = A k_a(T, x)[T - T_w] \tag{7.10}$$

with a nonconstant heat transfer coefficient

$$k_a(T, x) = a\,\frac{\kappa(T)}{d}\,\mathcal{R}^b(T, x) \tag{7.11}$$

which is a function of the thermal conductivity $\kappa(T)$ and the Reynolds number $\mathcal{R}(T, x)$ and depends on the fluid temperature T and the position x of the piston. The formula also contains the inside diameter d of the cylinder and the constants a and b as fixed, design-dependent parameters. Basically, the thermal conductivity κ incorporates the temperature dependence of the heat conduction into the modeling while the Reynolds number $\mathcal{R}$ introduces the influence of flow patterns on the convective part of the heat transfer.

A summary of the empirical values of these properties is found in Table 7.1 (also see [21]). For simplicity we have left out higher-order effects and assumed air as the predominant component in the working fluid. The parameters a and b are constants. The value of the convective constant a is typically in the range between 0.35 and 0.8, and depends on the conditions of convection in the cylinder. Low values of a correspond to "quiet" combustion and high values of a were measured for combustion chambers which are specially optimized for fast combustion. Here we have chosen $a = 0.628$ such that a sinusoidal piston motion modeled with the Annand expression (7.10) gives the same work output as the model with the linear heat conduction law (7.9) for $k_n = 1305\,\mathrm{W\,m^{-2}\,K^{-1}}$.

If we evaluate the definitions of Table 7.1 and insert them into expression (7.11) we obtain the Annand heat transfer coefficient as a function of the fluid temperature T and piston position x:

$$k_a(T, x) = 0.873\,a T^{0.321} x^{-0.7} d^{-1.7} \bar{v}^{0.7} N^{0.7}\,. \tag{7.12}$$

The heat transfer coefficient depends on given constant design parameters $a, d, \bar{v}$, as well as on the mole number N which is variable in time according to (7.4). Note that the numbers and the parameters are given in SI units so that all other variables and parameters in (7.12) should also be in SI units.

7.2.4 Radiative Heat Leak

If the combustion process has not started, the air–fuel mixture is quasi-transparent for heat radiation and the heat leakage out of the cylinder is, for all practical purposes, entirely conductive and convective.

During the power stroke, when the mixture burns, solid incandescent carbon particles appear as intermediate combustion products and radiate heat toward the cooler cylinder wall. The carbon particles are in strong convective contact with their environment and thus are often assumed to have the same temperatures as the working fluid. Since the carbon content in Diesel fuels is high, the predominant sources of radiative energy are the carbon particles. Chemical luminescence, which corresponds to a much higher effective temperature, is relatively weaker and is often ignored. Even if only the radiation from the carbon particles at gas temperature is considered, exact calculations of the radiative heat transfer is very difficult because it involves the computation of integrals with respect to cylinder volume, surface area, and wavelength thus requiring detailed data about the temperatures, concentration, and geometry of the carbon particles as well as their spectral absorbtivity during the combustion process.

Instead we follow the work of Annand [15], and use the well known expression for black-body radiation to estimate the radiative heat transfer

$$q_r = Ac\sigma(T^4 - T_w^4). \tag{7.13}$$

The radiative constant c incorporates all the unknown factors into a single number. During the compression stroke, where the radiative heat transfer is negligible, c is set to zero. The value for c during the power stroke has been obtained from experimental data and is typically in the range between 0.04 to 0.32 for Diesel engines [15]. For our model we assume $c = 0.1$.

In summary, the heat leak through the cylinder walls in the Diesel engine has two components, one due to conduction/convection, q_c, as well as one due to radiation, q_r:

$$q_{\text{leak}} = q_c + q_r. \tag{7.14}$$

In the following analysis the Diesel engine will be modeled for both, "Newton-type" and "Annand-type" heat transfer, with and without radiative effects in order to analyze how these contributions affect the performance of the engine.

7.3 Optimization

One of the very important points before performing an optimization is to define clearly the optimization goal, i.e., the property which should be extremalized, the controls, the boundary conditions, and the constraints of the model.

The literature on heat engine optimization offers quite a variety of optimization goals including maximum (average) power output, maximum efficiency, minimum entropy production, minimum loss of availability and maximum net revenue [22], [23], [24], [1], [25], [26], [27]; see also [28], [29], [30] for an overview.

In the following analysis the experimental accessible net output of mechanical work during the compression and power stroke is used as an optimization objective, i.e., the functional

$$W = \int_0^{t_{\text{tot}}} \left(\frac{NRT}{x} v - \alpha v^2 \right) dt \tag{7.15}$$

is maximized where the losses in mechanical work due to friction are also included. Since the amount of fuel and the cycle time are fixed, the path yielding maximum work also maximizes the average power output and the thermodynamic efficiency of the engine.

The piston position x is chosen as the control variable, meaning that the goal of the optimization is to find that piston path which maximizes the above functional (7.15).

The optimization is subject to the following boundary conditions and constraints:

- The total time t_{tot} for the compression and power stroke is fixed.
- The start and end positions of the piston are equal and fixed $x(0) = x(t_{\text{tot}}) = x_f$ and the piston must not go beyond this maximum value.
- The lowest accessible position x_0 of the piston leads to a maximal compression ratio of 1:16 for the engine.
- The wall temperature of the cylinder T_w as well as the initial temperature of the working fluid T_0, is given while the end temperature of the fluid is allowed to vary freely.

The values of these and further parameters are found in Table 7.1.

The temperature and piston position are state variables and are subject to the constraints

$$\dot{T} = \frac{h - NRTv/x - q_{\text{leak}} - \dot{\xi}[C(N_f - N_i) + N(C_f - C_i)]T}{NC}, \quad (7.16)$$

$$\dot{x} = v. \quad (7.17)$$

The state equation for the temperature (7.16) is derived from the energy balance equations where the rate of change in internal energy

$$\dot{U} = \frac{\partial}{\partial t}(TNC) \quad (7.18)$$

is related to the heating function h, the mechanical work $NRTv/x$, and the losses due to the heat leak q_{leak}. Note that the time-dependent expressions (7.4) and (7.5) for the mole number N and heat capacity C have also to be considered here.

7.3.1 Control Theory

After defining the problem, the actual optimization can be performed by using control theory, see, for example, [2], [9]. The Hamiltonian for our optimization problem is

$$H = \frac{NRT}{x}v - \alpha v^2 + \lambda_1 \dot{T} + \lambda_2 v + \mathcal{P}(x) \quad (7.19)$$

with the adjoint variables λ_1 and λ_2. An appropriate penalty function

$$\mathcal{P}(x,p,T) = -\alpha v^2 \left(\frac{x_0'}{x}\right)^{n_x} \quad (7.20)$$

ensures that the constraint on x is kept. The exponent n_x is typically in the range of 10 to 70 and specifies the "smoothness" of the constraint. Larger values of n_x correspond to a "sharper" constraint. The parameter x_0' is chosen such that the constraint on the minimum piston position x_0 is reached, yet never violated.

The constraint on x could also be implemented into the Hamiltonian by introducing an additional adjoint variable. This would give rise to a splitting of the piston path into separate branches [2]. The optimal path would then consist of pieces which would have to be connected such that the state variables are continuous at the joints and the overall work output is maximized.

Here we have chosen the approach with the penalty function in order to calculate the optimal solution as a single continuous trajectory with reduced computational effort.

Control theory provides differential equations for adjoint variables. The actual expressions depend on the models used for the heat leakage rate q_{leak}.

In case of a "Newton-type" heat transfer law with radiation we get

$$\dot{\lambda}_1 = -\frac{\partial H}{\partial T} = \frac{NRv}{x}\left(\frac{\lambda_1}{NC} - 1\right) + \frac{\lambda_1}{NC}\{(k_n + 4c\sigma T^3)\pi d(d/2 + x)$$
$$+ \dot{\xi}[C(N_f - N_i) + N(C_f - C_i)]\}, \quad (7.21)$$

$$\dot{\lambda}_2 = -\frac{\partial H}{\partial x} = -\frac{NRT}{x^2}v\left(\frac{\lambda_1}{NC} - 1\right) + \frac{\lambda_1}{NC}\pi d[k_n(T - T_w) + c\sigma(T^4 - T_w^4)]$$
$$- \frac{\alpha v^2 n_x}{x}\left(\frac{x_{\min}'}{x}\right)^{n_x} \quad (7.22)$$

for the adjoint variables. The Pontryagin maximum principle [31] requires that H is maximized with respect to v. Solving $\partial H/\partial v = 0$ with respect to the piston velocity v yields

$$v = \frac{1}{2\alpha[1 + \mathcal{P}(x)]}\left[\lambda_2 - \frac{NRT}{x}\left(\frac{\lambda_1}{NC} - 1\right)\right]. \quad (7.23)$$

In the case of an "Annand-type" heat transfer law with radiation we obtain for the adjoint variables

$$\dot{\lambda}_1 = \frac{NR}{x}v\left(\frac{\lambda_1}{NC} - 1\right) + \frac{\lambda_1}{NC}\left\{\left[k_a\left(1 + \frac{l - mb}{T}(T - T_w)\right) + 4c\sigma T^3\right]\right.$$
$$\left. \times \pi d(d/2 + x) + \dot{\xi}[C(N_f - N_i) + N(C_f - C_i)]\right\}, \quad (7.24)$$

$$\dot{\lambda}_2 = -\frac{NRT}{x^2}v\left(\frac{\lambda_1}{NC}-1\right)+\frac{\lambda_1}{NC}\pi d\left\{c\sigma(T^4-T_w^4)\right.$$
$$\left.+k_a(T-T_w)\left[1-\frac{b}{x}\left(\frac{d}{2}+x\right)\right]\right\}-\frac{\alpha v^2 n_x}{x}\left(\frac{x'_{\min}}{x}\right)^{n_x}, \quad (7.25)$$

where, according to (7.11), $k_a = a\kappa \mathcal{R}^b/d$. The expression for the speed v is the same as for the Newton case (7.23).

Note that in the above equations the radiative constant c is zero during the compression stroke and jumps to its finite value at the beginning of the power stroke when $t \geq t_z$. Radiative effects can be, if necessary, switched off completely by setting c to zero during both the compression and the power stroke.

By inserting the expression (7.23) for v into the equations of the adjoint variables, (7.21) and (7.22) or (7.24) and (7.25), and considering the equations of state (7.16) and (7.17), we find a closed system of four nonlinear differential equations which has to be solved to obtain the optimal piston path [5], [2], [9].

The boundary conditions of a given end position for the piston, $x(t_{\text{tot}}) = x_f$, and a freely varying end temperature $T(t_{\text{tot}})$ which is equivalent to $\lambda_1(t_{\text{tot}}) = 0$ are achieved by adjusting the initial values of the adjoint variables λ_1 and λ_2, according to the common shooting method of numerical mathematics (for example, see [32]).

7.3.2 Stochastic Optimization

In principle, mechanical feasibility requires restrictions on the acceleration of the piston. In order to investigate the effects of a bounded acceleration on the optimal solutions we have alternatively employed a Monte Carlo method which allows us to easily include various constraints into the modeling. Our approach starts with an arbitrary nonoptimal path $x(t)$. This path is modified to $x'(t)$ by random choice but with full consideration of all constraints in position, velocity, and acceleration of the piston. As illustrated in Figure 7.2, changes with gain in mechanical work are kept while changes with loss in work are rejected.

For example, if the constraint on the piston acceleration is in the form $|\ddot{x}(t)| \leq a_{\max}$ one can use the following algorithm:

- For a random integer i with $0 < i < t_{\text{tot}}/h$, where h is the width of the time step, change the value $x_i = x(i\,h)$ according to

$$x'_i = (2z-1)\cdot a_{\max}\, h^2\, \frac{x_{i-1}+x_{i+1}}{2}$$

using a real random number z in $[0\ldots 1]$.

7. Optimal Piston Paths for Diesel Engines

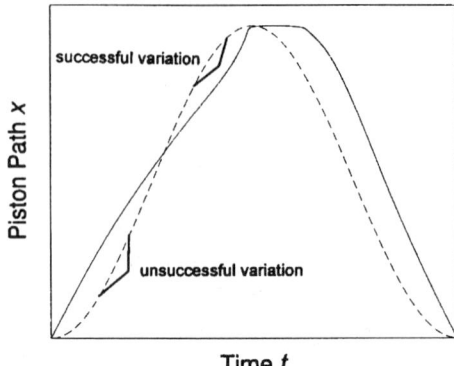

FIGURE 7.2. Statistic optimization of piston movement where a path to be optimized (dashed line) is varied until no further improvements in the respective work output is achieved and the optimal path (solid line) is found.

- If the restrictions on the acceleration for the adjacent points,

$$\left|\frac{x'_{i-2} + x'_i - 2x'_{i-1}}{h^2}\right| \quad \text{and} \quad \left|\frac{x'_i + x'_{i+2} - 2x'_{i+1}}{h^2}\right| \leq a_{\max},$$

or on the piston position, $x_0 \leq x(t) \leq x_f$, are violated, choose a new $z \in [0 \ldots 1]$ until all restrictions are fulfilled.
- If $W' > W$ then keep $x'(t)$,
  else reject $x'(t)$ and keep $x(t)$.

This procedure has to be repeated until no significant improvement in W can be achieved. In principle, there is the possibility of obtaining a local optimum. In order to increase the probability of finding a global optimum, the optimizations need to be performed a couple of times for different starting paths of the piston.

While this Monte Carlo method has the advantage that constraints, like those on the position and acceleration of the piston, can easily be included, it also has the drawback of a high computational effort, even if parallel computers are used to handle a lot of path variations simultaneously.

Nevertheless we have applied the Monte Carlo method to successfully model and optimize the Diesel engine [9]. We found that a bound on the acceleration has little influence on the overall results and only leads to a slight rounding of the piston path with a typical loss in output work of 1% [9]. Thus releasing the constraint on the acceleration hardly injures the advantage of the computationally much more effective optimal control method. Therefore we have used the optimal control method to obtain the following results.

7.4 Results

Piston paths for the simultaneously optimized compression and power stroke are calculated for engines with the parameters listed in Table 7.1. Most of the parameters are adapted from previous studies [4], [1], [2], [9] for reasons of compatibility. The results for the path optimized and conventional sinusoidal piston motion are compared for "Newton"- and "Annand"-type models of heat conduction and the role of the radiative heat transfer is discussed.

7.4.1 *Optimal Path*

Let us first assume a fixed ignition time, $t_z = 8.25$ ms, which is close to the top dead center at $t = 8.5$ ms of a conventional sinusoidal cycle and has been used in other studies [2], [9]. In the following we are comparing the path-optimized engine to an engine with sinusoidal piston motion and an Annand-type model for the heat leakage.

Optimized piston paths yielding maximum work output are markedly different from the sinusoidal piston motion of a conventional engine, as can be seen in Figure 7.3. The sinusoidal piston motion starts at the bottom dead center with slope zero corresponding to a zero velocity, accelerates, and reaches its maximum speed in the middle of the compression stroke. At the end of the compression stroke, near the top dead center, the velocity is decreased to zero again.

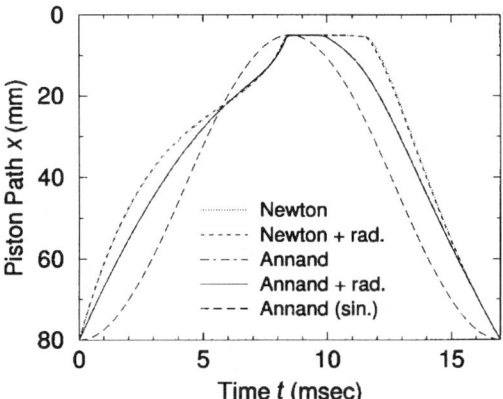

FIGURE 7.3. Comparison of a conventional sinusoidal piston motion to optimized trajectories with Newton and Annand models for the heat leak, with and without radiative effects calculated for an 8.25 ms fixed ignition time. Note that for the optimized Annand model, the trajectories with and without radiative heat transfer are almost identical.

186   7. Optimal Piston Paths for Diesel Engines

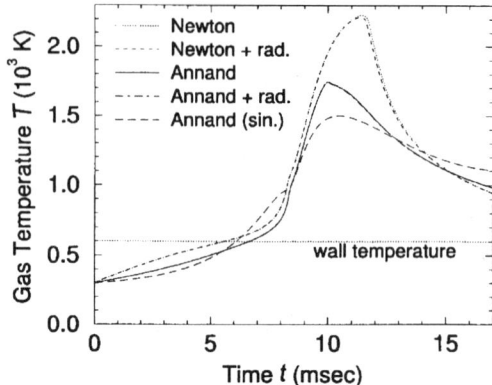

FIGURE 7.4. Temperature of the working fluid for different heat conduction models and an 8.25 ms fixed ignition time.

The optimized trajectories start with a finite velocity, followed by a small deceleration in the middle part of the compression stroke and an acceleration near the time of ignition. This behavior balances the effects of frictional losses, heat leakage, and compression work. If there were only frictional losses, the optimal path would be a straight line of constant velocity, since we used a frictional loss rate proportional to the square of the velocity.

Because of the heat leakage and the volume work of the piston, the optimal paths deviate from a straight line. Initially, the walls of the cylinder are hotter than the working fluid and the cold gas receives heat from the walls until the temperature of the working fluid becomes higher than the temperature of the wall and the direction of the heat flow switches. The heat losses and compression work are minimized if the switching point is reached a short time before ignition.

Figure 7.4 shows that in the optimized engines the temperature of the gas remains below the wall temperature until late in the compression stroke and rises rapidly after reaching the switching point. A modestly increased piston speed during the start phase of the compression stroke which reduces the temperature differences between the wall and the working fluid and lowers the initial heating of the working fluid by the warm walls.

The piston deceleration in the middle part of the compression stroke prevents the temperature of the gas rising too much. The acceleration toward the end of the compression stroke gets the piston to bottom center at the time of ignition.

The different shapes of the optimal compression curves for Newton- and Annand-type heat transfer models stem from differences in the heat conduction rate which is depicted in Figure 7.5. The Newton model shows a larger absolute magnitude for the heat transfer rate at the beginning and end of the compression stroke resulting in steeper optimal piston paths

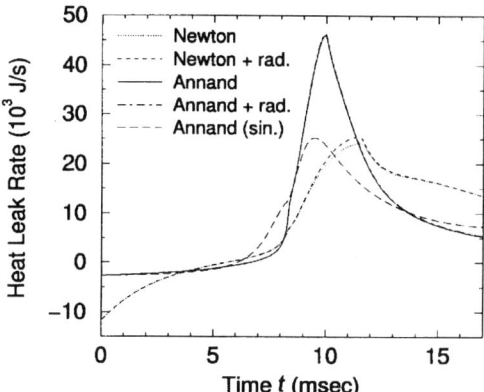

FIGURE 7.5. Heat leak rate for the optimized piston trajectories in the case of "Newton"- and "Annand"-type heat conduction laws with and without radiation compared with the "Annand-type" heat conduction for a conventional sinusoidal piston path.

during these parts of the stroke than for the Annand model.

Radiative effects are not present during the compression stroke, yet the inclusion of radiative heat transfer into the power stroke has some small effects on the compression stroke since both strokes interact via final and initial conditions. This again makes clear why it is important to optimize the compression and power strokes simultaneously. Optimizing the compression on its own would yield a path with a large, monotonously decreasing piston velocity to use the low pressure of the cold gas as long as possible. This, however, would result in a relatively low gas temperature at the time of ignition, and would decrease the possible gain of work during the power stroke.

The compression stroke is completed at ignition time $t_z = 8.25$ ms where the optimized path reaches its minimum after decelerating the piston to standstill at the highest possible rate. Subsequently, a zero-velocity phase is entered where the piston stays at its minimum position, before a phase with more or less constant piston velocity starts.

At first glance, this delayed motion seems bizarre since this increases the frictional losses as the piston has to move back to its maximal position in less time at higher velocities. The delayed motion additionally increases the maximum temperature of the burning working fluid causing higher losses from heat leakage. However, the higher temperatures also increase the maximum availability of the system during the power stroke and the losses are outweighed by gains in efficiency.

The zero-velocity phase is followed by a large acceleration and a volume increase that limits the temperature rise, even though the combustion

188   7. Optimal Piston Paths for Diesel Engines

within the cylinder continues. The zero-velocity phase is a consequence of the existence of a minimal piston position. Without this constraint the optimized system would move the piston to positions x smaller than x_0 to achieve even higher internal temperatures. This behavior is reminiscent of the work of Band, Kafri, and Salamon [33], who found that the optimal expansion stroke for their particular heating function was one which at first compressed the gas.

A plot of the temperature versus time, as in Figure 7.4, shows the temperature declining immediately after the zero-velocity phase. In this way the system minimizes the heat-conduction losses. Additionally, the piston moves quite close to a constant velocity trajectory which leads to decreased frictional losses. The optimization yields temperature profiles with higher peak temperatures and lower end temperatures than for the conventional sinusoidal piston motion. The low exhaust temperature of the optimized paths is equivalent to a decreased exergy content of the working fluid at the end of the power stroke. Note that this might not always be desirable in practice, if for example, the engine is fitted with a turbocharger which utilizes the residual exergy of the exhaust. Although further components are easily included in the model, this has to be done before the optimization. Therefore our discussion is restricted to precisely that system which has been introduced above.

Significant differences between the Newton- and Annand-type models in trajectories, temperature profile, and heat conduction are clearly apparent in the respective Figures 7.3, 7.4, and 7.5. The type of heat conduction obviously has strong effects on the optimal solution. The temperature for instance rises over $2200\,\mathrm{K}$ in the case of Newton-type conduction compared to only $1700\,\mathrm{K}$ for Annand-type conduction.

The Newton-type model assumes a heat conduction rate which is proportional to the temperature differences of the gas and cylinder wall, mathematically expressed through a constant heat transfer coefficient k_n. The heat transfer coefficient k_a of the more sophisticated Annand model is subject to large variations during the compression and power stroke, as can be seen in Figure 7.6.

The Annand-type heat transfer coefficient k_a increases for high gas temperatures and small volumes, leading to a sharp increase of the heat losses during the zero-velocity phase of the power stroke as depicted in Figure 7.5. In consequence, the zero-velocity phase of the power stroke is shortened for the Annand model and the gain in work output by optimization is less than for the Newton model. Both models have in common that the losses due to conductive heat transfer are significantly higher than the losses due to radiative heat transfer.

Table 7.2 displays a summary of the work output and loss terms for different cases of heat conduction. The relatively lowest influence on the performance of the model engines comes from the radiative heat transfer Q_r which contributes between 2% and 5% to the total heat losses. The domi-

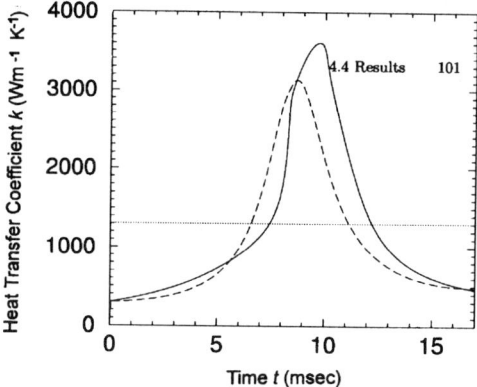

FIGURE 7.6. Constant heat conduction coefficient k_n for "Newton-type" heat conduction (dotted line) compared to the heat-conduction coefficient k_a for "Annand-type" heat conduction for an optimized (solid line) and a sinusoidal (dashed line) piston path.

TABLE 7.2. Work output W, losses due to friction W_f, conductive/convective Q_c, and radiative Q_r heat transfer in units of Joule are displayed for the sinusoidal (sin) and optimized (opt) trajectories with "Newton"- (N) and "Annand"-type (A) heat conduction, with and without radiative effects (R). The lengths of the bars are proportional to the magnitude of the respective values.

Case	W	W_f	Q_c	Q_r
N (sin)	265.1	21.1	144.5	0.0
N (opt)	319.4	24.8	133.3	0.0
NR (sin)	264.4	21.1	143.8	3.2
NR (opt)	317.4	24.7	132.2	4.7
A (sin)	265.1	21.1	121.2	0.0
A (opt)	289.8	20.3	150.7	0.0
AR (sin)	264.6	21.1	120.9	2.6
AR (opt)	288.9	20.3	149.8	2.6

nant conductive/convective heat losses Q_c are quite sensitive to variations in the path, especially in case of Annand-type heat transfer. Optimization of the work output effectively increases the heat losses; this is mainly caused by the higher temperatures during the power stroke. The type of heat conduction has contrary effects on the frictional losses W_f in the optimized models: In the case of Newton-type heat transfer the extended

190   7. Optimal Piston Paths for Diesel Engines

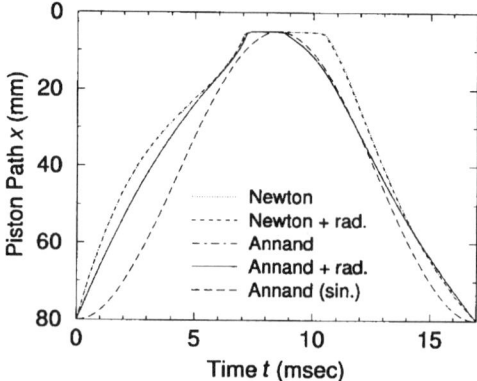

FIGURE 7.7. Trajectories for optimized ignition times. Note that the trajectories for the Annand model with and without radiative heat transfer are almost identical.

zero-velocity phase is responsible for higher frictional losses during the rest of the power stroke, while for Annand-type heat transfer the shorter zero-velocity phase in effect reduces the frictional losses. Friction dissipates 6% to 8% of the available volume work of the working fluid. The work output W is significantly improved by path optimization yielding a relative increase of 20.5% and 20.0% for the Newton model and 9.3% and 9.2% for the Annand model, without and with radiative heat transfer, respectively. The optimization with the more exact Annand model only achieves about half of the gains of the Newton model indicating that possible improvements by nonsinusoidal paths are still important but lower than previous studies suggested [1], [2], [9].

7.4.2 Optimal Time of Ignition

The next step of optimization is to find an optimal ignition time t_z^{opt} for the different models of the engine. This is done by systematically calculating the optimal path and the corresponding work output for various ignition times t_z. The ignition time which gives the largest work output is t_z^{opt}. In the following, optimal ignition times are also determined for sinusoidal piston paths to show possible improvements within conventional designs.

The optimized piston paths are depicted in Figure 7.7 and show only minor differences compared to the trajectories with fixed ignition time $t_z = 8.25$ ms. Apparent features are small shifts of the zero-velocity phases toward earlier times which correspond to shifts in the optimal ignition time to about 7 ms. The durations of the zero-velocity phases almost remain the same while the piston paths during the compression stroke have become a

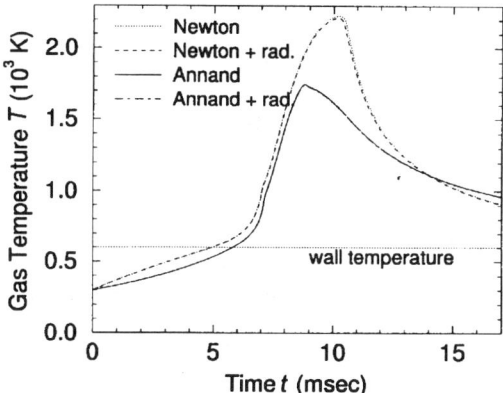

FIGURE 7.8. Temperatures of the working fluid for the path-optimized case with optimized ignition time t_z.

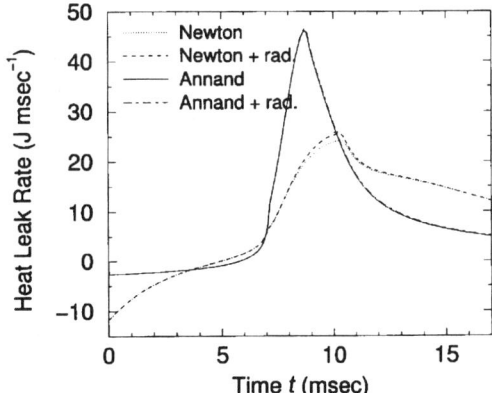

FIGURE 7.9. Heat leak rate for the path-optimized case with optimized ignition time t_z.

little more linear compared to their counterparts in Figure 7.3. This reduces the friction losses by a few percent.

As a result of the simulations, the maximum, minimum, and exhaust temperatures of the working fluid, as well as the heat conduction rate are little affected by the optimization of the ignition time as can be seen from Figures 7.8 and 7.9. The total heat losses, however, are enhanced by the shift of the ignition time because the switching points, where the temperature of the working fluid rises beyond the temperature of the wall, are reached earlier than for the case $t_z = 8.25\,\mathrm{ms}$.

7. Optimal Piston Paths for Diesel Engines

Table 7.3 demonstrates that for path-optimized cases the gains in work output by optimizing the ignition time are only about 0.2%. Apparently, the potential increases in efficiency due to the longer duration of the power stroke and smaller frictional losses are almost canceled by increased heat losses. Over a wide range of ignition times, the work output of the path optimized engine is nearly constant.

TABLE 7.3. Optimal ignition time t_z^{opt} in milliseconds, work output W, losses due to friction W_f, conductive/convective Q_c and radiative Q_r heat leakage in units of Joule are displayed for different types of models of heat conduction similar to Table 7.2. The parenthesized numbers refer to the percentage of relative increase to the results obtained for an 8.25 ms fixed ignition time.

Case	t_z^{opt}	W	W_f	Q_c	Q_r
N (sin)	7.17	296.2 (11.7)	21.1 (0.0)	149.6 (3.5)	0.0
N (opt)	7.00	320.0 (0.2)	23.7 (−4.4)	150.3 (12.8)	0.0
NR (sin)	7.19	294.7 (11.5)	21.1 (0.0)	148.3 (3.1)	4.3 (34)
NR (opt)	7.02	318.0 (0.2)	24.5 (−4.2)	148.7 (12.5)	12.7 (5)
A (sin)	7.47	282.9 (6.7)	21.1 (0.0)	146.1 (20.6)	0.0
A (opt)	7.03	290.4 (0.2)	19.8 (−2.5)	160.6 (6.6)	0.0
AR (sin)	7.48	282.2 (6.7)	21.1 (0.0)	145.2 (20.1)	2.7 (4)
AR (opt)	7.04	289.6 (0.2)	19.8 (−2.5)	159.6 (6.5)	2.8 (8)

It is important to note that we have optimized the path *after* choosing an ignition time in order to find the optimal ignition time with the largest work output. If we would optimize the path first and then change the ignition time, the work output would be significantly lowered. In other words: The work output for a given path, optimized or sinusoidal, is still sensitive to deviations from the optimal ignition time.

In the case of a sinusoidal piston path the ignition time indeed plays an important role and the work output can be increased by about 12% for the Newton and 7% for the Annand model if the ignition occurs at about 7.2 ms and 7.5 ms, respectively. These findings are consistent with the fact that in real Diesel engines the time of the injection of the fuel, which is related to the time of ignition, needs to be carefully adjusted to achieve an optimal performance of the engine.

The improvements are primarily caused by higher peak temperatures of the working fluid, as the peak of the combustion heat production is brought into coincidence with the minimum position of the piston.

The temperature curves in Figure 7.10 are all for sinusoidal piston motions with different types of heat conduction. The shapes of temperature profiles are still quite similar to the corresponding sinusoidal case in Figure 7.4 which was obtained for $t_z = 8.25$ ms and the Annand model only.

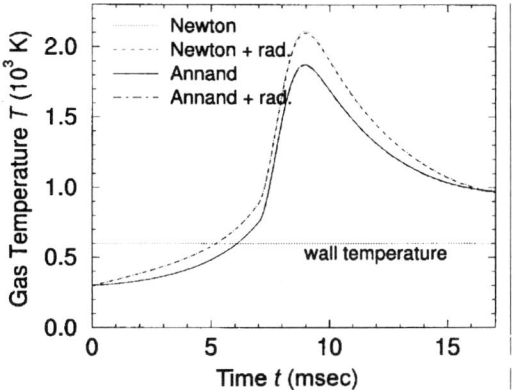

FIGURE 7.10. Temperatures of the working fluid for sinusoidal piston motion with optimized ignition time t_z.

Yet the t_z optimization results in an increase of the peak temperatures by approximately 400 K to over 2000 K and by 200 K to over 1700 K while the exhaust temperatures are decreased by about 90 K to 1050 K and by 130 K to 1100 K for the Newton and Annand models, respectively. The lower exhaust temperatures of the optimized models are also a consequence of the increased heat losses. The differences between models with and without radiation are minor.

A comparison between the effective heat transfer k_a of the Annand model for the sinusoidal and path-optimized case is shown in Figure 7.11. The coincidence of the two curves is surprising because the optimal ignition times are 0.3 ms apart.

The Annand heat transfer coefficient can be understood as the key parameter for the performance of the engine. At the one hand, large heat transfer coefficients indicate small volumes, high temperatures, and high pressures which serve as good initial conditions for a potentially high output of work. On the other hand, large heat transfer coefficients are connected to momentarily large heat loss rates. The peaking of heat transfer coefficients in Figure 7.11 balances both effects, creating high exergies for efficient power production while at the same time keeping the period of increased heat losses short.

In summary, the optimal allocation of the ignition time has decreased the relative advantages of the path-optimized engine. Relative to the conventional sinusoidal case, the work output has increased through path optimization by 8.0% and 7.9% for the Newton model and by 2.7% and 2.6% for the Annand model without and with inclusion of radiative heat transfer, respectively.

194   7. Optimal Piston Paths for Diesel Engines

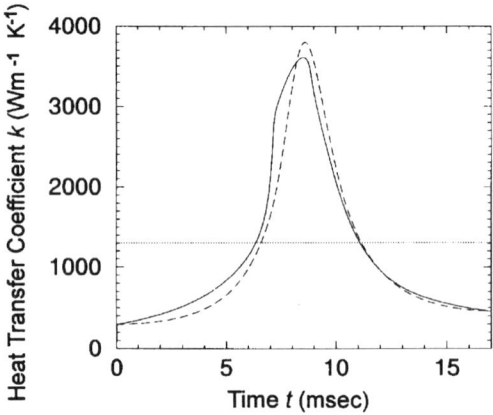

FIGURE 7.11. Heat transfer coefficients are displayed for optimized ignition times t_z in the case of a path-optimized (solid line) and sinusoidal piston motion (dashed line), both for Annand-type heat conduction. The constant heat transfer coefficient for the Newton-type heat conduction is drawn as a dashed line. The cases with and without radiation are almost identical.

7.5 Conclusion

In the endeavor for advances in the designs of Diesel engines we have addressed the question of finding an optimal path for the engine's piston during the compression and power stroke to achieve maximal work output for a given fuel input. Our model of a Diesel engine accounts for temporal variations of the heat produced in the combustion process and includes major irreversibilities such as friction and heat leakage by conductive, convective, and radiative heat transfer. Two kinds of descriptions for the combined conductive/convective heat transfer have been investigated, a simple linear dependence and a more sophisticated empirical expression due to Annand.

The piston trajectory for both, the compression and power stroke of the models were optimized simultaneously with a control theory method and the results were compared to the conventional engines with sinusoidal piston motion. The conductive/convective heat leakage was found to be the dominating loss mechanism in the model engines; friction losses and radiative heat transfer are of smaller importance and contribute only a few percent to the total losses. The path optimization decreases the friction losses and increases the peak temperatures thus causing larger heat losses. The losses are outweighed by gains in exergy of the working fuel due to the increased temperatures. The optimization yields a significantly higher work output compared to the conventional design.

The ignition times were optimized by systematically repeating the calculations for various ignition times. It was demonstrated that not much improvement can be achieved for the path-optimized models but the work output of the conventional engine can be increased by a few percent for the respective optimal time of ignition.

All simulations show that the model for heat leakage, linear or due to Annand, has a strong influence on the results. The more sophisticated Annand-type expression for the heat conduction rate generally leads to higher heat losses and lowers the possible gains through optimization. It is still not clear what impact even more detailed descriptions of processes such as the combustion would have on the results of the modeling. The differences in the results obtained for two different formulas of heat leakage indicate that a more precise understanding of the processes in the engine could lead to new interesting results and is therefore desirable.

Further studies could consider a simultaneous modeling of the entire cycle with intake, compression, power, and exhaust strokes, and possibly include additional components like turbochargers. They could also employ a more precise description of the combustion process, and changes of the working gas properties, as well as nonuniform temperatures of the cylinder's wall. Another important question concerns the role of constraints, such as the minimum volume and maximum temperatures of the working gas.

With today's ever-increasing computer power and advances in algorithms, the theoretical modeling of Diesel engines can certainly be improved to provide an even better guide for the improvement of applications in practice.

Acknowledgments

The authors thank Professor Dr. Ing. Heinz Herwig from the Technische Universität Chemnitz for a compilation of valuable literature, notes on the works of Annand and Woschni, and for fruitful discussions. J. M. Burzler acknowledges a scholarship from the Bischöfliche Studienstiftung Cusanuswerk.

7.6 References

[1] M. Mozurkewich and R. S. Berry: Optimal paths for thermodynamic systems: The ideal Otto cycle, *J. Appl. Phys.* **53**, 34–42, (1982).

[2] K. H. Hoffmann, S. J. Watowich, and R. S. Berry: Optimal paths for thermodynamic systems: The ideal Diesel cycle, *J. Appl. Phys.* **58**, 2125, (1985).

[3] M. H. Rubin: Optimal configuration of a class of irreversible heat engines II, *Phys. Rev. A* **19**, 1277, (1979).

[4] C. F. Taylor: *The Internal-Combustion Engine in Theory and Practice*, volumes 1 and 2, MIT, Cambridge MA, 1977.

[5] S. J. Watowich, K. H. Hoffmann and R. S. Berry: Intrinsically irreversible light-driven engine, *J. Appl. Phys.* **58**, 2893–2901, (1985).

[6] S. J. Watowich and R. S. Berry: Optimal current paths for model electrochemical systems, *J. Phys. Chem.* **90**, 4624–4631, (1986).

[7] A. E. Bryson and Y. C. Ho: *Applied Optimal Control*, Wiley, New York, 1975.

[8] V. M. Alekseev, V. M. Tikhomirov, and S. V. Fomin: *Optimal Control*, Nauka, Moscow, 1979.

[9] P. Blaudeck and K. H. Hoffmann: Optimization of the power output for the compression and power stroke of the Diesel engine, *Proceedings of the International Conference ECOS'95*, Istambul, 1995, vol. 2, p. 754.

[10] W. Nusselt: Der Wärmeübergang in den Verbrennungskraftmaschinen, *Z. des Vereins deutscher Ingenieure* **67**, 692 and 708, (1923), in German.

[11] G. Eichelberg: Temperaturverlauf und Wärmespannung in Verbrennungsmotoren, *Forsch. Ing. Wes.*, 1923, in German.

[12] V. D. Overbye, J. E. Bennethum, O. A. Uyehara, and P. S. Myers: Unsteady heat transfer in engines, *Trans. Soc. Automot. Engrs.* **69**, 461, (1961).

[13] G. Woschni: Die Berechnung der Wandverluste und der thermischen Belastung der Bauteile von Dieselmotoren, *Motortechnische Zeitschrift* **31**, 491–499, (1970), in German.

[14] G. Woschni: Experimental investigation of the heat transfer in internal combustion engines with insulated combustion chamber walls, 1989, Chapter 50, pp.53–65.

[15] W. J. D. Annand: Heat transfer in the cylinder of reciprocating internal combustion engines, *Proc. Inst. Mech. Eng.* **177**, 973–990, (1963).

[16] F. Angulo-Brown, J. A. Rocha-Martinez, and T. D. Navarrete-Gonzalez: A non-endoreversible Otto cycle model: Improving power output and efficiency, *J. Phys. D: Appl. Phys.* **29**, 80–83, (1996).

[17] C. Wu and D. A. Blank: The effects of combustion on a work-optimized endoreversible Otto cycle, *J. Inst. Energy* **65**, 86–89, (1992).

[18] C. Wu and D. A. Blank: Optimization of the endoreversible Otto cycle with respect to both power and mean effective pressure, *Energy Conversion and Management* **34**, 1255–1259, (1993).

[19] D. A. Blank and C. Wu: The effect of combustion on a power optimized endoreversible Diesel engine, *Energy Conversion and Management* **34**, 493–498, (1993).

[20] W. Kleinschmidt: Die Wärmeübertragung in aufgeladenen Dieselmotoren aus neuerer Sicht, 5. *Aufladetechnische Konferenz*, Augsburg, 1993.

[21] W. Pflaum and K. Mollenhauer: *Wärmeübergang in der Verbrennungskraftmaschine*, Springer-Verlag, Wien, 1977.

[22] B. Andresen, R. S. Berry, A. Nitzan, and P. Salamon: Thermodynamics in finite time I. The step-Carnot cycle, *Phys. Rev. A* **15**, 2086–2093, (1977).

[23] B. Andresen, P. Salamon and S. R. Berry: Thermodynamics in finite time: extremals for imperfect heat engines, *J. Chem. Phys.* **66**, 1571, (1977).

[24] P. Salamon, A. Nitzan, B. Andresen and R. S. Berry: Minimum entropy production and the optimization of heat engines, *Phys. Rev. A* **21**, 2115, (1980).

[25] J. H. Van Gerpen and H. N. Shapiro: Second law analysis of Diesel engine combustion, in: *Analysis and Design of Advanced Energy Systems: Computer-Aided Analysis and Design, Proc. ASME Winter Annual Meeting*, Boston, Dec. 13–18, 1988 (eds. M. J. Moran, R. A. Bajura, and G. Tsatsaronis), vol. 3–3, pp. 53–65, 1988.

[26] N. M. Al-Najem and J. M. Diab: Energy–exergy analysis of a Diesel engine, *Heat Recovery Systems & CHP* **12**, 525–529, (1992).

[27] V. A. Mironova, A. M. Tsirlin, V. A. Kazakov, and S. R. Berry: Finite-time thermodynamics: Exergy and optimization time-constrained processes, *J. Appl. Phys.* **76**, 629–636, (1994).

[28] A. Bejan: Entropy generation minimization: The new thermodynamics of finite-size devices and finite-time processes, *J. Appl. Phys.* **79**, 1191–1218, (1996).

[29] K. H. Hoffmann, J. M. Burzler, and S. Schubert: Endoreversible thermodynamics, *J. Non-Equilib. Thermodyn.* **22**, 311–355, (1997).

[30] K. H. Hoffmann, S. Schubert, and J. M. Burzler: Endoreversible dynamics, *J. Non-Equilib. Thermodynam.*, to appear.

[31] L. S. Pontryagin, V. G. Boltyanski, R. V. Gamkrelidze, and E. F. Mishtchenko: *The Mathematical Theory of Optimal Processes*, Interscience, New York, 1962.

[32] W. H. Press, S. A. Teukolsky, W. T. Vetterling, and B. P. Flannery: *Numerical Recipes in C*, Cambridge University Press, Cambridge, 1992.

[33] Y. B. Band, O. Kafri, and P. Salamon: Maximum work production from a heated gas in a cylinder with piston, *Chem. Phys. Lett.* **72**, 127–130, (1980).

8
Qualitative Properties of Conductive Heat Transfer

H. Farkas
I. Faragó
P. L. Simon

ABSTRACT. In Section 8.1 we overview the classical theory of heat conduction formulated in thermodynamic terms. The dissipative character of pure heat conduction is manifested in the heat conductional inequality, the maximum principle, and other related properties (Section 8.2). These properties can be stated in general, far beyond the linear theory. The classical heat equation yields an infinite velocity of propagation. The hyperbolic heat equation has been proposed to overcome this paradox. However, the maximum principle is not valid for a hyperbolic equation; a satisfactory theory involving the maximum principle, as well as finite propagation, is not yet known (Section 8.3). The required basic properties may be used as postulates in searching for a new theory. An attempt of this kind is outlined. For a homogeneous one-dimensional-medium it is demonstrated in Section 8.4 that a theory for stationary heat conduction can be derived. The solution of the initial-boundary value problem for the classical linear heat conduction equation of parabolic type has several special characteristic properties, like contractivity in time, nonoscillatory behavior, exponential convergence, and others. In Section 8.5 we list some such properties. As is typical, the continuous problem cannot be solved analytically. In Section 8.6 some numerical processes are applied. This means that we define the approximate solution at the discrete points of a mesh. Obviously, the basic question is the convergence, that is, when refining the mesh, the numerical solution should be convergent to the solution of the original continuous problem. It is no less important to require the preservation of the discrete analogues of the basic qualitative properties of the continuous solution mentioned the above. In Section 8.7 the exact conditions for the preservation of some qualitative properties are established. There are damped traveling wave solutions of the sourceless parabolic heat equation. In Section 8.8 we list all the "shape-preserving signal forms," that is, all the signal forms that can propagate inside the body without distortion, after a transient period. The solutions of this "shape-preserving" type have an attractive property: the asymptotic solutions belonging to different initial conditions tend to solutions of that kind.

200   8. Qualitative Properties of Conductive Heat Transfer

8.1 Theoretical Background

8.1.1 *Fourier's Differential Equation*

The mathematical theory of heat conduction [1] usually restricts us to the investigation of the heat equation (in other words, Fourier's differential equation, diffusion equation):

$$\frac{\partial T}{\partial t} - D \, \Delta T = f, \qquad (8.1)$$

where T is temperature, t is time, $D > 0$ is the heat diffusivity coefficient, Δ is the Laplace operator, and f is the source. Equation (8.1) is a linear parabolic differential equation with a constant coefficient D. In the following we embed this equation into a very general physical frame.

8.1.2 *Balance of Internal Energy*

The thermodynamical theory of heat conduction is based on more general physical models and laws. In this theory (8.1) is derived under some simplifying assumptions. The limits of validity have great importance from a physical point of view. Now we start from the general basic equations formulated in thermodynamics. The balance equation of the internal energy is [2]:

$$\frac{\partial \rho_u}{\partial t} + \text{div } q = \sigma_u, \qquad (8.2)$$

where ρ_u is the density of internal energy, q is the heat current density, and σ_u is the heat production density (production density of the internal energy, in other words, source).

8.1.3 *Material (Constitutive) Equations*

The differential equation of heat conduction is derived from (8.2) by substituting the material equations into it. These material (or in other words: constitutive) equations enable us to express the quantities $\partial \rho_u / \partial t$, div q, and σ_u in terms of the variable T and its time and spatial derivatives. Using the static state equation we get the relation

$$\frac{\partial \rho_u}{\partial t} = \rho c \frac{\partial T}{\partial t},$$

where $\rho > 0$ is (mass) density and $c > 0$ is specific heat. This relation is valid only if no work is involved in the conduction process, that is, thermal expansion can be neglected.

Fourier's law reads as

$$q = -\lambda \text{ grad } T,$$

where $\lambda > 0$ is heat conductivity. This relation is applied not only to isotropic solids, but also to isotropic fluids if thermal convection is neglected.

8.1.4 Transport Equation. Initial and Boundary Conditions

Under these assumptions we have the differential equation of heat conduction in the form

$$\rho c \frac{\partial T}{\partial t} - \text{div}(\lambda \text{ grad } T) = \sigma_u. \tag{8.3}$$

In the general case the quantities ρ, c, and λ are functions of the independent variables $\mathbf{r}$ (space variable) and t (time), as well as the temperature T. The differential equation must be satisfied in the points of a region V. Special attention will be paid to the cases:

- time-independent medium (the listed quantities do not depend explicitly on time);
- homogeneous medium (the listed quantities do not depend explicitly on space); and
- linear medium (the listed quantities do not depend explicitly on T).

For a time-independent homogeneous linear medium equation (8.3) takes the form (8.1) with the notation

$$D = \frac{\lambda}{\rho c}.$$

The initial condition for a typical heat conductional problem is

$$T(\mathbf{r}, t_0) = T_0(\mathbf{r}) \quad \text{for all} \quad \mathbf{r} \in V,$$

where the initial temperature distribution $T_0(\mathbf{r})$ is a prescribed function of space.

The general form of the linear boundary condition is

$$a(\mathbf{r}, t) T(\mathbf{r}, t) + b(\mathbf{r}, t) \frac{\partial T}{\partial n} = g(\mathbf{r}, t) \quad \text{for all} \quad t > t_0, \quad \text{for all } \mathbf{r} \in \partial V,$$

where a, b, and g are assumed to be given functions, $\partial T / \partial n$ is the normal component of $\text{grad } T$. This is called the boundary condition of the third kind which includes the special boundary conditions of the first ($b = 0$) and the second ($a = 0$) kind.

8.1.5 Heat Conduction in Irreversible Thermodynamics

The entropy balance is

$$\frac{\partial \rho_s}{\partial t} + \text{div } J_s = \sigma,$$

where ρ_s is the entropy density

$$d\rho_u = \frac{d\rho_s}{T},$$

J_s is the entropy current density

$$J_s = \frac{q}{T},$$

and σ is the entropy production density

$$\sigma = \frac{\sigma_u}{T} + q \operatorname{grad}\frac{1}{T}.$$

The most fundamental quantity in Onsagerian irreversible thermodynamics of continua is σ. As a general rule, this quantity is the sum of the product of thermodynamic fluxes and forces. The linear conductivity law of Onsagerian thermodynamics reads as

$$q = L \operatorname{grad}\frac{1}{T}.$$

This formula is essentially equivalent to Fourier's law with

$$\lambda = \frac{L}{T^2}.$$

In linear Onsagerian theory the conductivity may depend on the state, in our case on T. This case is termed as a quasilinear case [2], since it leads to quasilinear differential equations of the second order. For the quasilinear case we can work with different temperature scales ("pictures," see later). Nonlinear generalization of Onsagerian thermodynamics considers the flux q as a nonlinear function of the thermodynamic force grad($1/T$).

8.1.6 Variational Principles

A model described with differential equations may be formulated equivalently with variational principles. The solutions of the original differential equation are extremals of an appropriate functional. If the differential operator is self-adjoint, then the variational formulation is straightforward as, e.g., in classical mechanics.

The heat conduction equation however, is not self-adjoint, that is why difficulties arise with variational formulations [3], [4]. Several variational principles have been constructed which overcome these difficulties by imposing additional variational subsidiary conditions (Gyarmati's governing principle of dissipative processes [2]), explicit duplication of the variables (Rosen's restricted variational principle [5], Prigogine and Glansdorff's local potential [6], [7]), adding the adjoint problem (Morse and Feshbach [8]), introducing background variables (Biot [9]), and so on.

These variational formulations require space enlargement or the use of potentials. Both these operations can be accomplished by the use of Lagrange multipliers, yet some of these variational approaches follow other, particular ways.

The variational method developed by Gyarmati was based on the sum of two dissipation functions Φ (depending on thermodynamical fluxes) and Ψ (depending on thermodynamical forces). Gyarmati's functionals were basically developed for three-dimensional space, and in some forms the variations were subjected to the conservation laws of the process. Gyarmati's variational principle was applied to many special fields [2], and it was extended or reformulated involving nonlinear cases [10] and time integrals [11].

Working in four-dimensional space–time of independent thermodynamic fields, Sieniutycz [12, Chapters 10 and 11], [13] has minimized the four-volume-based integral over the dissipative Lagrangian $L = \Phi + \Psi$, to which the conservation or balance laws were explicitly adjoined by the Lagrange multipliers λ_i. In his approach, the Lagrange multipliers are fields (functions of x and t) that are independent of the fields of the thermodynamic variables appearing in $L = \Phi + \Psi$. Using this model, Sieniutycz has shown [12], [13], that the Euler–Lagrange equations of the balance-constrained variational problem are extensions of the standard transport and balance equations beyond the local equilibrium, in the sense that all Euler–Lagrange equations (which embrace the Fourier law of heat conduction, the equation of change for T, and the energy balance) are mutually independent at disequilibrium. If, however, the local equilibrium assumption is a posteriori made, as an extra physical postulate, then all these equations become dependent, that is, the equation of change for T follows as the usual combination of the energy balance and Fourier's law. This approach has also been extended [12] to hyperbolic systems (governed by the Cattaneo equation [14]) and chemically reacting systems [13] which cause thermal sources. For the relation between this approach and Biot's approach [9], as well as with traditional variational tools, see [12], [13], [15].

8.1.7 *Stationary Case*

The stationary temperature distribution $T(\mathbf{r})$ satisfies the equation

$$\text{div}(\lambda \text{ grad } T) = 0 \text{ (more general case)}$$

or

$$\Delta T = 0 \text{ (simple case).}$$

The last equation is the Euler–Lagrange equation of the variational problem

$$\mathcal{I} = \int (\text{grad } T)^2 \, dV = \text{minimum} \tag{8.4}$$

(Dirichlet's principle).

It can be proved that the functional $\mathcal{I}$ in (8.4) decreases monotonically and tends to its minimal value as t tends to infinity. This property remains valid for the case if λ depends on **r**.

8.1.8 *Temperature Scales: Pictures, Kelvin's Transformation*

Consider an arbitrary monotonous function of T:

$$\Gamma = \Gamma(T).$$

The material laws, the thermodynamical quantities, the variational principles, and the transport equation can be rewritten in this "Γ-picture" [16] belonging to this scale. Three special cases were worked out by Gyarmati [2]:

- the Fourier picture: $\Gamma = T$;
- the energy picture: $\Gamma = \ln(T)$; and
- the entropy picture $\Gamma = 1/T$.

For a time-independent homogeneous medium a new temperature scale can be introduced by Kelvin's transformation

$$\Gamma(T) = \int_{T_0}^{T} \lambda(\tau) \, d\tau.$$

Using this Γ temperature, Fourier's law reduces to the form

$$q = -\text{grad } \Gamma.$$

In other words, in this Γ-picture, the conductivity coefficient is constant, namely 1. Therefore, Dirichlet's principle will be valid in the form

$$\mathcal{I} = \int (\text{grad } \Gamma)^2 \, dV = \text{minimum}.$$

8.2 Consequences of the Second Law

Heat conduction is a dissipative process. It is irreversible in the thermodynamic sense: the initial state can be restored only if external work is utilized. The second law of thermodynamics is reflected in the dissipative character of heat conduction.

8.2.1 *Heat Conductional Inequality*

There are several possible formulations of the Second Law. The most often-cited is the principle of increasing entropy: the entropy of any adiabatically isolated system increases. As a consequence, the entropy production density must be positive definite. Although it refers only to the total entropy production, if the different processes are separable from each other, the corresponding contributions to the entropy production must be positive separately. As a consequence, the entropy production term belonging to the heat conduction process must be positive, that is,

$$q \text{ grad } \frac{1}{T} > 0, \tag{8.5}$$

i.e.,

$$q \text{ grad } T < 0$$

holds for any pure heat conduction process [17], [18]. For an anisotropic material obeying the linear law of conduction

$$q = \mathbf{L}(-\text{grad } T)$$

inequality (8.5) is equivalent to the requirement that the conduction matrix $\mathbf{L}$ must be positive definite [2], [17].

8.2.2 Maximum Principle

It was well known that all the solutions of the differential equation (8.1) satisfy the following maximum principle:

$$\max_{(\mathbf{r},t) \in \overline{V} \times [t_0,\tau]} T(\mathbf{r},t) = \max \left\{ \max_{\mathbf{r} \in \overline{V}} T_0(\mathbf{r}); \max_{(\mathbf{r},t) \in \partial V \times [t_0,\tau]} T(\mathbf{r},t) \right\},$$

that is, the temperature takes its maximum at the initial instant or at the boundaries [19], [20], [21]. A similar statement is valid for the minimum.

The maximum principle enables us to prove uniqueness for the linear equation in a very simple way. Namely, if T_1 and T_2 are two solutions of (8.1) which satisfy the same initial and boundary conditions, then their difference satisfies zero initial and boundary conditions, and consequently, its maximum and minimum is zero, that is, the difference of two solutions is identically zero.

The maximum principle is a straightforward consequence of the Second Law. To see this, recall the Kelvin–Planck statement [22]: the natural direction of heat flow is the downward direction, that is, without any external work, heat flows from the places with higher temperature to places with lower temperature.

Although the essence of the maximum principle is obvious, the conventional mathematical proof needs some sophisticated mathematics [20] and uses the linear equation (8.1). It was possible to prove the maximum principle for a very general, nonlinear case. This proof [21] used Gyarmati's variational principle. The proof was an indirect one: supposing an internal maximum we were able to construct a piecewise continuous function (by cutting the small neighborhood off the top) yielding a more optimal value for the variational functional than the supposed function.

Nevertheless, from a physical point of view the maximum principle does not need a mathematical proof. We can simply state that pure heat conduction should satisfy the maximum principle [23]. Otherwise, the Second Law would be violated.

There are several different formulations of mathematical theorems belonging to the maximum principle. One of them is the strong maximum principle [24]. If at a fixed instant T has a local spatial maximum in an internal point, then the time derivative is negative at that point. A consequence of the strong maximum principle is the following statement: the number of local extrema must not increase in time.

8.3 The Velocity of Propagation

The differential equation (8.1) implies a paradox property: the infinite velocity of heat [20]. One of the simplest manifestations of this property can immediately be seen from the solution

$$T(\mathbf{r}, t) = \frac{C}{(\sqrt{t})^3} \exp\left(-\frac{|\mathbf{r}|^2}{4Dt}\right). \tag{8.6}$$

This solution describes the propagation of heat in an infinite space generated by a heat source at the origin at the initial state $t = 0$. The effect of this source produces a temperature different from zero at arbitrary distant regions.

The infinite velocity of heat is unacceptable from a physical point of view. The theory of relativity requires that the velocity of any process–by which one can send signals–should not surpass the velocity of light. Furthermore, since the heat conduction is carried out by molecules, one can expect that the velocity in heat conduction cannot surpass the velocity of molecules. Maxwell [25] derived a hyperbolic heat equation based on kinetic consideration. Later, several other authors derived differential equations of the same type [14], [26]:

$$\tau \frac{\partial^2 T}{\partial t^2} + \frac{\partial T}{\partial t} - D\Delta T = f, \tag{8.7}$$

The occurrence of an additional inertial term $\tau \partial^2 T/\partial t^2$ can be derived from a generalized linear law of conduction

$$\tau \frac{\partial q}{\partial t} + q = -\lambda \text{ grad } T.$$

These generalized linear laws and the corresponding hyperbolic transport equations were widely studied and established on a thermodynamical base,

8.3. The Velocity of Propagation

see extended thermodynamics [27], [28] and the wave approach of thermodynamics [29]. In the limiting case $\tau \to 0$ the solution of the hyperbolic equation (8.7) tends to the solution of the parabolic heat equation (8.1) [20]. The hyperbolic heat equation removes the paradox of infinite velocity. On the other hand, the involved inertial term makes the process not purely dissipative. Consequently, some qualitative properties will not hold for the hyperbolic case. The heat conductional inequality and the maximum principle are not valid now. Even the positiveness of the absolute temperature is at stake: after a certain time the temperatures may assume negative values, even the initial and boundary temperatures are positive [23].

To describe real heat conductional processes we should decide which properties are decisive. For the vast majority of practical cases, the original Fourier equation predicts the propagation velocity with fairly good accuracy, e.g., the practically observed rather slow propagation is in good agreement with the distribution (8.6), since the exponential term approaches zero very quickly. Nevertheless, there are experimental observations, where the inertial term is relevant, see, e.g., the second sound in liquid He [30].

From a theoretical point of view, a unified generalization would be satisfying only if it could account for such qualitative properties as maximum principle, and at the same time it would yield a finite propagation velocity. Presently such a unified linear theory is not known.

For the nonlinear heat equation there are mathematical results which yield a finite propagation velocity without removing the dissipative character of the equation [31], [32], [33]. In these works, the velocity is defined as the velocity of the free boundary, more precisely the temperature at the initial instant it is assumed to be zero outside a bounded region, *the support is bounded* and the propagation velocity is defined as the velocity of the support boundary. In [33] one can find exact conditions for the temperature dependencies in order to get finite propagation velocity. It is found, e.g., that if the heat conductivity coefficient is proportional to the mth degree of the temperature ($m > 1$), then the support boundary moves with finite velocity. This case, however, is the case of the "degenerate parabolic heat equation". The finite velocity is due to the fact that heat conductivity is zero at the boundary. From a physical point of view the propagation velocity is determined by the processes happening at the boundary, and zero conductivity slows down the propagation to finite velocity. If the conductivity is "regular" (that is, remains between positive bounds) the propagation velocity is infinite for quasilinear cases, too.

8.4 System Theory Approach

8.4.1 *Heat Conduction and Dynamical Systems Theory*

Dynamical systems theory is a modern interdisciplinary branch of science [34], [35]. This theory applies sophisticated mathematics to model physical, chemical, biological, and even social phenomena. The dynamical systems theory provides us with a common terminology to describe the formal properties of behavior of essentially different systems.

A dynamical system is defined as a deterministic, autonomous system. This means that the initial state uniquely determines the evolution of the state in the future and in the past. Systems described by parabolic partial equations can be considered as semidynamical systems, since only their future is determined by the initial state, the past is not. For example, the solution of (8.1) exists for $t > 0$, but it maybe does not exist for $t < 0$.

The other problem is that heat conduction is generally not an autonomous process. The boundary conditions and the sources can be described in the frame of control theory.

8.4.2 *Principle of Superposition*

Consider the following heat conduction problem:

$$\frac{\partial T}{\partial t} - D\,\Delta T = f, \tag{8.8}$$

$$T(\mathbf{r}, t_0) = T_0(\mathbf{r}) \text{ for all } \mathbf{r} \in V, \tag{8.9}$$

$$a(\mathbf{r}, t)T(\mathbf{r}, t) + b(\mathbf{r}, t)\frac{\partial T}{\partial n} = g(\mathbf{r}, t) \text{ for all } t > t_0, \text{ for all } \mathbf{r} \in \partial V. \tag{8.10}$$

Here D is a given positive constant, $a(\mathbf{r}, t)$, $b(\mathbf{r}, t)$, $f(\mathbf{r}, t)$, $T_0(\mathbf{r})$, $g(\mathbf{r}, t)$ are given functions. The given data D, $a(\mathbf{r}, t)$, $b(\mathbf{r}, t)$ characterize the system, the functions $f(\mathbf{r}, t)$, $T_0(\mathbf{r})$, $g(\mathbf{r}, t)$ can be considered as inputs, while the solution to the problem (8.8), (8.9), (8.10) is the output which is determined uniquely by the inputs for any specified system. This problem is a linear one: the output depends linearly on the input vector (f, T_0, g). Therefore, the superposition principle holds: the superposition of inputs yields the superposition of the corresponding outputs.

The principle of superposition enables us to reduce the original problem to simpler ones: the solution of the original problem is the sum

$$T = T_S + T_I + T_B,$$

where T_S is a source-generated field (with zero initial and boundary conditions); and T_I is the solution of the initial value problem (with zero source and zero boundary conditions); T_B is the solution of the boundary value problem (with zero source and zero initial condition).

Further applications of the superposition principle enable us to extend the scalar fields T_S, T_I, T_B as a sum (or integral) of the solutions of even simpler problems. As an illustration, the source $f(\mathbf{r},t)$ can be expanded as superposition of the unit instantaneous point sources $\delta(\mathbf{r}-\mathbf{s},t-\tau)$:

$$f(\mathbf{r},t) = \int_0^\infty \int_V f(\mathbf{s},\tau)\delta(\mathbf{r}-\mathbf{s},t-\tau) \, ds \, d\tau.$$

According to the superposition principle the solution of the problem generated by the source $f(\mathbf{r},t)$ is

$$T_S(\mathbf{r},t) = \int_0^t \int_V f(\mathbf{s},\tau) G(\mathbf{r},\mathbf{s},t,\tau) \, ds \, d\tau,$$

where G is the Green function belonging to the source-generated problem, that is, the solution belonging to the unit instantaneous (at the instant τ) point source (located at the point $\mathbf{s}$), while the initial and boundary values of temperature are assumed to be zero. In other words, G gives the response at $(\mathbf{r},t)$ to the unit source acting at $(\mathbf{s},\tau)$.

This technique can be applied to the initial value problem, and also to the boundary value problem. Moreover, the solutions of these can also be expressed in terms of the Green function G [1], [20], [36].

We remark that the superposition function and the Green function techniques can be applied to any linear problem, not only for problems belonging to the differential equation (8.8). In the frame of rational thermodynamics [17], [18] the transport equation is given by integro-differential equations involving not only functions but functionals, too (nonlocal systems with memory).

As we saw above (Section 8.3), the parabolic and hyperbolic heat differential equations have principal disadvantages: the former yields unrealistic infinite propagation velocity, while the latter disobeys the maximum principle. The Green function approach could perhaps help us in future. Our guess is that it is possible to establish such a reasonable linear model which would match both the requirement of finite propagation velocity and the maximum principle. The wanted model could be defined by a Green function satisfying the appropriate requirements.

8.4.3 *A Postulatory Approach to Stationary Heat Conduction*

Now we outline a postulatory approach to heat conduction. This approach was worked out in the homogeneous one-dimensional case to stationary heat conduction [23], [37]. This postulatory approach involves only two primary concepts: the space coordinate x and the temperature T. Our aim is to derive the differential equation of the problem and to define the heat current q as a secondary concept. In general, it is a much easier task to

210   8. Qualitative Properties of Conductive Heat Transfer

measure x and T than to measure q; this fact may also justify such an approach.

Our treatment contains the following steps:

Step 1. Formulation of our postulates. We assume that the stationary temperature field $T(x)$ is uniquely determined by the temperatures at the boundaries. In the one-dimensional case this means that our problem is a two-point problem. The postulates are imposed on the possible temperature fields.

Step 2. The equation of the set of possible temperature fields will contain two parameters. One of these refers to the space displacement invariance (the homogeneous case), the other one plays a role in the definition of heat current (in the next step).

Step 3. From the two-point problem we derive a one-point problem by involving the space derivative T' of T. The other parameter can be interpreted as "empirical heat current."

Step 4. Using the additivity property at branching, we derive the true ("absolute") heat current.

The relation between the empirical and absolute scales is the same as used in thermodynamics [38]: the empirical temperature (or entropy) has an arbitrary scale assigned to isotherms (or adiabatics, resp.), while the absolute scale is selected uniquely from the empirical scales by imposing some additional general requirements. The empirical heat current enables us to define classes in the set of possible fields (the heat current is constant for a class), while the absolute heat current defines a natural measure (distance) between the different classes.

Our postulates obviously hold for any medium obeying the Fourier law with bounded positive thermal conductivity λ:

$$0 < k < \lambda(T) < K$$

which depends on T continuously, therefore, our approach is a possible generalization of the Fourier law [23].

Postulates

The following postulates are aimed at modeling the one-dimensional stationary heat conduction in an infinite homogeneous medium.

Postulate 1. There exists a differentiable five-variable function F with the properties:

- the value of F is the stationary temperature $T(x)$ at the point x in the interval $[a, b]$:
$$T(x) = F(a, b, T_a, T_b; x);$$

- $F(a, b, T_a, T_b; a) = T_a$ and $F(a, b, T_a, T_b; b) = T_b$;

- if $[c, d]$ is a subinterval of $[a, b]$, then

$$F(a, b, T_a, T_b; x) = F(c, d, F(a, b, T_a, T_b; c), F(a, b, T_a, T_b; d); x)$$

for any x in $[c, d]$.

In the following, sometimes we will concentrate on the x dependence, and the first four variables of F will be considered parameters. The graph $(x, T(x))$ will be referred to as temperature characteristic.

Remark that the requirements involved in Postulate 1 resemble those used in the definition of the dynamical system [34].

Postulate 2 (Equilibrium). If the boundary temperature values are equal $T_a = T_b$, then the temperature is constant:

$$F(a, b, T_a, T_a; x) = T_a$$

for any x in $[a, b]$.

Postulate 3 (Strict Monotonicity). If $T_b < T_b'$, then

$$F(a, b, T_a, T_b; x) < F(a, b, T_a, T_b'; x)$$

for all x in $(a, b]$ and, moreover,

$$F_x(a, b, T_a, T_b; a) < F_x(a, b, T_a, T_b'; a),$$

where F_x denotes the (right-hand) derivative of the temperature with respect to the space coordinate x.

This property expresses that the inside temperature, as well as the initial slope, are strictly monotonically increasing functions of the right boundary temperature value. Dual requirements can be formulated for the left boundary, and really, we assume the validity of the dual properties:
If $T_a < T_a'$, then

$$F(a, b, T_a, T_b; x) < F(a, b, T_a', T_b; x)$$

for all x in $[a, b)$ and, moreover,

$$F_x(a, b, T_a, T_b; b) > F_x(a, b, T_a', T_b; b).$$

Instead of this postulate we could have assumed that two temperature characteristics intersect each other transversally [35], because transversality and Postulate 2 imply monotonicity.

Postulate 4 (Homogeneity).

$$F(a, b, T_a, T_b; x) = F(a + d, b + d, T_a, T_b; x + d) \quad \text{for all} \quad d \in \mathbf{R}.$$

This postulate expresses the homogeneity of the medium: its behavior is invariant under an arbitrary spatial displacement (d).

Postulate 5 (Controllability). (i) Let $[a, b]$ be a subinterval of $[c, d]$. Given any values T_a, T_b, there exist T_c and T_d such that

$$F(c, d, T_c, T_d; a) = T_a \quad \text{and} \quad F(c, d, T_c, T_d; b) = T_b.$$

(ii) Let $[T_a, T_b]$ be a subinterval of $[T_c, T_d]$. Given any values a, b there exist c and d such that

$$F(c, d, T_c, T_d; a) = T_a \quad \text{and} \quad F(c, d, T_c, T_d; b) = T_b.$$

This postulate guarantees the continuation of the temperature characteristics: any characteristic is a part of an infinite one, defined for all x due to (i), and taking all values of T due to (ii).

For simplicity, we made no restriction to the region of the admissible temperature values. It is possible (e.g., by use of another temperature scale) to reformulate our postulates to a bounded region of admissible temperature values.

Consequences

Here we recall some consequences of the above postulates. These consequences can be derived by simple mathematics [23], [37].

Postulate 2 implies the maximum principle in this case, that is,

$$\min\{T_a, T_b\} \leq T(x) = F(a, b, T_a, T_b; x) \leq \max\{T_a, T_b\}$$

for any x in $[a, b]$.

Two characteristics must be identical if they have two points in common. This property is a consequence of Postulate 1.

Postulate 5 enables us to introduce infinite characteristics. More formally, there exists a two-parameter family of curves in the plane (x, T) defined by a differentiable function f:

$$T(x) = f(H, x + d), \quad -\infty < x < +\infty, \tag{8.11}$$

such that any finite part of these infinite characteristics (that is, their restriction to any finite interval in x) is a finite characteristic and any finite characteristic is a part of an infinite one.

The homogeneity of the medium (that is, invariance under spatial displacement) implies that one of the parameters (d) is displacement: horizontal displacement of a characteristic by any d yields another infinite characteristic.

Without loss of generality we can impose the following requirements:

$$f(H, 0) = 0 \quad \text{and} \quad f(H, L) = H,$$

where L is an arbitrary positive constant length. If we use this convention, the physical meaning of the other parameter H is the temperature value at $x = L$. The family of characteristics going through the origin is characterized by the parameter H. For $H > 0$, these are strictly monotonically increasing, for $H < 0$, these are strictly monotonically decreasing, for $H = 0$ it is constant, namely

$$f(0, x) = 0.$$

Differentiating (8.11) with respect to x, we obtain

$$T'(x) = f_x(H, x + d). \tag{8.12}$$

Due to strict monotonicity, the function f is invertible with respect to the second variable

$$x + d = f^{-1}(H, T).$$

Inserting this into (8.12) we conclude that the characteristics are solutions of the one-parameter family of differential equations

$$T' = g(H, T). \tag{8.13}$$

Therefore, we obtained a relation between the values of T' and T. This relation is valid along a characteristic (belonging to the parameter H).

From Postulate 3 it follows that if two different characteristics have a common point, then their slope must be different at the joint point. Due to this property T' is a strictly monotone function of H, and hence H can be expressed from (8.13). Consequently, the differential equation of characteristics can be written in the form

$$h(T, T') = \text{constant} \ (= H). \tag{8.14}$$

Empirical and Absolute Heat Current

Let us compare our results with the formulas of the thermodynamical theory of heat conduction. It is obvious that the balance equation of the internal energy in our case (one-dimensional, sourceless, stationary) is

$$\frac{dq}{dx} = 0,$$

that is, $q = $ constant. This is the same form as (8.14). However, it does not mean immediately that the parameter H can be interpreted as the wanted heat current q. Recall that the definition of H implies some arbitrariness (partly because L was arbitrary). Nevertheless, it follows that q must be a function of H since $q = $ constant and $H = $ constant determine the same family of characteristics, that is,

$$q = q(H).$$

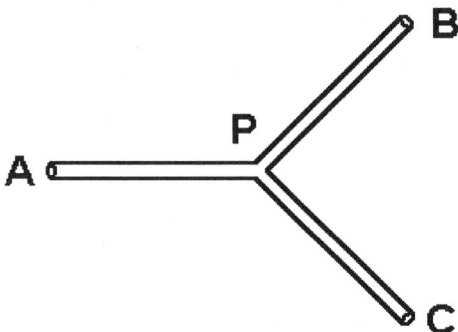

FIGURE 8.1. Branching.

The parameter H can be interpreted as the "empirical heat current," and its value can be determined from the temperature characteristics. But its scale is arbitrary in the same manner as that of empirical temperature and empirical entropy.

We can define the scale function $q(H)$ by using the additivity property for the true (absolute) heat current. To do this, let us consider three rods made of the same homogeneous material and having the same geometrical data. Let the three rods (A, B, C) have a common point P (Figure 8.1). The temperatures at the free ends are $T_A < T_B = T_C$. For this arrangement, the heat currents in the three rods should be

$$q_A = 2q_B = 2q_C. \tag{8.15}$$

It can be seen that this requirement guarantees the unity of the scale q apart from a constant multiplier [23], [37]. The parameter values H can be determined separately for the three rods. From the relation between them we can determine the absolute heat current. The existence of the absolute scale satisfying (8.15) is obvious from a physical point of view, however, the mathematical proof for its existence is still lacking.

Examples

Example 1. Let us assume that the temperature characteristics are linear, that is,

$$f(H, x + d) = (x + d)\frac{H}{L}.$$

So the empirical heat current is the slope, apart from a constant factor, and, furthermore,

$$T' = \frac{H}{L}. \tag{8.16}$$

Let us further assume that in the branching arrangement of Figure 8.1 we find that the slope in the rod B is just half of that in rod A:

$$H_A = 2H_B. \tag{8.17}$$

Since the absolute heat current q is a function of H, therefore applying (8.15), we get

$$q(2H_B) = 2q(H_B). \tag{8.18}$$

The only continuous solution of this functional equation is a homogeneous function, that is, the absolute heat current q is the same as H, apart from a constant factor, and we obtain the constitutive law from (8.16):

$$q = -KT',$$

with a positive constant K.

Example 2. The temperature characteristics are linear just as in Example 1, but let us assume that we have at the branching point

$$H_A = 2^{1/3} H_B$$

instead of (8.17). Now the following functional equation holds:

$$q(2^{1/3}H) = 2q(H).$$

Let us introduce a new function $p(H^3) = q(H)$. For this function p we get the functional equation of the type (8.18), therefore in this case the absolute heat current q is proportional to H^3. It can be seen immediately that the constitutive law is

$$q = -K(T')^3,$$

with a positive constant K.

Example 3. Let us assume that the temperature characteristics are hyperbolas

$$T(x) = f(H, x + d) = \frac{T_0}{1 + L(T_0 - H)/(x + d)H}. \tag{8.19}$$

It is obvious that by a suitable transformation $(\Theta = T/(T - T_0))$ this example can be reduced to the case of Example 1.

However, now we show a direct systematic way to introduce the heat current for this example. In this case, the admissible region of T is bounded below, $T > T_0$, so Postulate 5 does not hold. As we will see, this fact does not prevent us from deriving the absolute heat current.

Following the way detailed above we calculate the derivative $T'(x)$ from (8.19):

$$T'(x) = \frac{H(T_0 - T)^2}{T_0 L(T_0 - H)}. \tag{8.20}$$

216   8. Qualitative Properties of Conductive Heat Transfer

Let us further assume that in the branching arrangement of Figure 8.1 we find that between the slope in rod B and that in rod A, the following relation holds:
$$H_B = \frac{H_A T_0}{2T_0 - H_A}.$$

Since the absolute heat current q is a function of H, therefore applying (8.15), we get
$$q(H) = 2q\left(\frac{HT_0}{2T_0 - H}\right).$$

In order to solve this functional equation we introduce the new function $g(H/(H - T_0)) = q(H)$ (this formula was obtained by choosing appropriately the coefficients in a rational algebraic function of first degree). For this function g we get the functional equation of the type (8.18), therefore in this case the absolute heat current q is proportional to $H/(H - T_0)$:

$$q(H) = c \frac{H}{H - T_0}. \tag{8.21}$$

Using the formulas (8.20) and (8.21) we obtain the wanted constitutive relation
$$q = \frac{-KT'}{(T_0 - T)^2}$$
with a positive constant K. This quasilinear constitutive relation becomes linear in the temperature scale $\Theta = T/(T - T_0)$.

8.5 Properties of the Solution of the Linear Heat Equation

As we saw above, the temperature distribution function must satisfy some general requirements imposed by the physical nature of heat conduction. Now we list some of the qualitative properties that may have interest from a mathematical point of view.

Consider the following one-dimensional parabolic initial-boundary value problem:

$$\frac{\partial T}{\partial t} = D \frac{\partial^2 T}{\partial x^2}, \quad x \in (0, \pi), \ t > 0, \tag{8.22}$$
$$T(x, 0) = T_0(x), \quad x \in (0, \pi), \tag{8.23}$$
$$T(0, t) = T_1, \ T(\pi, t) = T_2, \quad t > 0, \tag{8.24}$$

where the following consistency conditions are assumed to be valid:
$$T_0(0) = T_1, \quad T_0(\pi) = T_2.$$

8.5. Properties of the Solution of the Linear Heat Equation

Here T_0 is the given initial heat distribution, $D = \text{constant} > 0$, the boundary condition in (8.24) is assumed to be constant and we applied a scaling in space to transform the interval $(0, L)$ to $(0, \pi)$.

The solution of this model $T(x,t)$ is defined on the domain $\Omega = [0, \pi] \times \mathbb{R}_0^+$ and has the following properties:

Property 5.1. If $T_0 \geq 0$, $T_1 \geq 0$, and $T_2 \geq 0$, then $T \geq 0$ on Ω.

Property 5.2. If T_0 is a monotonically increasing (decreasing) function then, for any fixed $t^* > 0$, the function $T(x,t^*)$ is also of the same kind.

Property 5.3. If T_0 is concave then, for any fixed $t^* > 0$, the function $T(x,t^*)$ is also concave.

Property 5.4. If T_0 is concave then, for any fixed $x^* \in (0, \pi)$, the function $T(x^*,t)$ is monotonically decreasing. Moreover, this property is valid only in the case when T_0 is concave.

Property 5.5. For any t_1, t_2 such that $0 \leq t_1 \leq t_2$ the inequality

$$\|T(\cdot, t_1)\| \geq \|T(\cdot, t_2)\|$$

holds where $\|\cdot\|$ denotes the L_2-norm over the interval $[0, \pi]$.

Property 5.6. By letting $t \to \infty$, the function $T(x,t)$ tends to the solution of the corresponding elliptic problem (to the so-called stationary solution). Moreover, the convergence takes place without oscillation and the convergence rate in the L_2-norm is the same as the convergence rate of the function $\exp(-\lambda_1 t)$ to zero. (Here $-\lambda_1$ denotes the greatest (negative) eigenvalue of the elliptic operator $D\,(\partial^2 T/\partial x^2)$.)

Property 5.7. Let us define the function

$$E(t) = \int_0^\pi T(x,t)\,dx. \tag{8.25}$$

If $T_0 \geq 0$ and $T_1 = T_2 = 0$, then $E(t)$ is a monotonically decreasing function. This yields the *energy decreasing property* of the solution.

Property 5.8. If we define the function

$$F(x) = \int_0^\infty T(x,t)\,dt, \tag{8.26}$$

then the definition of the function $F(x)$ is correct, that is, the improper integral on the right-hand side always exists and is finite. This is called *finitely summable in the time property* of the solution.

For the details and proofs we refer the reader to [39], [40], [41], [42].

218   8. Qualitative Properties of Conductive Heat Transfer

8.6 Numerical Solution of the Linear Heat Equation

In this section we deal with the basic numerical methods of problem (8.22), (8.23), (8.24).

8.6.1 *Solution of the Problem by the Fourier Method*

Our goal is to determine the unknown function of the problem (8.22), (8.23), (8.24).

We assume, for simplicity, that the given function T_0 is sufficiently smooth, such that there exists a unique solution $T : \Omega \to \mathbf{R}$. Considering the simple *homogeneous boundary condition*

$$T(0,t) = T(\pi,t) = 0, \quad t > 0. \tag{8.27}$$

the solution can be easily determined by use of the Fourier method (i.e., [41], [43]) as follows:

$$T(x,t) = \sum_{m=-\infty}^{+\infty} a_m \exp(imx - m^2 Dt), \tag{8.28}$$

where

$$a_m = \frac{1}{2\pi} \int_{-\pi}^{\pi} T_0(x) \exp(-imx) \, dx. \tag{8.29}$$

(In (8.29) the function T_0 is extended to the interval $[-\pi, 0]$ by the definition $T_0(x) = -T_0(-x)$.) Consequently, in order to determine the value of the solution at some fixed point (x,t), we have to summarize an infinite series of the form (8.28) with the coefficients from (8.29). Clearly this approach is not convenient for practical computation, even for the simplest model. Therefore, as usual, we apply numerical methods. (For simplicity, we shall further restrict our numerical considerations to the homogeneous boundary conditions.)

8.6.2 *Finite Difference Method*

The general idea of the finite difference method is the following. First, we replace the continuous domain of the original problem Ω with a *discrete mesh*. This means that we define the set of mesh points by

$$\Omega_{h,\tau} = \{(x_n, t_j), \, x_n = nh, \, n = 0, 1, \ldots, N+1, \, t_j = j\tau, \, j = 0, 1, \ldots, \}, \tag{8.30}$$

where the step-size of the space discretization is defined by $h = \pi/(N+1)$ and $\tau > 0$ means the time discretization parameters. Such a mesh $\Omega_{h,\tau} \subset \Omega$ is called *uniform*.

8.6. Numerical Solution of the Linear Heat Equation

Then we replace the continuous problem with the discrete problem, namely, the differential equation is replaced by the finite difference equations and the boundary and initial conditions are approximated, as well. Mathematically, this means the approximation of the continuous solution at the mesh points.

For instance, if we apply the well-known explicit method to the approximation, then we obtain the discrete problem of the form

$$\frac{u_n^{j+1} - u_n^j}{\tau} = D \frac{u_{n+1}^j - 2u_n^j + u_{n-1}^j}{h^2}, \quad n = 1, \ldots, N, \ j = 0, 1, \ldots \quad (8.31)$$

$$u_n^0 = T_0(x_n), \quad n = 1, 2, \ldots, N, \quad (8.32)$$

$$u_0^j = u_{N+1}^j = 0, \quad j = 1, 2, \ldots, \quad (8.33)$$

where u_n^j means approximation to the value $T(x_n, t_j)$.

Consequently, using the above two steps, we can define an approximation to the true solution at the mesh points of a *fixed* mesh $\Omega_{h,\tau}$. Let us denote by $u_{h,\tau}$ this numerical solution. The basic question of the numerical methods can be formulated as follows. Refining the mesh we construct a sequence of the meshes. On each mesh we define a numerical solution. Under which condition does the sequence of the numerical solutions converge to the solution?

That is, our aim is to define the conditions under which the relation $\Omega_{h,\tau} \to \Omega$ implies $u_{h,\tau} \to T$. This is the question of *convergence*. At a glance it seems there is no additional condition: by $\tau \to 0$ and $h \to 0$ the relations (8.31), (8.32), (8.33) "converge" to the relations (8.22), (8.23), (8.27). So, we can expect that the solutions are also convergent.

Unfortunately, that is not the case. Let us consider the case when, for the sequence of the meshes $\Omega_{h,\tau}$, the ratio

$$q = \frac{D\tau}{h^2} \quad (8.34)$$

is constant. We examine two cases:

- $q = 5/9$; and
- $q = 5/11$.

Doing a numerical experiment we can see that for the first case the numerical solutions show incorrect behavior on each mesh: as the time approaches infinity the numerical solution strongly oscillates and the amplitudes grow to infinity! Therefore, obviously, in this case there is no convergence. However, for the second case the numerical solution corresponds well to the continuous solution and, by refining the mesh, we have convergence, too. (For the details see [44].)

By direct verification we can easily check that the solution of the discrete problem (8.31), (8.32), (8.33) has the form

$$u_j^n = \sum_{m=-\infty}^{+\infty} a_m (\xi(m))^n \exp(imjh), \qquad (8.35)$$

where

$$\xi(m) = 1 - \frac{2D\tau}{h^2}(1 - \cos mh) \qquad (8.36)$$

and a_m is defined by (8.29).

Comparing (8.28) and (8.35) we can see that the approximation property of the discrete problem (8.31), (8.32), (8.33) depends on the approximation of the expression $\exp(-m^2 D\tau)$ by $\xi(m)$. The Taylor series of these expressions are

$$\xi(m) = 1 - m^2 D\tau + \tfrac{1}{12} m^4 D\tau h^2 - \cdots,$$
$$\exp(-m^2 D\tau) = 1 - m^2 D\tau + \tfrac{1}{2} m^4 D^2 \tau^2 - \cdots.$$

Obviously, by decreasing the step-sizes h and τ, the above numbers are arbitrarily close. This is called *approximation*, that is, it means that in the case $h, \tau \to 0$, the value $\xi(m)$ approximates $\exp(-m^2 D\tau)$. On the other hand, let us notice that for any *fixed* h, τ, and sufficiently large integers $m \in N$, they can be very different! Moreover, the numerical solution remains bounded only in the case

$$\max |\xi(m)| \leq 1. \qquad (8.37)$$

Otherwise, the *global error* $|u_j^n - u(jh, n\tau)|$ does not remain bounded. We say that the numerical method is *numerically stable* if the global error is bounded.

That is, by any approaching $h, \tau \to 0$, we always have the approximation, but the numerical stability, and, consequently, the convergence, is only under a stricter condition, namely, under condition (8.37).

Theorem 6.1. *Assume that the Fourier series of the function T_0 is absolutely convergent. Then the discretization (8.31), (8.32), (8.33) is numerically stable if and only if condition (8.37) is satisfied.*

Proof. We prove first a simple relation for the Fourier coefficients. The expansion into Fourier series of the function T_0 has the form

$$T_0(x) = \sum_{m=-\infty}^{+\infty} a_m \exp(imx), \qquad (8.38)$$

where a_m are defined in (8.29). Due to our assumption the series (8.38) are absolutely convergent, that is, the sum $\sum_{m=-\infty}^{+\infty} |a_m|$ is finite.

8.6. Numerical Solution of the Linear Heat Equation

Now we prove the statement of the theorem. Obviously the continuous solution $T(x,t)$ of the form (8.28), (8.29) is bounded on Ω. Therefore the global error is bounded if and only if the numerical solution u_j^n is bounded. If the condition (8.37) is satisfied then by virtue of (8.35) we have the estimation

$$|u_j^n| = |\sum_{m=-\infty}^{+\infty} a_m(\xi(m))^n \exp(imjh)| \leq \sum_{m=-\infty}^{+\infty} |a_m||\xi(m)|^n \leq \sum_{m=-\infty}^{+\infty} |a_m|.$$

Since the right-hand side is bounded, therefore the first part of the theorem is proved.

To prove the necessity of condition (8.37) let us choose $u_0(x) = \sin mx$ for some fixed $m \in \mathbb{N}$. Clearly, in this case,

$$u_j^n = (\xi(m))^n \sin mjh,$$

which does not remain bounded in the case of

$$|\xi(m)| > 1.$$

□

In practice, to prove the convergence is a difficult task. Usually, the investigation of the approximation and the stability is easier. Therefore the *basic theorem of the numerical investigation* is very useful, which states that, in the case of approximation, stability is a necessary and sufficient condition of the convergence.

As an easy computation shows, condition (8.37) implies the condition

$$q = \frac{D\tau}{h^2} \leq \frac{1}{2}, \qquad (8.39)$$

which is consequently the necessary and sufficient condition of the convergence of the explicit scheme (8.31), (8.32), (8.33).

The condition of stability (8.39) is very restrictive in the practical computation: by refining the mesh the decrease of the space discretization step h requires too many new time levels. Therefore it is desirable to get out of such a condition. To this aim, we define a class of numerical methods, called θ-methods.

Assume that $\theta \in [0,1]$ is a fixed given parameter. We consider the discretization of the form

$$\frac{u_n^{j+1} - u_n^j}{\tau} = \theta D \frac{u_{n+1}^{j+1} - 2u_n^{j+1} + u_{n-1}^{j+1}}{h^2} + (1-\theta) D \frac{u_{n+1}^j - 2u_n^j + u_{n-1}^j}{h^2},$$

$$n = 1, 2, \ldots, N, \quad j = 0, 1, \ldots, \qquad (8.40)$$

$$u_n^0 = T_0(x_n), \quad n = 1, 2, \ldots N, \qquad (8.41)$$

$$u_0^j = u_{N+1}^j = 0, \quad j = 1, 2, \ldots. \qquad (8.42)$$

Obviously, in the case $\theta = 0$, the θ-method results in the explicit method considered above. For $\theta = 1$ and $\theta = 0.5$ the method is called the implicit Euler method and the Crank–Nicolson method, respectively.

The numerical solution of the discrete problem (8.40), (8.41), (8.42) has the form (8.35), where now

$$\xi(m) = \frac{2 - (1-\theta)q(1 - \cos mh)}{2 + \theta q(1 - \cos mh)}. \tag{8.43}$$

As we already know, the method is numerically stable (and consequently convergent) if and only if condition (8.37) is satisfied. An easy computation shows that this condition is equivalent to the condition

$$q \leq \frac{1}{2(1 - 2\theta)} \quad \text{for } \theta \in [0, 1),$$

$$q \text{ is any number} \quad \text{for } \theta \in [0.5, 1]. \tag{8.44}$$

That is, for $\theta \in [0.5, 1]$, the θ-method is convergent without any restriction with respect to the number q. Such methods are called *unconditionally stable methods*. If there is any restriction for the number q, then the method is called a *conditionally stable method*. (For more details about stability and convergence, see [45] and [46].)

We note that the discrete problem (8.40), (8.41), (8.42) yields a system of linear algebraic equations at each time level. Namely, introducing the notation $u^j \in \mathbb{R}^N$ for the unknown values at the time level $t = t_j$, the problem can be rewritten in algebraic form

$$(I - \theta q Q)u^{j+1} = (I + (1-\theta)q Q)u^j, \quad j = 0, 1, 2, \ldots, \tag{8.45}$$

with given u^0. Here I is the unit matrix and $Q = \text{tridiag}(1, -2, 1)$ denotes the tridiagonal matrix in $\mathbb{R}^{N \times N}$, respectively.

8.6.3 Galerkin Finite Element Method

Another approach to obtaining the discretization is the *Galerkin finite element method*. The basic idea is the following. We seek the weak solution of the problem (8.22), (8.23), (8.27), that is, such a function $T(x, t)$, which belongs to the Sobolev space $H_0^1(0, \pi)$ [47] for any fixed $t \geq 0$ and satisfies the equation

$$\int_0^\pi \left(\frac{\partial T}{\partial t} - D \frac{\partial^2 T}{\partial x^2} \right) v(x) \, dx = 0$$

for all $v \in H_0^1(0, \pi)$ and fixed t. Using integration by parts this equation is equivalent to the equation

$$\int_0^\pi \left(\frac{\partial T}{\partial t} v + D \frac{\partial T}{\partial x} \frac{dv}{dx} \right) dx = 0 \tag{8.46}$$

8.6. Numerical Solution of the Linear Heat Equation

for all $v \in H_0^1(0, \pi)$ and fixed t. This serves as the basic equation of the discretization.

Further, we apply the Galerkin principle. First, we define an N-dimensional subspace $S_h \subset H_0^1(0, \pi)$. Here h denotes the parameter of the space, usually defined by the relation $h = 1/N$. The relation $h \to 0$ means that the dimension of the spaces S_h approaches infinity, as well. Then we seek the function $T_h(x, t)$ such that:

- $T_h \in S_h$ for all fixed $t \geq 0$; and
- the equation

$$\int_0^\pi \left(\frac{\partial T_h}{\partial t} v_h + D \frac{\partial T_h}{\partial x} \frac{d v_h}{d x} \right) dx = 0 \qquad (8.47)$$

holds for all $v_h \in S_h$ and fixed t.

Assume that the set of functions $\phi_1, \phi_2, \ldots, \phi_N$ forms a basis in S_h. Then the unknown discretization T_h has the form

$$T_h(x, t) = \sum_{n=1}^{N} \alpha_n(t) \phi_n(x), \qquad (8.48)$$

where, on the basis of (8.47), the unknown coefficient functions $\{\alpha_n, n = 1, 2, \ldots, N\}$ satisfy the conditions

$$\int_0^\pi \left(\sum_{n=1}^{N} \frac{d\alpha_n}{dt} \phi_n \phi_k + D\alpha_n \frac{d\phi_n}{dx} \frac{d\phi_k}{dx} \right) dx = 0 \qquad (8.49)$$

for all $k = 1, 2, \ldots, N$.

Clearly (8.49) yields a linear system of ordinary differential equations of first order. In order to determine the solution we add the initial condition using the initial function T_0 from (8.23). If T_0 is a continuous function, we take

$$\alpha_n(0) = T_0(x_n), \quad n = 1, 2, \ldots, N. \qquad (8.50)$$

Now (8.49) and (8.50) result in a Cauchy problem that can be rewritten in matrix form

$$M \frac{d\alpha}{dt} + Q_G \alpha = 0, \qquad (8.51)$$

where M and Q_G are given matrices in $\mathbb{R}^{N \times N}$ with the elements

$$m_{n,k} = \int_0^\pi \phi_n \phi_k \, dx$$

and

$$q_{n,k} = \int_0^\pi D \frac{d\phi_n}{dx} \frac{d\phi_k}{dx} dx,$$

respectively. For the solution of the Cauchy problem (8.51), (8.50) we apply the numerical method, namely, the θ-method:

$$M \frac{\alpha^{j+1} - \alpha^j}{\tau} + Q_G(\theta \alpha^{j+1} + (1-\theta)\alpha^j) = 0. \tag{8.52}$$

We remark that (8.52) yields a system of linear algebraic equations. In order to solve them, we want to turn it into the simplest possible form. Therefore it is suitable to apply the well-known linear spline basis functions (linear finite element basis) for the functions $\phi_1, \phi_2, \ldots, \phi_N$. In this case, the solvable system has a simple structure, namely, the matrix of the system is tridiagonal with equal elements in the same diagonals. (This type of matrix is called the *uniformly continuant matrix*.) For the above choice of basis functions we have $\phi_n(x_i) = \delta_{i,n}$, where $\delta_{i,n}$ is the Kronecker symbol. Due to (8.48), $T_h(x_n, t) = \alpha_n(t)$, therefore for this case the components of the vector α^j are the approximations of the solution at the jth time level of the mesh $\Omega_{h,\tau}$. Moreover, in (8.52) the matrices have the form

$$M = \frac{h}{6} \text{tridiag}(1, 4, 1), \quad Q_G = \frac{D}{h} \text{tridiag}(1, -2, 1).$$

(For the details, see [48] and [49].)

8.7 Properties and Their Preservation for the Discretization

As we saw, the numerical solution of the heat equation leads to the solution of the sequence of systems of linear algebraic equations of the special form

$$X_1 u^{j+1} = X_2 u^j, \quad j = 0, 1, 2, \ldots, \tag{8.53}$$

with given u^0, where the unknown vector $u^j \in \mathbb{R}^N$ yields the approximation to the solution $T(x, j\tau)$ at the mesh points x_n ($n = 1, 2, \ldots, N$). Here X_1 and X_2 are given uniformly continuant tridiagonal matrices of the form

$$X_1 = I + \alpha Q, \quad X_2 = I - \beta Q, \tag{8.54}$$

where we have used the notations

$$\alpha = -\sigma + \theta q, \quad \beta = \sigma + (1-\theta)q, \tag{8.55}$$

with certain given values of σ. The number q and matrix Q are defined in (8.34) and (8.45), respectively. (In the sequel the dependence of the matrices and vectors on dimension N will not be denoted.) Obviously, the case $\sigma = 0$ corresponds to the *finite difference method* and the case $\sigma = \frac{1}{6}$ results in the *Galerkin finite element method*. Regarding the numerical

8.7. Properties and Their Preservation for the Discretization

methods, in addition to the convergence, it is no less important to require the preservation of the discrete analogues of the basic qualitative properties of the continuous solution mentioned above at a certain fixed numerical solution (or at all of them). As we will see, it results in new conditions for the choice of the parameters. They appear to be stronger, in general.

8.7.1 Qualitative Properties of the Numerical Solution

First of all we formulate the *discrete qualitative properties* to the discretization method defined by (8.53).

Property 7.1. The method is called l_2-*contractive* if the relation

$$\|u^{j+1}\| \leq \varrho\|u^j\|$$

holds for any $j \in \mathbb{N}$ with some fixed $\varrho \in [0,1)$, where $\|\cdot\|$ denotes the usual Euclidean norm in $\mathbb{R}^N$.

Property 7.2. Clearly the l_2-contractivity yields the estimation

$$\|u^j\| \leq \varrho^j\|u^0\|$$

for any $j \in \mathbb{N}$. Besides this geometrical convergence to the limit vector (called in the literature *linear convergence*), the discretization method usually should also have an exponential convergence of the form

$$\|u^j\| \leq \exp(-\lambda_1^{(n)}qj)\|u^0\|$$

for all $j \in \mathbb{N}$, which is called *regular exponential convergence*. (Here $-\lambda_1^{(n)}$ denotes the greatest (negative) eigenvalue of the discrete elliptic operator (matrix) DQ.)

Property 7.3. The discretization method is called *nonnegative* if, for all nonnegative vectors u^0, the iteration process (8.53) results in a nonnegative vector sequence $\{u^j, j \in \mathbb{N}\}$, i.e., $u^0 \geq 0$ implies the relation $u^j \geq 0$ for all $j \in \mathbb{N}$.

Property 7.4. A vector $u \in \mathbb{R}^n$ is called *concave* (or *convex*) if the relations

$$Qu \geq 0$$

or

$$Qu \leq 0$$

hold. The discretization method (8.53) is called *shape preserving* if the concavity or convexity of the initial vector implies the same for every vector obtained by the iteration.

Property 7.5. The discretization method is said to be *conserving monotonicity with respect to the space variable* if the relation $(u^j)_1 \leq (u^j)_2 \leq$

$\ldots \leq (u^j)_N$ (or $(u^j)_1 \geq (u^j)_2 \geq \cdots \geq (u^j)_N$) implies the same for the components of vector u^{j+1}, too.

Property 7.6. The discretization method (8.53) is called *coordinatewise monotonically decreasing* if relation

$$u^{j+1} \leq u^j$$

holds for all $j = 0, 1, 2, \ldots$.

Property 7.7. For the vector sequence $\{u^j, j \in \mathbb{N}\}$, produced by the discretization (8.53), let us define the numbers $E_d(j)$ by the formula

$$E_d(j) = \sum_{n=1}^{N} u_n^j, \quad j = 1, 2, \ldots .$$

The discretization method (8.53) preserves the *energy decreasing property* if in the case of any nonnegative vector u^j the relation $E_d(j+1) \leq E_d(j)$ holds.

Property 7.8. Assume that the number $F_d(n)$ is defined by the formula

$$F_d(n) = \sum_{j=1}^{\infty} u_n^j, \quad n = 1, 2, \ldots, N,$$

The discretization method (8.53) preserves the *finitely summable-in-time property* if $F_d(n)$ is finite for all $n = 1, 2, \ldots, N$.

8.7.2 Conditions for the Preservation of Qualitative Properties

In the following we formulate the conditions under which the convergent numerical methods preserve the above qualitative properties. To this end we rewrite the method (8.53) in a more suitable form

$$u^{j+1} = X u^j, \quad j = 0, 1, 2, \ldots, \tag{8.56}$$

where we used the notation

$$X = X_1^{-1} \cdot X_2.$$

1. The method (8.56) is convergent if and only if the eigenvalues of the matrix X (the so-called *iteration matrix*) in absolute value are less than 1. Since the corresponding eigenvalues can be defined *exactly*, we are therefore able to formulate the *exact bound of the convergence* which guarantees the

8.7. Properties and Their Preservation for the Discretization

convergence for all norms, independent of the dimension of the matrix X [50]:

$$q \leq \frac{1-4\sigma}{2(1-2\theta)}, \quad \text{when } \theta \in [0, 0.5),$$
$$q \text{ is arbitrary} \quad \text{when } \theta \in [0.5, 1]. \tag{8.57}$$

Consequently, for any σ, the schemes with $\theta \in [0, 0.5)$ are conditionally stable and for the values $\theta \in [0.5, 1]$ they are unconditionally stable. This restriction for the considered discretization methods yields the following bounds: the finite difference method is convergent for all dimensions if and only if the condition

$$q \leq \frac{1}{2(1-2\theta)}$$

is satisfied. For the Galerkin finite element method the corresponding condition is

$$q \leq \frac{1}{6(1-2\theta)}.$$

We remark that the above bounds are valid for all dimensions. For any fixed dimension these bounds can be slightly increased, but they are very close to (8.57), already for comparatively low dimensions. As a consequence of the l_2-norm property, under the above convergence's condition, the method is l_2-contractive, too. That is, the considered discretization methods can be convergent and contractive in the l_2-norm only at the same time. We mention, however, that this is true *only* for the l_2-norm and for the equation of the form (8.22)! For instance, under the bounds (8.57) the method is also convergent in the maximum norm, but, it is not contractive. Moreover, for the equation, also having a conductive term, we can define such a discretization method which is convergent, but not contractive, in the l_2-norm.

2. The detailed analysis of the eigenvalues of the matrix X also gives the possibility to formulate the *exact* condition of the regular exponential convergence [51]: the discretization method (8.56) is regularly exponentially convergent if and only if the condition

$$q \leq \begin{cases} \frac{1-2\sigma}{2(1-\theta)} & \text{when } \theta \in [0, \frac{1}{2}+\sigma], \\ \frac{2\sigma}{2\theta-1} & \text{when } \theta \in (\frac{1}{2}+\sigma, 1], \end{cases}$$

holds. For the special methods we obtain the following conditions. The finite difference method is regularly exponentially convergent for all dimensions if and only if the condition

$$q \leq \frac{1}{2(1-\theta)}, \quad \theta \in [0, 0.5],$$

is satisfied. For the Galerkin finite element method the condition is

$$q \leq \begin{cases} \dfrac{1}{3(1-\theta)} & \text{when } \theta \in [0, \tfrac{2}{3}], \\ \dfrac{1}{3(2\theta-1)} & \text{when } \theta \in (\tfrac{2}{3}, 1]. \end{cases}$$

We emphasize that for the finite difference method the regular exponential convergence can be guaranteed *only* for the values $\theta \in [0, 0.5]$! For the other values of θ we lose this property, that is, the rate of convergence is lower, as well.

3. One of the most important qualitative properties is the nonnegativity preservation of the discretization method. Obviously, the discretization method (8.56) is nonnegativity preserving if and only if the matrix X is nonnegative, that is, its elements are nonnegative. Due to the special structure of the matrices X_1 and X_2 in (8.54), we can represent the matrix X in the *explicit form*:

$$X = \begin{cases} -\dfrac{\beta}{\alpha}I + \dfrac{\alpha+\beta}{\alpha}X_1^{-1}, & \text{if } \alpha \neq 0, \\ I - \beta Q, & \text{if } \alpha = 0. \end{cases} \qquad (8.58)$$

Since the inverse of the uniformly tridiagonal matrices can be computed exactly, therefore, taking into consideration the definition of the parameters (8.55), we can formulate the *exact* bound of the nonnegativity preservation. (For the details, see [50], [52].) The discretization method (8.56) with $\theta \in (0,1)$ is nonnegativity preserving for all dimensions if and only if the condition

$$\frac{\sigma}{\theta} \leq q \leq \frac{\tfrac{1}{2}-\sigma}{1-\theta}, \quad \theta \geq 2\sigma, \qquad (8.59)$$

is satisfied. For the case $\sigma = \theta = 0$ the condition is $q \leq 0.5$. When $\theta = 1$, then the condition is $q \geq \sigma$.

For the considered discretization methods condition (8.59) yields the following bounds. The finite difference method is nonnegativity preserving for all dimensions if and only if the condition

$$q \leq \begin{cases} 0.5 & \text{when } \theta = 0, \\ \dfrac{1}{2(1-\theta)} & \text{when } \theta \in (0,1), \end{cases}$$

holds. For the method $\theta = 1$ (the so-called *backward Euler method*) there is no restriction. In the Galerkin finite element method for the choice of the number q the bound is

$$\frac{1}{6\theta} \leq q \leq \frac{1}{3(1-\theta)}, \quad \theta \in [\tfrac{1}{3}, 1].$$

8.7. Properties and Their Preservation for the Discretization

This bound shows that for the values $\theta \in [0, \frac{1}{3})$ the Galerkin finite element method is not nonnegativity preserving with any choice of q. Moreover, as the bound (8.59) implies, for the case $\sigma \neq 0$, there always exist both *lower* and *upper* bounds to the choice of q. That is, to the preservation of nonnegativity the number q cannot be too small. For instance, in the Galerkin finite element method with $\theta = 0.5$ we have the restriction $\frac{1}{2} \leq q \leq \frac{2}{3}$.

Finally, we remark that (8.59) is the bound of the preservation of the nonnegativity for *all* dimensions. With an increase of the dimension the upper bound can also be increased. However, we are able to formulate the necessary conditions of the preservation of the nonnegativity [50]. For the finite difference method this restriction has a comparatively simple form: if and only if the relation

$$q < \frac{1 - \sqrt{1-\theta}}{\theta(1-\theta)}, \quad \theta \in (0,1),$$

holds, then there exists an N_0 such that the finite difference method is nonnegativity preserving for all dimensions $N \geq N_0$.

4. Under the condition of nonnegativity preservation the discretization method (8.53) is shape preserving, too. That is, the nonnegativity preserving property plays a basic role in the preservation of the convexity or concavity of the initial function. (For the details, see [50], [53].)

5. The same result is true for the discretization method in the preservation of the property of conserving monotonicity with respect to the space variable: under the condition of nonnegativity preservation this property is also preserved [54].

6. The condition of the coordinatewise monotonically decreasing property is interesting. If the inverse of the matrix X_1 is nonnegative (that is, X_1 is a *monotone matrix*), then the discretization method (8.53) is coordinatewise monotonically decreasing. Due to (8.58), the monotonicity of X_1 is a necessary condition of the nonnegativity of the matrix X. Therefore, under the condition of nonnegativity preservation, the coordinatewise monotonically decreasing property is preserved, too [53]. On the other hand, we require this property *only* on the concave vectors. As an easy example shows [50], it results in the requirement of the nonnegativity of the matrix X_1. Consequently, the matrix X_1 should be monotone and nonnegative at the same time. Clearly, it is possible *only* in the case when X_1 is a diagonal matrix. Therefore, let us assume that $\sigma > 0$, $\theta \geq 2\sigma$, and $q = \sigma/\theta$. Then the discretization method is monotonically decreasing on the concave vectors, only.

7. For the energy decreasing property we can point out the basic role of the nonnegativity property, too [55]. Namely, if the discretization method (8.53) is nonnegative, then it is energy decreasing, too.

8. It is remarkable that for the preservation of the finitely summable-in-time property the weakest possible condition is already enough: under the condition of convergence (that is, (8.57)) the method preserves it.

Finally, we make some remarks:

- In this chapter we considered the case when the number q is constant. However, this number can be dependent on the number of the space division N, too. In this case, for the critical values of the numbers q_N, we can formulate some *sufficient* condition, too [50].

- Basically we did not consider the qualitative properties connected with the *maximum norm*. For the details, we refer the reader to [44], [55], [56].

- In this chapter we did not analyze the discretization methods to the heat equations with convection. For such problems the qualitative properties and their preservation are more complicated [45], [46], [49].

- The qualitative properties and their preservation can be investigated for (8.22) with some source function on the right-hand side, too. (See [50], [51], [52].) For this case, the bounds remain the same as for the homogeneous case.

- The problem of the choice of *optimal mesh* is considered in [57]. Here under the optimal mesh we should understand the meshes on which the discretization method results in maximal accuracy, within fixed arithmetic operation. This problem is closely related to the *schemes with higher accuracy*, which is particulary investigated in [58].

8.8 Temperature Waves

8.8.1 *Shape Preserving Property*

The essence of the so-called *temperature waves* [1] can be illustrated with a very simple example. Let the temperature at the end of a semi-infinite rod vary sinusoidally in time. Then the temperature of any internal point depends on time–after the decay of the transient terms–also sinusoidally with the same period, and with an exponentially decreasing amplitude and linearly increasing phase delay. Since the phase delay depends on the period, too, in the case of a general (i.e., nonharmonic) periodic signal, dispersion takes place, that is, the signal will be distorted inside the body. So, sinusoidal signals inserted at the boundary are preserved, while the nonharmonic signal are forms are not preserved generally. This fact offers an experimental test of the Fourier differential equation: this equation fails if the sinusoidal signal distorts.

On the other hand, the shape preserving property shown by sinusoidal signals, also occurs in the case of linear signals [59, p. 17].

A question arises here: Is there any other time-dependence for which the shape preserving property is valid? We formulated this problem in mathematical terms [60].

Let us consider the following boundary value problem:

$$\frac{\partial T}{\partial t} - D\,\Delta T = f, \tag{8.60}$$

$$a(\mathbf{r},t)T(\mathbf{r},t) + b(\mathbf{r},t)\,(\partial T/\partial n) = Q(\mathbf{r})\Phi(t) \text{ for all } t > 0, \text{ for all } \mathbf{r} \in \partial V, \tag{8.61}$$

where $V \subset \mathbf{R}^3$ is a bounded domain with $\partial V \in C^{2+\alpha}$ for some $\alpha \in (0,1)$, $a, b, Q \in C^{1+\alpha}(\partial V)$, and $D > 0$. We assume that Q and $a^2 + b^2$ do not vanish on ∂V.

Now we introduce two essential definitions.

Definition 1. We call a solution $T \in C^2(\mathbf{R}_+ \times V) \cap C([0,+\infty)) \times \overline{V})$ of (8.60) a *boundary following solution* (BFS) if there are space functions $T_0, \Psi \in C^2(V) \cap C(\overline{V})$, and $\Phi \in C^2(\mathbf{R})$ such that the solution has the form

$$T(\mathbf{r},t) = T_0(\mathbf{r})\Phi(t - \Psi(\mathbf{r})). \tag{8.62}$$

We call a function $\Phi : \mathbf{R} \to \mathbf{R}$ a *shape preserving signal form* (SPSF) if there exists a BFS of the form (8.62).

Note that the BFS satisfies a special initial condition.

8.8.2 Classification of SPSFs

Now we determine the general form of the SPSFs. More exactly, we shall give a class of functions that contain the shape preserving signal forms (that is, we establish a necessary condition for a function to be an SPSF). Later we shall give sufficient conditions for functions of these classes to be an SPSF.

In order to get the form of the SPSFs, let us insert (8.62) into (8.60). We get

$$-D\,\Delta T_0(\mathbf{r})\Phi(t - \Psi(\mathbf{r}))$$
$$+[T_0(\mathbf{r}) + 2D(\nabla T_0(\mathbf{r})\cdot\nabla\Psi(\mathbf{r})) + DT_0(\mathbf{r})\,\Delta\Psi(\mathbf{r})]\Phi'(t-\Psi(\mathbf{r}))$$
$$- DT_0(\mathbf{r})(\nabla\Psi(\mathbf{r})\cdot\nabla\Psi(\mathbf{r}))\Phi''(t-\Psi(\mathbf{r})) = 0.$$

Fixing the vector $\mathbf{r}$, and introducing the new variable $u = t - \Psi(\mathbf{r})$, we get a second-order linear ordinary differential equation with constant coefficients for the function Φ:

$$A\Phi(u) + B\Phi'(u) + C\Phi''(u) = 0, \tag{8.63}$$

where

$$A = -D\, \Delta T_0(\mathbf{r}),$$
$$B = T_0(\mathbf{r}) + 2D(\nabla T_0(\mathbf{r}) \cdot \nabla \Psi(\mathbf{r})) + DT_0(\mathbf{r})\, \Delta\Psi(\mathbf{r}),$$
$$C = -DT_0(\mathbf{r})(\nabla\Psi(\mathbf{r}) \cdot \nabla\Psi(\mathbf{r})).$$

Thus if Φ is an SPSF, then it is the solution of (8.63), hence it has one of the following forms:

I. $\Phi(u) = c_1 e^{\mu_1 u} + c_2 e^{\mu_2 u}$, where $\mu_1 \neq \mu_2$,   (8.64)

II. $\Phi(u) = c_1 e^{\mu u} + c_2 u e^{\mu u}$,   (8.65)

III. $\Phi(u) = e^{\mu u}(c_1 \cos\omega u + c_2 \sin\omega u)$,   (8.66)

where $c_1, c_2, \mu_1, \mu_2, \mu \in \mathbf{R}$ and $\omega > 0$ are constants.

Now we give sufficient conditions for a function Φ in the first and second class to be an SPSF.

In case I, let us substitute the function (8.64) into (8.62). We get

$$T(\mathbf{r}, t) = c_1 T_1(\mathbf{r}) e^{\mu_1 t} + c_2 T_2(\mathbf{r}) e^{\mu_2 t}, \qquad (8.67)$$

where

$$T_i(\mathbf{r}) = T_0(\mathbf{r}) e^{-\mu_i \Psi(\mathbf{r})} \quad \text{for } i = 1, 2. \qquad (8.68)$$

Let us write (8.67) into (8.60) and fix the value of $\mathbf{r}$. Using the independence of the functions $e^{\mu_1 t}$ and $e^{\mu_2 t}$ we get the following boundary value problems for the functions T_i:

$$D\, \Delta T_i(\mathbf{r}) - \mu_i T_i(\mathbf{r}) = 0 \quad \text{for } i = 1, 2,$$
$$a(\mathbf{r}) T_i(\mathbf{r}) + b(\mathbf{r}) \frac{\partial T_i}{\partial n} = Q(\mathbf{r}) \quad \text{for } i = 1, 2 \text{ for all } \mathbf{r} \in \partial V.$$

If $\mu_1 > \mu_2 \geq 0$, then according to Theorem 6.31 in [61] this system has unique solutions T_1 and T_2. Due to the maximum principle and the boundary point lemma [62] these are positive in $\overline{V}$. These functions yield the BFS because from (8.68) for the functions Ψ and T_0 we get

$$\Psi(\mathbf{r}) = \frac{1}{\mu_1 - \mu_2} \ln \frac{T_2(\mathbf{r})}{T_1(\mathbf{r})},$$
$$T_0(\mathbf{r}) = T_1(\mathbf{r}) e^{\mu_1 \Psi(\mathbf{r})}.$$

Thus we have proved that if $\mu_1 > \mu_2 \geq 0$, then the function $\Phi(u) = c_1 e^{\mu_1 u} + c_2 e^{\mu_2 u}$ is an SPSF for every $c_1, c_2 \in \mathbf{R}$ and each of these Φ's determines a BFS.

In case II, let us substitute the function (8.65) into (8.62). We get

$$T(\mathbf{r}, t) = T_1(\mathbf{r}) e^{\mu t} + T_2(\mathbf{r}) t e^{\mu t}, \qquad (8.69)$$

where

$$T_1(\mathbf{r}) = T_0(\mathbf{r})e^{-\mu\Psi(\mathbf{r})}(c_1 - c_2\Psi(\mathbf{r})), \qquad (8.70)$$
$$T_2(\mathbf{r}) = T_0(\mathbf{r})c_2 e^{-\mu\Psi(\mathbf{r})}. \qquad (8.71)$$

Similarly, as in the previous case, write (8.69) into (8.60). We get the following boundary value problems for the functions T_i:

$$D\,\Delta T_1(\mathbf{r}) - \mu T_1(\mathbf{r}) - T_2(\mathbf{r}) = 0,$$
$$D\,\Delta T_2(\mathbf{r}) - \mu T_2(\mathbf{r}) = 0,$$

$$a(\mathbf{r})T_1(\mathbf{r}) + b(\mathbf{r})\frac{\partial T_1}{\partial n} = Q(\mathbf{r}) \quad \text{for all } \mathbf{r} \in \partial V,$$
$$a(\mathbf{r})T_2(\mathbf{r}) + b(\mathbf{r})\frac{\partial T_2}{\partial n} = c_2 Q(\mathbf{r}) \quad \text{for all } \mathbf{r} \in \partial V.$$

If $\mu \geq 0$, then according to Theorem 6.31 in [61], there exists unique solutions to T_1 and T_2 this system. Due to the maximum principle and the boundary point lemma [62] $c_2 > 0$ implies $T_2 > 0$ and $c_2 < 0$ implies $T_2 < 0$. One can assume that $c_2 \neq 0$, because $c_2 = 0$ belongs to the first case. Hence the functions T_1 and T_2 give the BFS because, from (8.70) and (8.71) for the functions Ψ and T_0, we get

$$\Psi(\mathbf{r}) = \frac{c_1}{c_2} - \frac{T_1(\mathbf{r})}{T_2(\mathbf{r})},$$
$$T_0(\mathbf{r}) = \frac{1}{c_2} T_2(\mathbf{r})e^{\mu\Psi(\mathbf{r})}.$$

8.8.3 *Asymptotic Behavior; Stability*

Heat conduction is a dissipative process, therefore we expect that the effect of the initial condition diminishes in time. To sustain a process forever we need an everlasting stimulation at the boundaries. The BFSs have an attractive property (some kind of stability), namely, as we will prove below, a solution belonging to an arbitrary initial condition is superposed from a transient term and a BFS, the latter belongs to a peculiar initial condition.

Let us assume that the boundary condition (8.61) is one of the classes listed in Section 8.8.2. Let $T^*(\mathbf{r},t)$ be a solution of (8.60) with an arbitrary initial condition and let $T(\mathbf{r},t)$ be a BFS of (8.60) belonging to that boundary condition. Then $T^{**} := T^* - T$ satisfies (8.60) with the homogeneous boundary condition, therefore

$$\lim_{t\to\infty} T^{**}(\mathbf{r},t) = 0.$$

Nevertheless, this does not necessarily mean that T^{**} is the transient term, since it may occur that the corresponding BFS T also decays in time. If the coefficient μ in the function Φ in (8.64), (8.65), (8.66), corresponding to the given boundary condition, is nonnegative then T does not tend to 0 as t tends to infinity. Consequently, in this case, T^{**} can be termed transient while the BFS T is "asymptotic."

If $\mu < 0$, then T also tends to 0. Therefore the partition into a transient and an asymptotic term is generally not straightforward. The crucial problem here is the comparison of the two decays: the decay of the function Φ (characterized by the value of μ) and the natural decay due to conduction (characterized by the largest eigenvalue $\lambda_{\max}$ of the homogeneous boundary value problem $D \, \Delta T = \lambda T$). If $0 > \mu > \lambda_{\max}$ (resp., $0 > \mu_i > \lambda_{\max}$), then

$$\lim_{t \to \infty} \frac{T^{**}(\mathbf{r},t)}{T(\mathbf{r},t)} = 0.$$

Consequently, the term "asymptotic" is justified for T in this case too.

8.9 References

[1] H. S. Carslaw, and J. C. Jaeger: Conduction of Heat in Solids, Oxford University Press, London, 1959.

[2] I. Gyarmati: *Non-Equilibrium Thermodynamics; Field Theory and Variational Principles*, Springer-Verlag, Berlin, 1970.

[3] B.A. Finlayson, and L.E. Scriven: On the search for variational principles, *Internat. J. Heat Mass Transfer* **10**, 799–821, (1967).

[4] H. Farkas: On the phenomenological theory of heat conduction, *Internat. J. Engng. Sci.*, 1035–1053, (1975).

[5] P. Rosen: Use of restricted variational principles for the solutions of differential equations, *J. Chem. Phys.* **25**, 336–338, (1953).

[6] P. Glansdorff, and I. Prigogine: *Thermodynamic Theory of Structure Stability and Fluctuations*, Wiley, New York, 1971.

[7] R. Donelly: *Non-Equilibrium Thermodynamics, Variational Techniques and Stability*, University of Chicago Press, Chicago, IL, 1965.

[8] P. M. Morse, and H. Feshbach: *Methods of Theoretical Physics*, McGraw-Hill, New York, 1953.

[9] M. A. Biot: *Variational Principles in Heat Transfer*, Oxford University Press, Oxford, 1970.

[10] H. Farkas, and Z. Noszticzius: On the non-linear generalization of the Gyarmati principle and theorem, *Ann. Phys.* **27**, 341–348, (1971).

[11] J. Verhás: *Thermodynamics and Rheology*, Akadémiai Kiadó, Budapest, 1997.

[12] S. Sieniutycz: *Conservation Laws in Variational Thermo-Hydrodynamics*, Kluwer Academic, Dordrecht, 1994.

[13] S. Sieniutycz: Variational thermomechanical processes and chemical reactions in distributed systems, *Internat. J. Heat Mass Transfer* **40**, 3467–3485, (1997).

[14] C. Cattaneo: Sur une forme de l'équation éliminant le paradoxe d'une propagation instantanée, *C. R. Hebd. Seanc. Acad. Sci.* **247**, 431–433, (1958).

[15] S. Sieniutycz: Extended conservation laws from Hamilton's principle for nonequilibrium fluids with heat flow, *Periodica Polytechnica Ser. Phys. Nucl. Sci.* **2**, 61–83, (1994).

[16] H. Farkas: The reformulation of the Gyarmati principle in a generalized "Γ-picture", *Z. Phys. Chem.* **239**, 124–132, (1968).

[17] C. Truesdell: *Rational Thermodynamics*, McGraw-Hill, New York, 1969.

[18] D. Coleman, and V.J. Mizel: Thermodynamics and departures from Fourier's law of heat conduction, *Arch. Rat. Mech. Anal.* **13**, 245–261, (1963).

[19] A. Tychonoff: Théorèmes d'unicité de la chaleur, *Receuil Mathématique (Sbornik)* **42**, 199–216, (1935).

[20] I. Rubinstein, and L. Rubinstein: *Partial Differential Equations in Classical Mathematical Physics*, Cambridge University Press, Cambridge, 1994.

[21] H. Farkas: A new proof and generalization of the maximum principle of heat conduction, *J. Non-Equilib. Thermodyn.* **7**, 355–362, (1982).

[22] M. W. Zemansky: *Heat and Thermodynamics*, McGraw-Hill, New York, 1943.

[23] H. Farkas: Generalization of the Fourier law, Periodica Polytechnica ME **24**, 291-308, (1980).

[24] L. Nirenberg: A strong maximum principle for parabolic equations, *Comm. Pure Appl. Math.* **6**, 167–177, (1953).

[25] J. C. Maxwell: On the dynamical theory of gases, *Phylos. Trans. Roy. Soc.* **157**, 49, (1867).

[26] P. Vernotte: Les paradoxes de la théorie de l'équation de la chaleur, *Comptes Rend.* **246**, 3154, (1958).

[27] D. Jou, J. Casas-Vãzquez, and G. Lebon: *Extended irreversible thermodynamics*, Springer-Verlag, New York, 1996.

[28] I. Müller, and T. Ruggeri: *Extended Thermodynamics*, Springer Tracts of Natural Philosophy 37, Springer-Verlag, New York, 1993.

[29] I. Gyarmati: On the wave approach of thermodynamics and some problems of non-linear theories, *J.. Non-Equilib. Thermodyn.* **2**, 233–260, (1977).

[30] I. Müller: Thermodynamics of simple mixtures of fluids with application to second sound and liquid helium, *Banach Center Publications* **15**, 543–592, (1995).

[31] A. S. Kalashnikov: Some problems of the qualitative theory of nonlinear degenerate second-order parabolic equations, *Russion Math. Surveys* 42, 169–222, (1987), translation of *Uspekhi Mat. Nauk* **42**, 135–176, (1987).

[32] R. Kersner: Filtration with absorption: Necessary and sufficient condition for the propagation of perturbations to have finite velocity, *J. Math. Anal. Appl.* **90**, (1982) 463–479.

[33] B.H. Gilding, and R. Kersner: The characterization of reaction–convection–diffusion processes by travelling waves, *J. Differential Equations* **124**, 27–79, (1996).

[34] V. V. Nemytskii, and V. V. Stepanov: *Qualitative Theory of Differential Equations*, Princeton University Press, Princeton, 1960.

[35] J. Guckenheimer, and Ph. Holmes: *Nonlinear Oscillations, Dynamical Systems and Bifurcations of Vector Fields*, Springer-Verlag, New York, 1983.

[36] C. V. Pao: *Nonlinear Parabolic and Elliptic Equations*, Plenum Press, New York, 1992.

[37] H. Farkas: Properties of stationary temperature fields, *Magyar Fizikai Folyóirat* **23**, 157–163, (1975), (in Hungarian).

[38] H. A. Buchdahl: *The Concepts of Classical Thermodynamics*, Cambridge University Press, Cambridge, 1966.

[39] L. Bers, F. John, and M. Schechter: *Partial Differential Equations*, Interscience, New York, 1964.

[40] I. Faragó, and T. Pfeil: Preserving concavity in initial-boundary value problems of parabolic type and its numerical solution, *Period. Math. Hungar.* **30,** 135–139, (1995).

[41] A. Friedmann: *Partial Diffferential Equations of Parabolic Type*, Prentice-Hall, Englewood Cliffs, 1964.

[42] J. Smoller: *Shock Waves and Reaction–Diffusion Equations*, Springer-Verlag, Berlin, 1982.

[43] A. A. Samarskii, and A. N. Tychonoff: *Equations of the Mathematical Physics*, Nauka, Moscow, 1967 (in Russian).

[44] R. D. Richtmyer, and K. W. Morton: *Difference Methods for Initial-Value Problems*, Interscience, New York, 1967.

[45] A. A. Samarskii: *Theory of the Difference Schemes*, Nauka, Moscow, 1977 (in Russian).

[46] V. Thomée: Finite difference methods for linear parabolic equations, in: *Handbook of Numerical Analysis* (eds. P. G. Ciarlet and J. L. Lions), North-Holland, Amsterdam, 1990.

[47] R. A. Adams: *Sobolev Spaces*, Academic Press, New York, 1975.

[48] G. Strang, and G. Fix: *An Analysis of the Finite Element Method*, Prentice-Hall, Englewood Cliffs, 1973.

[49] V. Thomée: Galerkin finite element methods for parabolic problems, *Lecture Notes in Math.* 1054, 1984.

[50] I. Faragó: One step methods of solving a parabolic problem and their qualitative properties, *Publ. Appl. Anal.* **3,** , 1–30, (1996).

[51] I. Faragó: Regular and exponential convergence of difference schemes for the heat-conduction equation, *Comput. Math. Appl.*, Pergamon Press (accepted).

[52] I. Faragó: Nonnegativity of the difference schemes,*Pure Math. Appl.* **6,** , 47–60, (1996).

[53] I. Faragó, and P. Tarvainen: Two-stage algebraic models with symmetric tridiagonal Toeplitz matrices and their qualitative analysis, *Period. Math. Hungar.* **35,** 177–192, (1997).

[54] I. Faragó: Qualitative properties of the numerical solution of linear parabolic problems with nonhomogeneous boundary conditions, *Comput. Math. Appl.* **26,** 143–150, (1996).

[55] R. Horváth: Some qualitative properties of the numerical solution of the heat conduction equation, Eötvös L. University, M.Sc. thesis, 1994 (in Hungarian).

[56] R. Horváth: Maximum norm contractivity in the numerical solution of the heat equation, *Appl. Numer. Math.* (accepted).

[57] I. Faragó, and R. Horváth: An optimal mesh choice in the numerical solution of the heat equation, *Comput. Math. Appl.* (accepted).

[58] I. Faragó: On the mesh for difference schemes of higher accuracy, in: *Finite Element Method, Superconvergence* (eds. M. Krizek and P. Neittanmaaki), Kluwer, Amsterdam, 1998, pp. 127-133.

[59] P. J. Schneider: *Temperature Response Charts*, Wiley, New York, 1963.

[60] H. Farkas, and I. Mudri: Shape preserving time-dependencies in heat conduction, *Acta Phys. Hungar.* **55**, 267–273, (1984).

[61] D. Gilbarg, and N. S. Trudinger: *Elliptic Partial Differential Equations of Second Order*, Springer-Verlag, New York, 1977.

[62] M. H. Protter, and H. F. Weinberger: *Maximum Principles in Differential Equations*, Prentice-Hall, Englewood Cliffs, 1967.

9
Energy Transfer in Particle–Surface Collisions

Z. Herman

ABSTRACT. Energy transfer in collisions of single neutral and ionized particles with surfaces is reviewed. Collision energy domains (thermal to MeV) are only briefly characterized, as the main emphasis is on energy transfer in collisions of slow particles. Basic features of energy transfer in surface collisions of thermal and slightly hyperthermal neutral atoms, molecules, and clusters are described. Results on surface collisions of ions (hyperthermal energy region from 1 eV to about 1 keV) are discussed separately for atomic, simple molecular, polyatomic, and cluster ions. Neutralization processes of ions at surfaces are outlined. The important problem of energy transfer and dissociation in polyatomic ion–surface interactions is treated in more detail.

9.1 Introduction

Surface collisions of neutral and ionized particles, atoms, molecules, and clusters are of great interest to many areas of science and technology. Studies of surface collisions phenomena span a broad range of basic and applied research stretching from molecular physics to applied molecular science. Characterizing structure, morphology, chemical, and physical properties of well-defined surfaces; chemical reactions at surfaces and heterogeneous catalysis; gas–surface interactions in environmental problems; energy transfer in ion interaction with walls of thermonuclear devices; modification of surfaces by particle bombardment; elucidation of the structure of large molecular ions and clusters by their interaction with surfaces; surface deposition of molecules and clusters; surface thin film synthesis; these are but few examples of areas of research of prime importance in physics, chemistry, electronics, metallurgy, biophysics, and other disciplines. Energy transfer in collisions with surfaces represents one of the central questions in particle–surface interactions. The incident energy of particles may change from thermal to MeV and different phenomena prevail in the particular collision energy regions. These phenomena can be used in characterizing both the surface and the incident particle.

240   9. Energy Transfer in Particle–Surface Collisions

In this chapter we will review some of the phenomena connected with energy transfer in surface collisions of slow single particles. The projectile of thermal energy will be mostly neutral atoms, molecules, and clusters. Hyperthermal projectiles will be mostly charged particles (ions).

The subject of gas–surface interactions and energy transfer has been treated in many reviews and books. Of the multitude of available literature, let us mention here only some of the more recent reviews, collections of papers, and monographs on neutral particle–surface interactions [1]–[3] and ion collisions with surfaces [4]–[6].

9.1.1 *Collision Energy Domains of Neutral and Ion Projectiles*

Two basic parts of the problem of particle collisions with surfaces may be distinguished, namely collisions of neutral particles and collisions of ions. Detailed information on energy transfer in collisions of thermal and slightly hyperthermal (≤ 1 eV) neutral atoms and molecules can be obtained from scattering studies. In collisions of light thermal energy atoms, the de Broglie wavelength of the projectile is comparable to or smaller than the lattice constant of the solid, and quantum mechanical effects cannot be neglected. The main phenomena are diffraction (elastic scattering) and particle–(surface) phonon interaction (inelastic scattering). Structural information on the surface can be obtained, too. For collisions of heavy atoms with metal surfaces the importance of quantum effects is usually smaller, and classical molecular dynamics may be used to describe the nature of collisions and energy transfer. In addition, an important phenomenon of energy transfer is trapping of the particle on the surface and chemical reactions at surfaces.

Ion scattering phenomena, on the other hand, usually serve to explore the wide range of hyperthermal incident energies: from a few eV to about 1000 eV (hyperthermal energy range, slow ions), 1 keV to about 10 keV (low-energy range), 10–100 keV (medium energy range), and the MeV region (high-energy range). From the point of view of scattering, the de Broglie wavelength is usually less than 10^{-3} Å and thus quantum mechanical effects may usually be neglected. Various phenomena are typical for these energy ranges. Some of them are, however, not pertinent to the subject of this chapter (energy transfer in collisions of slow particles) and will be thus only briefly mentioned. Starting with the highest collision energies, Rutherford back-scattering is typical for the high (MeV) energy regime. Translational energy of projectiles is so high that the nucleus interacts with the bare nuclei of the target in the repulsive Coulomb potential field and back scattering depends on the atomic number of the projectile and target. Besides the classical example of Rutherford's determination of the effective cross section of the atomic nucleus, the method has been widely used in elemental analysis of the target material when bombarded by light projectiles of MeV energies.

In the medium incident energy range (10–100 keV) the scattering cross section is partly sensitive to electron screening of the nuclei and studies provide data for detailed analysis of the solid material, surface structures, and/or quantitative information on species present on the surface. In the low-energy range (1–10 keV) electron screening of nuclei is very important and scattering probability from the surface atomic monolayer becomes significant. The cross sections are generally high, but multiple scattering events may complicate data analysis.

In the lowest energy range (1 eV–1 keV) the projectiles tend to interact with several target atoms simultaneously and the interpretation of the results to determine the surface structures may be difficult. However, in this very low collision energy region attractive forces between the projectile and target atoms or molecules become increasingly more important, and thus studies provide information on energy transfer to (molecular) projectiles, on chemical reactions at surfaces, and with it directly connected technologically important aspects of chemical modification of surfaces, and also structural information on more complicated projectiles.

9.2 Neutral Particle–Surface Energy Transfer

When an atom or a molecule of thermal or slightly hyperthermal energy collides with a surface, various scattering processes can occur (Figure 9.1). In studying energy interchange between a particle and a surface, one would like to know the velocity and internal state distributions of the scattered species in dependence on the incident angle, translational energy, and internal state of the projectile impinging on a clean, well-defined surface.

9.2.1 Translational Energy Transfer

Translational energy exchange with the surface can be conveniently studied in beam experiments using non–reactive atoms impinging on smooth surfaces such as rare gas atoms on oriented metal surfaces. For collisions of slightly hyperthermal heavy atoms on metal surfaces (e.g., 0.14 eV Xe atoms on Pt(111) [7]) the processes identified were [8] a direct inelastic scattering channel, where the atom is scattered close to the specular angle with velocities only slightly broadened and shifted to lower values (in comparison with the incident particle); the atom is directly scattered exchanging with the surface both the parallel and normal components of motion. The direct inelastic type of scattering exhibits two regimes depending on the

242   9. Energy Transfer in Particle–Surface Collisions

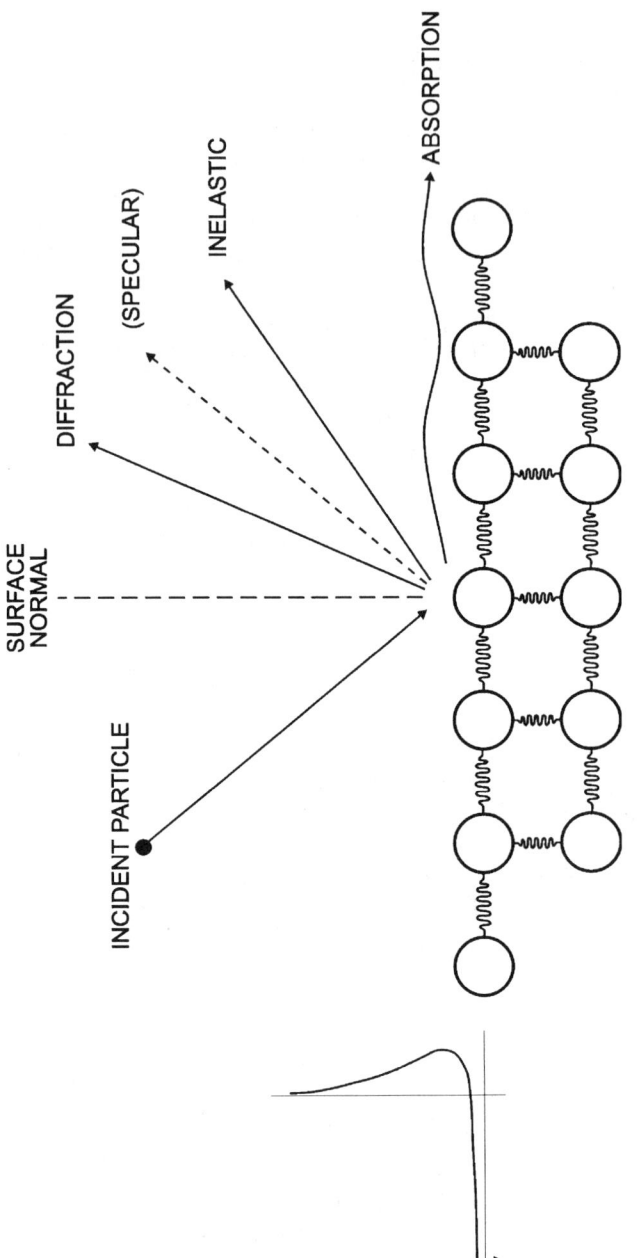

FIGURE 9.1. Schematics of scattering processes in particle–surface interactions.

energy of the incident particle. For incident energies in the thermal region the timescale of the projectile approach is comparable with periods of surface modes and the scattering is dominated by the surface motion (thermal scattering). For fast hyperthermal projectiles the motion of the surface atoms is slow in comparison with the particle motion and the interaction occurs with single surface atoms only (structure scattering).

At scattering angles close to surface normal, another type of scattering prevails. This can be characterized as the trapping–desorption channel in which the atom sticks to the surface, completely equilibrating its energy with it, and then it desorbes; its velocity distribution corresponds to the Maxwell–Boltzmann distribution pertaining to the surface temperature.

More detailed information could be obtained in collisions of slow light particles (usually He) with surfaces characterized by pronounced potentials, like oriented surfaces of alkali halide crystals [1]. As mentioned earlier, the de Broglie wavelength of such an incident particle is of the order of 1 Å and is thus comparable to the lattice dimensions; quantum scattering effects then become important and the scattering is dominated by diffraction phenomena. Scattering experiments also showed, however, characteristic weak energy-loss peaks which could be ascribed to the particular phonon dispersion curves characterizing the longitudinal and transversal modes of the phonons. Experimental results and theoretical models agree in the description of the inelastic scattering act, as formation or annihilation of a single surface phonon or a bulk phonon close to the surface.

9.2.2 Rotational Energy Transfer

Studies of collisions of molecular projectiles with surfaces made it possible to obtain information on rotational and vibrational energy exchange with the surface. Most frequently, diatomic molecules have been used and rotational distributions were measured using laser spectroscopy methods. Rotational excitation and deexcitation of molecular projectiles is directly connected with translational energy of the projectile and energy of the surface modes. Distribution of rotational states of molecules scattered off surfaces indicates that the two earlier-mentioned scattering channels, direct inelastic and trapping–desorption, can be used to describe the rotational energy transfer, too. In the trapping–desorption scattering, the distribution of rotational states reflects the Boltzmann distribution at a temperature close to the surface temperature, indicating energy equilibration while the particle was interacting with the surface. The direct-inelastic scattering leads to a rotational state distribution that can be characterized by an effective temperature which is, however, not directly related to the temperature of the surface [9].

9.2.3 *Vibrational Energy Transfer*

Vibrational energy transfer in collisions of simple molecules with surfaces is somewhat less well understood than rotational energy transfer. Various models have been used to describe vibrational excitation and deexcitation of various molecules in collisions with surfaces. Measurements of energy thresholds for excitation showed that some vibrational modes in molecules can be excited in a process that may be described by a simple mechanical model (ammonia on Au(111) [10]). In other cases, surface temperature rather than translational energy of the projectile determines an effective vibrational energy transfer (NO on Ag(111) [11]). Other models relate the efficient vibrational excitation to strong chemisorption forces acting during a molecule approach to the surface and to particle–surface interactions during adsorption of the particles on surfaces [4].

9.2.4 *Energy Exchange in Cluster–Surface Collisions*

Collisions of large neutral clusters with surfaces represent another important class of particle surface collisions as they underline a broad field of application spanning from particle-beam epitaxy to cluster-impact nuclear fusion. Experimental data exist on collisions of large clusters ($n = 10^2$–10^3) of noble gases, water, ammonia, and other molecules with a variety of surfaces (e.g., oriented carbon or metal planes) and theoretical simulations have been carried out. A thermochemical model has been developed [12] which fits well the available data. The model treats the surface collision as gliding of the incident cluster along the surface and thermal evaporation of small fragments during this time. The local temperature of evaporation in the case of argon was found to be close to its critical temperature [9]. This appears to be the dominant cluster decomposition channel at surface temperatures of about 1000 K. At much lower temperatures, the gliding of the clusters seems to be slowed down and the clusters lose momentum parallel to the surface before complete evaporation [13].

Theoretical simulations [14], [15] predict that collisions of large clusters with surfaces may lead to intracluster shock waves creating internal heating of clusters to temperatures of the order of 10^5 K on a sub-picosecond time scale. This may possibly be used to initiate specific chemical processes in particles embedded in clusters.

9.3 Slow Ion–Surface Energy Exchange

In this section energy exchange in collisions of hyperthermal atomic, molecular, and cluster ions with surfaces will be treated. For this purpose, hyperthermal means an energy region of 10–1000 eV. This in turn means velocities of light projectiles of the order of 10^6–10^7 cm s^{-1}. The interaction time depends on the trajectory length and this is a function of the ion incident angle. A reasonable estimate of the interaction length is about 10 Å (which means interaction times of the order of 10^{-14}–10^{-15} s). Thus rotational and solid state phonon effects can be regarded as not being important. Vibrational energy exchange has to be taken into consideration as well as electronic excitation and electron exchange between the projectile ion and the surface. The electron transition times are of the order of 10^{-15} s and thus the ion–surface interaction may be treated within the Born–Oppenheimer approximation. An interesting aspect is that large incident angles Φ (measured from the surface normal, i.e., glancing collisions with respect to the surface plane) can make the perpendicular energy, defined as $E_{pp} = E_0 \cos^2 \Phi$ (where E_0 is the incident ion energy) quite small and close to the surface chemical reaction region.

9.3.1 Neutralization of Ions at Surfaces

Low-energy atomic and molecular ions are neutralized at surfaces with high probability, and loss of charge is an important process in ion–surface interactions. The neutralization process is in many cases described reasonably well by the Hagstrum model [16], developed for neutralization of atomic ions of ionization energies equal to or greater than twice the work function of the surface. The neutralization process then occurs as an electron transfer from the solid to the approaching ion and an excitation of an Auger electron from the solid. If the ionization energy of the incoming particle is close to the energy levels of the solid (e.g., alkali atoms colliding with metal surfaces) an important charge transfer mechanism is the resonant transfer of electrons. A variety of theoretical models [17], [18] has been developed for systems that undergo resonant charge; they include, to a various degree, the problem of a multitude of states in the atom and/or the surface, the particle spin, etc. For molecular ion collisions, excited states of the molecule are accessible by resonant processes and dissociative neutralization must often be considered. Very often, however, various charge transfer mechanisms operate at the same time, even in systems as superficially simple as collisions of hydrogen molecular ions with metal surfaces [19].

Ion survival probability in collisions of slow alkali metal ions with metal surfaces varies greatly, as it depends on the overlap of energy levels of the projectile and the surface. Qualitative agreement between experiments and theoretical models has been achieved [17]. Measurements of scattered molecular ions yields [20] show that the ion survival probability increases appreciably with decreasing ion velocity and increasing incident angle (measured from the surface normal). The ion yields are low for collisions with metal surfaces (10^{-2}–10^{-3}). They increase for metal surfaces covered by an adsorbate layer or self-assembled monolayer, and for surfaces of semiconductors or insulators.

9.3.2 *Collisions of Atomic Ions with Surfaces*

A valuable insight into the translational energy exchange in slow ion collisions with surfaces was obtained in studies of slow (energies 10–500 eV) collisions of alkali metal ions with oriented metal surfaces [17]. Measurements of translational energies of 24.5 eV Na^+ ions scattered from Cu(001) showed characteristic energy losses of 40%, 50%, and 62% of the incident energy, decreasing in intensity. These could be well reproduced by model trajectory calculations in which a surface–particle potential was used that contained both the repulsive and attractive part. The energy losses correspond to different particle trajectory and interactions with atoms of the surface. One type of trajectory corresponds to scattering along a chain of surface atoms in which the projectile exchanges momentum with one atom only (quasi-single collision). Another type of trajectory describes a similar interaction along the chain of surface atoms, however, the projectile exchanges momentum with two neighboring atoms along the chain (quasi-double collision). The third type of trajectory describes a momentum transfer with surface atoms belonging to two neighboring chains (double zig-zag) and yet another concerns momentum exchange with three atoms of the two neighboring surface atom chains (triple zig-zag). These four types of trajectory describe very well the observed translational energy losses and indicate that an important characteristic of the translational energy transfer is a possibility of an interaction of the projectile with several surface atoms at once.

9.3.3 *Collisions of Simple Molecular Ions with Surfaces*

An important inelastic energy transfer process in collisions of molecular ions with surfaces leads to dissociation of the projectiles. Information obtained from velocity distribution measurements of both ionic and neutral products of interaction with the surface shows that dissociation with neutralization results from electron capture by the projectile into repulsive excited states of the neutral. Neutralization of the projectile is a dominant process in the interaction, sometimes accounting for something like 90% of the collision events [21]. Charge capture processes are even more significant in surface collisions of molecular species exhibiting electron affinity. Formation of negative ions may then result from the sequence neutralization of incoming positive ion–electron capture from the surface to form a negative ion–dissociation to negative, neutral (and positive) products [21], [22]. The importance of molecular ion–surface interaction research is exemplified by studies of grazing collisions of ions D_3^+ of keV energies with metal surfaces [23], directly related to the plasma–wall interaction in fusion reactors. The results showed, besides identifying ionic and neutral dissociation products, a remarkably large survival probability of the incident D_3^+ projectiles.

Models describing inelastic collisions of molecular species and dissociative interaction with surfaces are essentially extensions of models of atomic ion collisions. A remarkably useful is a simple model based on classical trajectories [24]. It describes a collision of an incident diatomic molecular ion in terms of collisions of the two constituent atoms with a surface particle resulting in deflections, and elastic energy losses characterized by the final velocity vectors of the two constituent atoms $\vec{v}_1$ and $\vec{v}_2$ (further collisions with surface particles in chain may follow). The condition for molecule survival is then $m/2(\vec{v}_1 - \vec{v}_2) \ll E_D$, where m is the mass of the atoms of the molecule and E_D is its dissociation energy. Despite its simplicity, the model was shown to fit experimental data reasonably well, if the attractive part was included into the interaction potential.

9.3.4 Collisions of Polyatomic Ions with Surfaces

Studies of polyatomic ion collisions with surfaces have turned out to be of special significance because of the following reasons:

- dissociation of polyatomic ions in collisions with surfaces (surface-induced-dissociation, SID) has been used by organic mass spectrometrists during the past 15 years as one of the tools to characterize the structural properties of organic cations;

- in collisions of slow polyatomic ions, chemical reactions with the surface material (surface-induced-reactions, SIR) have been observed; a new field of ion chemical reactivity studies has thus been opened; and

- surfaces of special properties have been prepared with self-assembled-monolayer (SAM) coverages which exhibit specific energy transfer or chemical reactivity behavior, studied conveniently by slow polyatomic ion impact.

Information on energy transfer in collisions of this type is, therefore, of great interest to researches in various fields of physics, chemistry, and material studies.

The choice of a surface in mass spectrometric studies, where the emphasis was on the characteristic dissociation pathways of the projectile, was for a long time dictated by practical means: namely by the requirement of its stability and sufficient ion reflectivity. Rather surprisingly, surfaces coated by adsorbed gases or contaminants turned out to be much more suitable [6] than clean, defined metal surfaces, where the neutralization of the incoming polyatomic ion projectile was usually quite high. Thus, a universal surface used in numerous SID studies was a polished stainless steel surface, covered under the operating mass spectrometry conditions (vacuum of 10^{-9}–10^{-6} Torr) by a layer of hydrocarbons (described as pump oil contaminants); recent experiments with SAM of hydrocarbons support this identification. Only recently, experiments using special UHV scattering machines have extended the studies to clean (oriented) metal and specific SAM surfaces (see, e.g., [25] and citations therein).

When a polyatomic ion of incident energy E_{inc} interacts with a surface, the basic energy partitioning relation is

$$E_{\text{inc}} = E'_{\text{int}} + E'_{\text{tr}} + E'_{\text{surf}}, \qquad (9.1)$$

where E'_{int} is the fraction of energy converted into internal excitation of the projectile, E'_{tr} is the translational energy of the product(s), and E'_{surf} is energy absorbed by the surface (primed quantities refer to products).

Early approximate estimations concerned the average energy transferred into the internal energy of the projectile. They were obtained from the energetics of observed specific projectile/fragment ion abundances and led to a conclusion that about 15% of E_{inc} was on the average converted to internal energy of the projectile [6]. More detailed information on incident translation-to-internal energy transfer has been obtained from the dissociation of "thermometer molecules" [26]. The procedure is based on using as a projectile a suitable molecular ion from which fragment ions are formed in a chain of consecutive dissociation processes of known critical energies and similar entropy requirements. Molecular ions like $Fe(CO)_5^+$ or $(C_2H_5)_4Si^+$ were shown to fit these requirements and thus useful data on the distributions of energy transferred into projectile internal energy, $P(E'_{\text{int}})$, and their dependence on incident energy could be obtained from mass spectrometric measurements [27]. They showed that about 10–15% of the incident energy is converted into internal energy in the 10–60 eV thermometer ion interaction with a stainless steel surface, with an approximately bell-shaped distribution of a full-width-at-half-maximum (FWHM) of about 3 eV.

In a recent important study [25] of $Si(CH_3)_3^+$ interaction (incident energy 10–70 eV, incident angle 45°) with a clean Au(111), Au with a hexathiolate SAM adsorbate, and NiO(111) surface, both the ion products and their kinetic energies were measured. From the data the average energy deposited as internal energy of the projectile was estimated. On average, about 18% of the incident energy was found to be transferred into internal energy in the case of the Au–SAM surface. The value of E'_{int} thus increased linearly with increasing incident energy, with small deviations to lower values above 30 eV. The fraction of energy transferred into internal energy in collisions of the ions with clean Au surface was found to be larger than 20% of the incident projectile energy. Data on the product ion translational distributions made it possible to estimate (from the energy balance) the average fraction of energy transferred to the surface. The results could be approximated by relations $E'_{\text{surf}} = E_{\text{inc}} - 5.1$ eV for the Au–SAM organic surface and by $E'_{\text{surf}} = 0.43 E_{\text{inc}} - 4.9$ eV for the clean Au surface. The results were in semiquantitative agreement with a classical impulsive excitation model [28].

In a set of experiments undertaken in our laboratory [29], we investigated dissociative and reactive scattering of the ethanol molecular ion

9.3. Slow Ion–Surface Energy Exchange

$C_2H_5OH^+$ from stainless steel surfaces with the aim of determining the distribution functions for energy transfer into projectile internal excitation, $P(E'_{int})$, product kinetic energy, $P(E'_{tr})$, and energy absorbed by the surface, $P(E'_{surf})$, in dependence on the projectile incident energy (10–30 eV) and its incident angle (40°–80°, measured from the surface normal). Product ion mass spectra, translational energy distributions, and angular distributions of the product ions were measured and analyzed to provide the above-mentioned distributions:,

- $P(E'_{int})$ was obtained by fitting the abundances of the dissociation product ions, as derived from the measured mass spectra, by abundances obtained from the break-down curves of the ethanol molecular ion (dependence of relative abundances of various fragment ions on the internal energy of the molecular ion) and a suitably chosen internal energy distribution curve;

- $P(E'_{tr})$ from measurements of product ion translational energy distributions (the velocity distributions of the product ions were the same as that of the undissociated projectile ion, indicating that the decomposition took place as a unimolecular dissociation of the molecular ion after the interaction with the surface); and

- $P(E'_{surf})$ from the difference of these two terms and the projectile ion incident energy (composed of the translational energy and its internal excitation).

The results are summarized in Figure 9.2 and Figure 9.3.

In Figure 9.2 we see the partitioning of energy for the impact of 23.4 eV ethanol molecular ions in dependence on the incident angle, Figure 9.3 shows the dependence of energy partitioning for an incident angle of 60° on the projectile incident energy. Several important conclusions can be drawn from these results. First of all, the dissociative collisions are strongly inelastic. The fraction of energy going into internal excitation of the projectile is independent of the angle of incidence and also (within the experimental error) independent of the incident ion energy, with the peak value of 6–8% of E_{inc}; this is somewhat smaller than the earlier data of 10% (18% for a SAM surface, see above). The fraction of energy remaining as the translational energy of the scattered ions is independent of the incident energy, but it increases substantially with increasing incident angle, being largest at close to the surface incident angles. There is a correlation between the peak values of the components of velocities parallel to the surface for the scattered and incident ions: their ratio, v'_{par}/v_{par}, was found to be independent of both the incident energy and the incident angle of the projectile ion [29]. The fraction of energy absorbed by the surface complementarily decreases with the incident angle, being largest for incident angles close to the surface normal. The absolute value of the energy transferred into internal energy

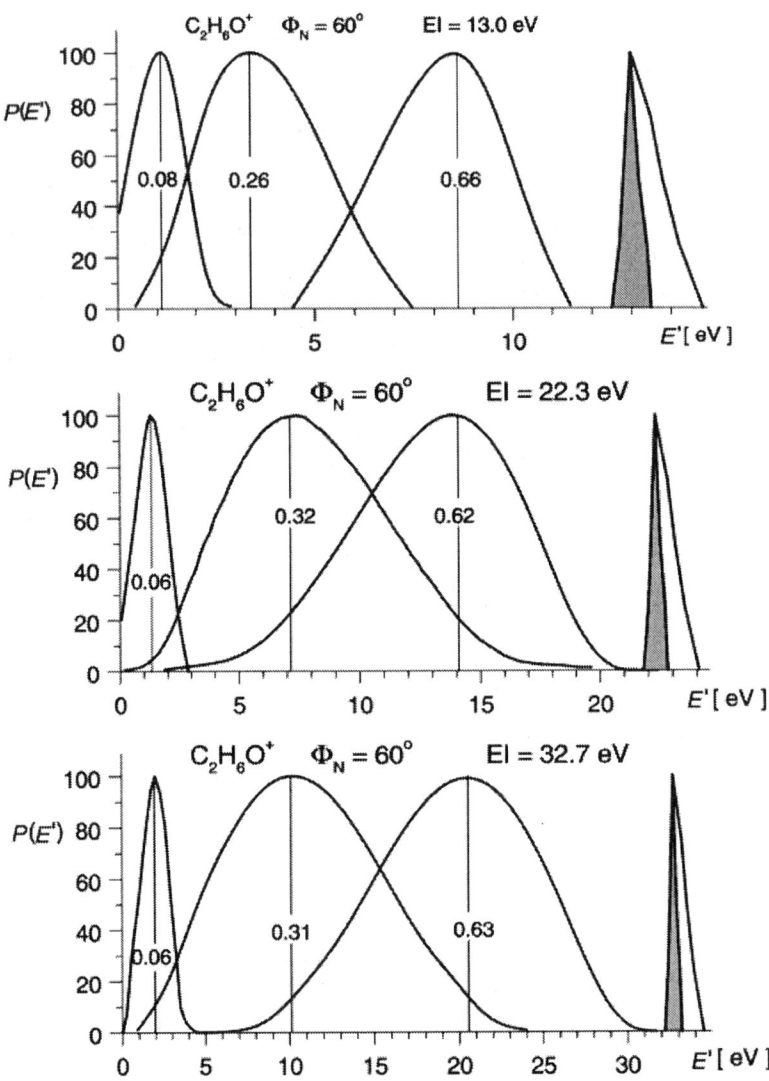

FIGURE 9.2. Energy partitioning into E'_{int}, E'_{tr}, and E'_{surf} in collisions of the ethanol molecular ion with a stainless steel surface in dependence on the projectile incident angle: 40°(A), 60°(B), and 80°(C). Projectile translational energy 22.3 eV (shaded); hatched: estimated internal energy distribution of the projectile ion [29].

9.3. Slow Ion–Surface Energy Exchange   251

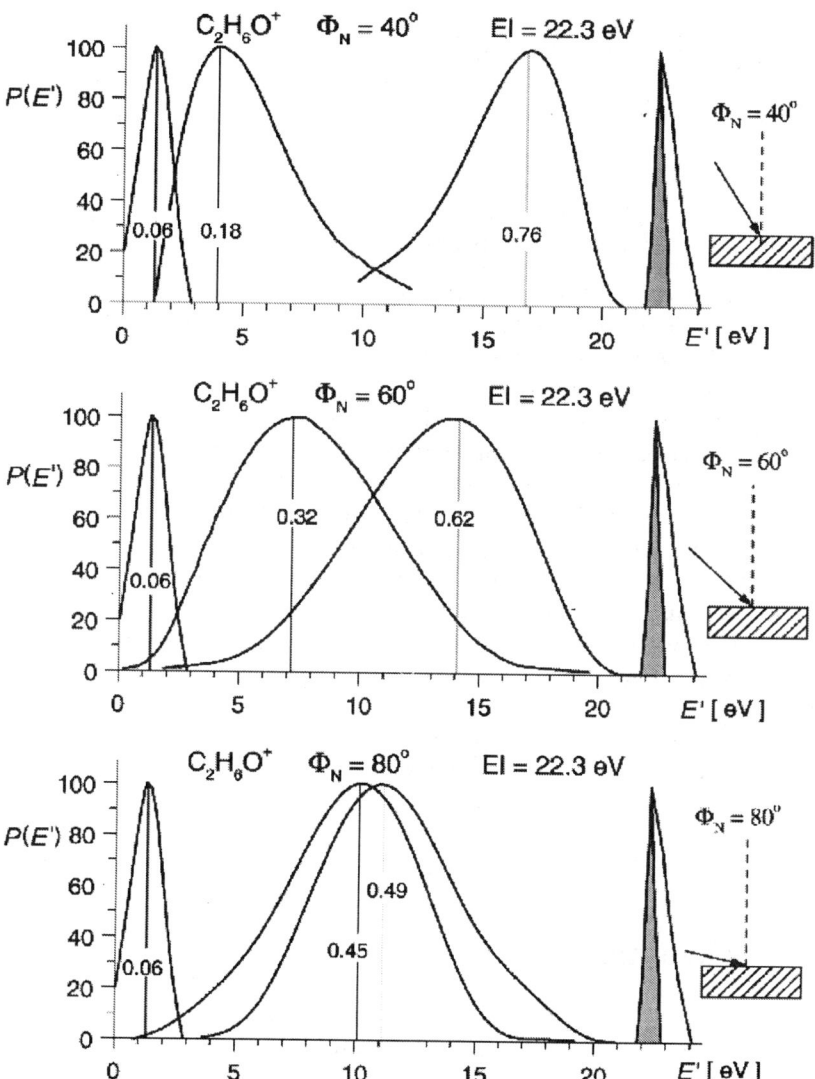

FIGURE 9.3. Energy partitioning into E'_{int}, E'_{tr}, and E'_{surf} in collisions of the ethanol molecular ion with a stainless steel surface in dependence on the projectile incident energy: 13.0 eV(A), 22.3 eV(B), and 32.7 eV. Projectile incident angle 60° [29].

of the projectile increased (in the studied incident energy region) linearly with increasing incident energy, as observed in other studies [6], [26]–[28].

Classical trajectory simulations [30], [31] of the collisions of polyatomic ions with surfaces suggest that short-range repulsive forces dominate the scattering for incident energies above 10 eV and vibrational energy transfer is important. Molecules of a surface adsorbate or SAM act as a deformable target and ions appear to be scattered in both single and multiple collisions from it.

9.3.5 *Collisions of Cluster Ions with Surfaces*

In comparison with collisions of neutral clusters with surfaces, surface interactions of ion clusters make it possible to achieve much higher impact velocities. This can lead to extensive or even complete dissociation of the impinging cluster into its constituents. At even higher impact velocities, energy dissipation by electronic excitation and impact ionization may be important [32], [33]. The phenomenon of shattering impact of fast clusters in collision with surfaces, predicted theoretically [34], may have important consequences. It indicates, on the one hand, a facile dissipation of energy in a compressed cluster and, on the other hand, it suggests that the dissipation of energy in the cluster may make the energy available to activate reactants embedded within the cluster ("burning air" in mixed nitrogen–oxygen clusters [35]). So far the theoretical prediction of shattering collisions was supported by results of experiments on the dissociation of fast ammonia clusters [36] and by the dissociation of ions enclosed in large clusters [37]. However, the first data on chemical reactions activated by energy transfer in cluster ion–surface collisions are beginning to appear [38].

Acknowledgment

Partial support of this work by the research grant No. 203/97/0351 of the Grant Agency of the Czech Republic is gratefully acknowledged.

9.4 References

[1] *Dynamics of Gas–Surface Interactions* (eds. G. Benedek and U. Valbusa), Springer-Verlag, Berlin, 1982.

[2] *Dynamics of Gas–Surface Interactions* (eds. C. T. Rettner and M. N. R. Ashfold), Royal Society of Chemistry, London, 1991.

[3] C. T. Rettner, D. J. Auerbach, J. C. Tully, and A. W. Kleyn: *J. Phys. Chem.* **100**, 13021, (1996).

[4] *Low Energy Ion-Surface Interactions* (ed. J. W. Rabelais), Wiley Series in Ion Chemistry and Physics, Wiley, Chichester, 1994.

[5] Polyatomic ion–surface interactions (ed. L. Hanley), *Internat. J. Mass Spectrom.* **174**, 1998.

[6] R. G. Cooks, T. Ast, and M. D. A. Mabud: *Internat. J. Mass Spectrom. Ion. Proc.* **100**, 209, (1990).

[7] J. E. Hurst, C. A. Becher, J. P. Cowin, K. C. Janda, L. Wharton, and D. J. Auerbach: *Phys. Rev. Lett.* **49**, 1175, (1979).

[8] F. O. Goodman and H. Y. Wachsman: *Dynamics of Gas–Surface Scattering*, Academic Press, New York, 1976.

[9] A. W. Kleyn, A. C. Lunz, and D. J. Auerbach: *Phys. Rev. Lett.* **47**, 1169, (1981).

[10] B. D. Kay, T. D. Raymond, and M. E. Coltrin: *Phys. Rev. Lett.* **59**, 2792, (1987).

[11] C. T. Rettner, F. Fabre, J. Kimman, D. J. Auerbach, and H. Morawitz: *Surf. Sci.* **192** 107, (1982).

[12] H. Vach, A. De Martino, M. Benslimane, M. Chatelet, and F. Pradere: *J. Chem. Phys.* **100**, 8526, (1994).

[13] P. U. Andersson, A. Tomsic, M. B. Andersson, and J. B. C. Petersson: *Chem. Phys. Lett.* **279**, 100, (1997).

[14] C. L. Cleveland and U. Landman: *Science* **257**, 355, (1992).

[15] U. Even, I. Schek, and J. Jortner: *Chem. Phys. Lett.* **202**, 303, (1993).

[16] H. D. Hagstrum: *Phys. Rev.* **96**, 325, 336, (1954).

[17] B. H. Cooper and E. R. Behringer: in [4, p. 263].

[18] E. B. Dahl, E. R. Behringer, D. R. Anderson, and B. H. Cooper: *Internat. J. Mass Spectrom.* **174**, 267, (1998).

[19] H. Müller and V. Kempter: *Internat. J. Mass Spectrom.* **174**, 285, (1998).

[20] Y. Murata, in: *Unimolecular and Bimolecular Reaction Dynamics* (eds. C. Y. Ng, T. Baer, and I. Powis), Wiley, New York, 1994, p. 428.

[21] W. Heiland: in [4, p. 313].

[22] W. R. Koppers, K. Tsumori, J. H. M. Beijersbergen, T. L. Weeding, P. G. Kistemaker, and A. W. Kleyn: *Internat. J. Mass Spectrom.* **174**, 11, (1998).

[23] W. Eckstein, H. Verbeek, and S. Datz: *Appl. Phys. Lett.* **27** 527, (1975).

[24] I. S. Bitenski, E. S. Parilis, and I. A. Wojciechowski: *Nucl. Instr. Meth.* **B47**, 246, (1990).

[25] S. B. Wainhaus, H. Lim, D. G. Schultz, L. Hanley: *J. Chem. Phys.* **106**, 10329, (1997).

[26] H. I. Kenttämaa and R. G. Cooks: *Internat. J. Mass Spectrom. Ion Proc.* **64**, 79, (1984).

[27] V. H. Wysocki, H. I. Kenttämaa, and R. G. Cooks: *Internat. J. Mass Spectrom. Ion Proc.* **75**, 181, (1987).

[28] A. Burroughs, S. B. Wainhaus, and L. Hanley: *J. Phys. Chem.* **103**, 6706, (1995).

[29] J. Kubista, Z. Dolejsek, and Z. Herman: *Europ. Mass Spectrom.* **4**, 311,(1998).

[30] D. G. Schultz, S. B. Wainhaus, L. Hanley, P. de Saint Claire, and W. L. Hase: *J. Chem. Phys.* **106**, 10337, (1997).

[31] S. D. M. Bosco and W. L. Hase: *Internat. J. Mass Spectrom.* **174**, 1, (1998).

[32] R. J. Beuhler and L. Friedman: *Chem. Rev.* **86**, 521, (1986).

[33] E. Hendell and U. Even: *J. Chem. Phys.* **103**, 9045, (1995).

[34] T. Raz, U. Even, and R. D. Levine: *J. Chem. Phys.* **103**, 5394, (1995).

[35] T. Raz and R. D. Levine: *Chem. Phys. Lett.* **246**, 405, (1995).

[36] W. Christen, U. Even, T. Raz, and R. D. Levine: *Internat. J. Mass Spectrom.* **174**, 35, (1998).

[37] H. Yasumatsu, U. Kalmbach, S. Koizumi, A. Terasaki, and T. Kondow: in: *Structure and Dynamics of Clusters* (eds. T. Kondow, K. Kaya, and A. Terasaki), Universal Academy Press, Tokyo, 1996, p. 485.

[38] C. Mair, T. Fiegele, F. Biasoli, J. H. Futrell, Z. Herman, and T. D. Märk: *Phys. Rev. Lett.* (submitted).

Part III

Energy in Geometrical Thermodynamics

10
Geometrical Methods in Thermodynamics

R. Mrugała

ABSTRACT. The aim of this chapter is to investigate some general aspects of classical thermodynamics from a geometrical point of view. Contact and Riemannian geometries are mainly employed, which correspond to the first and second laws of thermodynamics, respectively. These two structures have been defined on the so-called thermodynamic phase space (TPS). For a thermodynamic system having n degrees of freedom, TPS is a $(2n + 1)$-dimensional manifold. Its contact structure may be given by a nondegenerate Pfaff form θ, for instance, by $\theta = dU - T\, dS + P\, dV - \mu\, dN$ for $n = 3$. It is important that all variables in θ are treated as independent. In that respect, the developed formalism is similar to the symplectic formalism of classical Hamiltonian mechanics, with TPS and the contact form playing, in a sense, the role analogous to the phase space and symplectic form in mechanics.

Any real thermodynamic system, having degrees of freedom of the same type, is represented in TPS by a Legendre submanifold of the contact form.

With every function on TPS one can associate a contact vector field, or contact Hamiltonian equations, which generate a one-parameter group of continuous contact transformations. Under some conditions such transformations can be regarded as thermodynamic processes while in other cases regarded only as a one-parameter deformation of Legendre submanifolds. In the latter case, from a given thermodynamic system, but we may obtain a one-parameter family of systems with different constitutive relations. Numerous examples of contact vector fields of both types are presented.

The contact form also makes it possible to define on TPS three types of brackets which generalize the well-known Poisson bracket to the odd-dimensional case.

In the second part of this chapter it is shown how a Riemannian metric can be defined on TPS in a purely phenomenological way; and also how the contact and metric structures may be obtained in a natural way from the probability distributions of statistical mechanics. They are shown to be derived from the mean and variance of the microscopic entropy, respectively. As a consequence, the components of the metric tensor may be interpreted in terms of fluctuation theory. Finally, it is shown how the same metric may be obtained in another way using the notion of relative information.

10.1 Introduction

In the last two decades thermodynamics gained much by employing modern mathematical methods developed, and successfully used, earlier in other branches of physics. In particular, the methods of differential geometry, differential forms, and Lie groups, which play a fundamental role in classical mechanics, relativity, gauge fields, the theory of dynamical systems, and control theory, can be used with remarkable success in thermodynamics. More explicitly, in this chapter we shall concentrate on applications of the methods of contact and Riemannian (metric) geometries. Quite recently the concepts of the so-called Poisson and Jacobi geometries also found their place in thermodynamic formalism.

The basic object supporting all of these types of geometries in thermodynamics is the so-called thermodynamic phase space (TPS). For a thermodynamic system having n degrees of freedom, TPS is a $(2n+1)$-dimensional manifold endowed with a contact structure defined by the Gibbs one-form θ [1] in which all the thermodynamic parameters are treated as independent.

The contact structure of TPS allows us to associate two types of flows, and hence two types of vector fields, X_f and $\overline{X}_f$, to any smooth function f on TPS. The set of all X_f's forms a Lie algebra; this is not the case for $\overline{X}_f$'s. The fields X_f found numerous applications in thermodynamics. Formally they generate continuous transformations of TPS, and thus represent continuous symmetries of the thermodynamic formalism. More practically, for some special choices of f's, integral curves of X_f's represent thermodynamic processes. In the general case, X_f's generate one-parameter families of new thermodynamic systems. Thus they allow us to find new constitutive relations (equations of state) from the known ones. In this chapter we present some known and some new examples of X_f's of both types for various f's. In this context an interesting inverse problem arises, namely that of finding f for a given transformation of equations of state. The solution to this problem is not easy because it is equivalent to solving a system of coupled partial differential equations.

By means of X_f and $\overline{X}_f$ one can define three types of brackets on TPS; the Jacobi, Cartan, and Lagrange brackets. The meaning of these brackets in thermodynamics has not been fully recognized so far. However, some of them may be used for finding new invariants of some thermodynamic processes. Particularly useful is the Lagrange bracket defined by $(f,g) = X_f g$ for two differentiable functions f and g.

The contact structure of TPS also allows us to introduce a Riemannian metric G on TPS. So far, its thermodynamic meaning is not very clear. However, G may be reduced to the $(n+1)$-dimensional Gibbs space (not discussed here) and to the n-dimensional Legendre submanifolds $\mathcal{S}$ of TPS. Physically, various $\mathcal{S}$ represent states of different thermodynamic systems.

It is remarkable that G restricted to $\mathcal{S}$ is a metric equivalent to the Riemannian metrics considered previously by Weinhold, Ruppeiner, Janyszek, Mrugała, and others.

Both the contact and Riemannian structures allow us to compare not only different states of a given thermodynamic system, but we may also compare different states of different systems. There are numerous applications of the Riemannian metric on the Legendre submanifolds $\mathcal{S}$. It is known that the thermodynamic length of a curve describing a thermodynamic process is related to the efficiency and dissipation of the process. However, not only lengths represent useful thermodynamic quantities; it seems that local quantities are far more important, i.e., those depending on properties of a single state. For instance, the Riemannian scalar curvature may be related to the stability of thermodynamic systems. This in turn is related to the existing fluctuations around the state.

It is important that both these geometric structures (contact and metric) have been obtained in a purely phenomenological way as well as in a statistical way. In the latter case, the concepts of microscopic entropy, entropy and the relative (Kullback) entropy have been used.

Let us hope that the use of geometrical methods not only helps us to better understand and describe numerous thermodynamic phenomena, but also allows us to obtain new results for complex thermodynamic systems by means of the analysis of simple thermodynamic systems.

10.2 Contact Manifolds

In this and in the next three sections we present a few basic facts about contact geometry [2]–[8] and about some applications of this geometry in thermodynamics.

Definition 2.1. A differentiable $(2n+1)$-dimensional manifold M is said to be a *contact manifold* if it carries a global differential one-form θ such that

$$\theta \wedge (d\theta)^n \neq 0, \tag{10.1}$$

where $\wedge$ denotes the exterior product and $(d\theta)^n = d\theta \wedge \cdots \wedge d\theta$ (n times). The above condition means that θ is nondegenerate; it is called the *contact form*.

According to the Darboux theorem [2] there exist local canonical (contact) coordinates (x^0, x^i, p_i), $i = 1, \ldots, n$, in which θ has the simplest canonical form

$$\theta = dx^0 + p_i \, dx^i, \qquad i = 1, \ldots, n. \tag{10.2}$$

From now on we assume the summation convention, i.e., summation over repeated indices. The nondegeneracy condition (10.1) can be geometrically interpreted in several ways. The simplest way results from (10.2) from which we see that $\theta \wedge (d\theta)^n$ is the volume form on M.

The one-form θ defines on M a $2n$-dimensional distribution D, i.e., a field of tangent $2n$-dimensional hyperplanes D_m such that

$$D = \bigcup_{m \in M} D_m, \qquad D_m = \{X \in T_m M : \theta(X) = 0\}, \qquad (10.3)$$

where X denotes a vector field on M and $T_m M$ is a tangent space to M at a point $m \in M$. Locally, D can be given by $2n$ vector fields, e.g., by [8]

$$\mathcal{P}_k = \partial/\partial p_k, \qquad \mathcal{X}_k = \partial/\partial x^k - p_k \partial/\partial x^0, \qquad k = 1, \ldots, n. \qquad (10.4)$$

The distribution D is usually called a *contact distribution* or a *contact structure* on M [2].

Remark 2.1. Note that the contact structure D is not given by a unique one-form θ. If ρ is a nowhere vanishing function on M, the form $\rho\theta$ also satisfies the nondegeneracy condition (10.1) and defines the same field D of hyperplanes.

In thermodynamics the major role is played by the maximum dimensional integral submanifolds of the contact distribution D, by the so-called *Legendre submanifolds* denoted further on by $\mathcal{S}$. The name given to $\mathcal{S}$ is justified by the fact that the well-known-in-thermodynamics Legendre transformations preserve $\mathcal{S}$, i.e., they map any Legendre submanifold onto itself (cf. Theorem 2.2 below). From the thermodynamic point of view, the dimension of any Legendre submanifold coincides with the number of thermodynamic degrees of freedom.

The theorem below shows that—geometrically—condition (10.1) means that D is maximally nonintegrable, i.e., the dimension of its integral manifolds cannot exceed n.

Theorem 2.1. *Let (M, θ) be a $(2n+1)$-dimensional contact manifold. The maximal dimension of integral submanifolds of the contact distribution D (or, equivalently, of integral submanifolds of $\theta = 0$) is n.*

Proof. The existence of n-dimensional integral submanifolds is guaranteed because they may be given, for instance, by $n+1$ equations $x^l = C^l$, $l = 0, 1, \ldots, n$, where C^l are arbitrary constants, or by $n+1$ equations

$$x^0 = \Omega(x^1, \ldots, x^n), \qquad p_i = -\frac{\partial \Omega(x^1, \ldots, x^n)}{\partial x^i}, \qquad i = 1, \ldots, n. \qquad (10.5)$$

In the second part of the proof let us assume that $\mathcal{S}$ is a k-dimensional integral manifold of $\theta = 0$ and let $k > 0$. If $\varphi : \mathcal{S} \to M$ is an embedding of $\mathcal{S}$ into M, then $\varphi^*\theta = 0$ and $\varphi^*(d\theta) = d(\varphi^*\theta) = 0$, where φ^* is a pullback induced by φ. Let $v_1, \ldots, v_k, \ldots, v_{2n}, v_{2n+1}$ be linearly independent vector fields on M such that $v_1, \ldots, v_k$ are tangent to $\mathcal{S}$, whereas $v_1, \ldots, v_{2n}$ determine the contact distribution D, i.e., $\theta(v_i) = 0$ for $i = 1, \ldots, 2n$. It is obvious that $\theta(v_{2n+1}) \neq 0$ because otherwise θ would be identically equal to zero. Moreover, for $i,j \leq k$, one has $d\theta(v_i, v_j) = 0$ because $d\theta(v_i, v_j) = \frac{1}{2}\{v_i\theta(v_j) - v_j\theta(v_i) - \theta([v_i, v_j])\} = 0$. This must be so since the three vector fields, v_i, v_j, and their commutator (Lie bracket) $[v_i, v_j]$ are tangent to $\mathcal{S}$ by virtue of the Frobenius theorem [9], [10]. However, if one or both of i or j are bigger than k, then $d\theta(v_i, v_j)$ may be different from zero. In the expression

$$(\theta \wedge (d\theta)^n)(v_1, \ldots, v_{2n}, v_{2n+1})$$
$$= \sum_\sigma \text{sgn}(\sigma)\theta(\sigma(v_1))\, d\theta(\sigma(v_2), \sigma(v_3)) \ldots d\theta(\sigma(v_{2n}), \sigma(v_{2n+1})),$$

where σ runs through the set of all permutations of $2n+1$ elements, the only nonvanishing terms are those proportional to $\theta(v_{2n+1})$. However, for $k > n$ in each such term, there must occur at least one factor $d\theta(v_i, v_j)$ with $i, j \leq k$, which is zero. Hence, for $k > n$, one has $(\theta \wedge (d\theta)^n)(v_1, \ldots, v_{2n}, v_{2n+1}) = 0$. Therefore, the assumption $\dim \mathcal{S} = k > n$ is false since it contradicts the basic assumption (10.1) that θ is nondegenerate. $\square$

Corollary. *The theorem above, together with the Frobenius theorem, shows that the contact distribution is not involutive. In other words, on M there exists no such system of local coordinates $u^1, \ldots, u^{2n}, u^{2n+1}$ in which the contact distribution would be spanned by $2n$ vector fields $\partial/\partial u^1, \ldots, \partial/\partial u^{2n}$ [9], [10]. This fact may also be interpreted that on (M, θ) there is no such system of local coordinates in which D is given by a field of "parallel" hyperplanes.*

A local description of Legendre submanifolds in terms of a generating function ϕ is given by the following theorem [2], [11].

Theorem 2.2. *For any partition $I \cup J$ of the set of indices $\{1, \ldots, n\}$ into two disjoint subsets I and J, and for a function $\phi(p_I, x^J)$ of n variables p_i, $i \in I$, and x^j, $j \in J$, the $n+1$ equations*

$$x^i = \frac{\partial \phi}{\partial p_i}, \qquad p_j = -\frac{\partial \phi}{\partial x^j}, \qquad x^0 = \phi - p_i \frac{\partial \phi}{\partial p_i}, \qquad (10.6)$$

define a Legendre submanifold $\mathcal{S}$ of M^{2n+1}. Conversely, every Legendre submanifold of (M, θ) in a neighborhood of any point is defined by these equations for at least one of the 2^n possible choices of the subset I.

Proof. The proof that $n+1$ equations (10.6) define a Legendre submanifold is immediate since substitution of (10.6) into θ gives $\theta = 0$.

For the second part of the proof assume that a Legendre submanifold $\mathcal{S}$ can be locally parameterized by n variables (p_I, x^J). This assumption is true because an n-dimensional submanifold may be "vertical" at most with respect to n variables but never with respect to $2n$ variables (p, x). Thus $\mathcal{S}$ is given by $n+1$ equations ($i \in I$, $j \in J$)

$$x^0 = x^0(p_I, x^J), \qquad p_j = p_j(p_I, x^J), \qquad x^i = x^i(p_I, x^J). \qquad (10.7)$$

262   10. Geometrical Methods in Thermodynamics

Inserting these equations into $\theta = 0$ one gets

$$\left(\frac{\partial x^0}{\partial p_i} + p_{i'}\frac{\partial x^{i'}}{\partial p_i}\right)dp_i + \left(\frac{\partial x^0}{\partial x^j} + p_i\frac{\partial x^i}{\partial x^j} + p_j\right)dx^j = 0 \qquad (10.8)$$

$(i, i' \in I, j \in J)$. Since p_I and x^J are independent, the coefficients preceding dp_i and dx^j must vanish. From the vanishing of coefficients by dx^j we obtain

$$p_j = -\frac{\partial x^0}{\partial x^j} - p_i\frac{\partial x^i}{\partial x^j} = -\frac{\partial}{\partial x^j}\left(x^0 + p_i x^i\right) = -\frac{\partial \phi(p_I, x^J)}{\partial x^j}, \qquad (10.9)$$

where we have defined

$$\phi(p_I, x^J) = x^0(p_I, x^J) + p_i x^i(p_I, x^J). \qquad (10.10)$$

From the vanishing of coefficients by dp_i, after replacing x^0 by $\phi - p_i x^i$, we gather that

$$x^i = \frac{\partial \phi(p_I, x^J)}{\partial p_i}. \qquad (10.11)$$

The last three formulas prove the theorem. □

From (10.6) we see that $\phi(p_I, x^J)$ can be called a generating function for Legendre submanifolds because all the dependent variables p_J, x^I, and x^0 are defined entirely by ϕ and its derivatives. One can also readily see the connection of these solutions for S with Legendre transformations.

The first law of thermodynamics can be geometrically expressed in terms of Legendre submanifolds.

Postulate (First Law of Thermodynamics). Any equilibrium thermodynamic system is represented in an appropriate thermodynamic phase space (M, θ) by Legendre submanifolds of the equation $\theta = 0$.

We have used plural here because in almost all cases a thermodynamic system will not be represented by a single Legendre submanifold. Rather it will be represented by an n-dimensional surface composed of some fragments of various Legendre submanifolds, one piece for one phase. The total number of Legendre submanifolds in M is infinite. In fact, through every point of M there is an infinite number of Legendre submanifolds. Only some of them represent real thermodynamic systems.

From Theorem 2.2 we see that in the contact coordinates (cf. (10.2)) a given S can in principle be represented in equivalent ways by 2^n functions ϕ of n variables. These functions correspond to various thermodynamic potentials. Therefore, for a fixed ϕ the set of equations (10.6) may be interpreted as one fundamental relation and n equations of state [1].

There is also a dual one-dimensional *characteristic distribution* Ξ defined by a global *characteristic vector field* ξ such that

$$i_\xi\, d\theta = 0, \qquad i_\xi\, \theta = 1 \qquad (\text{or} \quad i_\xi\left(\theta \wedge (d\theta)^n\right) = (d\theta)^n), \qquad (10.12)$$

where i_ξ denotes the interior product (contraction) with ξ. In contact coordinates
$$\xi = \partial/\partial x^0. \tag{10.13}$$
Thus, $TM = D \oplus \Xi = \ker \theta \oplus \ker d\theta$, where D and Ξ are two complementary vector subbundles of TM.

The fields (10.4) and (10.13) satisfy the following commutation relations [8]
$$[\mathcal{X}_i, \mathcal{X}_j] = [\mathcal{P}_i, \mathcal{P}_j] = [\mathcal{X}_i, \xi] = [\mathcal{P}_i, \xi] = 0, \quad [\mathcal{X}_i, \mathcal{P}_j] = \delta_{ij}\xi. \tag{10.14}$$

The last of these commutators shows that the distribution D is not involutive.

10.3 Contact Transformations and Contact Vector Fields

In this section we shall consider a group of diffeomorphisms Λ of M which preserve its contact structure. Apart from finite diffeomorphisms we shall also consider their infinitesimal counterparts, i.e., vector fields (generators) associated to these diffeomorphisms.

Definition 3.1. A diffeomorphism $\lambda : M \to M$ is said to be a *contact diffeomorphism* if it preserves the contact distribution D of M, i.e., λ is such that
$$\lambda^* \theta = \rho\theta, \quad \lambda \in \Lambda, \tag{10.15}$$
where ρ is a nowhere vanishing function on M and λ^* is the pull-back map induced by λ.

Note that the new transformed form $\lambda^*\theta$ is a contact form because it is nondegenerate, $\rho\theta \wedge (d(\rho\theta))^n = \rho^{n+1}\theta \wedge (d\theta)^n \neq 0$. Thus λ preserves the contact structure but not the contact form. Diffeomorphisms with $\rho = 1$ also preserve the contact form and are called *strict contact transformations*.

Analogously, by a one-parameter group of *continuous contact transformations* we mean a subgroup of mappings $\lambda_t : M \to M$ of Λ which preserve the contact distribution D, i.e., λ_t are such that
$$\lambda_t^* \theta = \rho_t \theta, \quad \lambda_t \in \Lambda, \tag{10.16}$$
where again ρ_t is a nowhere vanishing function on M.

Let X be a generator of this one-parameter subgroup of Λ, that is, X is defined by the formula
$$(Xf)(m) = \left.\frac{d}{dt}\right|_{t=0} \lambda_t^* f(m) \equiv \left.\frac{d}{dt}\right|_{t=0} f(\lambda_t(m)), \quad \forall m \in M, \tag{10.17}$$
for any function f on M. Hence X is a vector field associated to λ_t.

In terms of X the definition (10.16) of λ_t is equivalent to

$$\mathcal{L}_X \theta \equiv \left.\frac{d}{dt}\right|_{t=0} \lambda_t^* \theta = \tau_t \theta, \qquad (10.18)$$

where $\tau_t = d\rho_t/dt$ and $\mathcal{L}_X$ denotes the Lie derivative. What really matters here is the fact that $\mathcal{L}_X \theta$ is a product of θ and a function τ_t on M (if this function is equal to zero we say that θ is invariant). This justifies the following definition.

Definition 3.2. A vector field X on M is said to be a *contact vector field* if it preserves the contact structure D or, equivalently, if

$$\mathcal{L}_X \theta = \tau \theta, \quad \text{i.e.,} \quad \mathcal{L}_X \theta \wedge \theta = 0. \qquad (10.19)$$

It is easy to prove that contact vector fields form a Lie algebra and that they do not belong to the contact distribution.

Let us now treat M as a one-dimensional principal fiber bundle with the fibers being the integral curves of ξ. Then θ becomes the connection form [9], [10], and D and Ξ the horizontal and vertical distributions, respectively. Any vector X may then be decomposed into the horizontal hX and vertical vX components,

$$X = vX + hX \quad \text{where} \quad vX = \theta(X)\xi, \quad hX = X - vX. \qquad (10.20)$$

This in turn allows one to introduce the notion of covariant differentiation on M. For a real-valued function f on M its covariant differential Df is defined by

$$Df(X) = df(hX) \quad \text{or} \quad Df = df - (\xi f)\theta \qquad (10.21)$$

for any vector field X on M.

Definition 3.3. By a *contact vector field* associated with a function f on M we mean a vector field X_f defined by

$$i_{X_f}\theta \equiv \theta(X_f) = f, \quad i_{X_f} d\theta = -Df. \qquad (10.22)$$

The above two equations define the horizontal $hX_f \equiv \overline{X}_f$ and the vertical vX_f components of X_f,

$$X_f = hX_f + vX_f = \overline{X}_f + f\xi. \qquad (10.23)$$

It is easy to check that in contact coordinates [11]–[13]:

$$X_f = \frac{\partial f}{\partial p_i}\frac{\partial}{\partial x^i} + \left(p_i \frac{\partial f}{\partial x^0} - \frac{\partial f}{\partial x^i}\right)\frac{\partial}{\partial p_i} + \left(f - p_i \frac{\partial f}{\partial p_i}\right)\frac{\partial}{\partial x^0}, \qquad (10.24)$$

$$\overline{X}_f = \frac{\partial f}{\partial p_i}\frac{\partial}{\partial x^i} + \left(p_i \frac{\partial f}{\partial x^0} - \frac{\partial f}{\partial x^i}\right)\frac{\partial}{\partial p_i} - p_i \frac{\partial f}{\partial p_i}\frac{\partial}{\partial x^0}. \qquad (10.25)$$

10.3. Contact Transformations and Contact Vector Fields 265

Remark. Note that X_f and $\overline{X}_f$ are combinations of the vector fields (10.4) and ξ,

$$X_f = (\mathcal{P}_i f)\mathcal{X}_i - (\mathcal{X}_i f)\mathcal{P}_i + f\xi, \qquad \overline{X}_f = (\mathcal{P}_i f)\mathcal{X}_i - (\mathcal{X}_i f)\mathcal{P}_i, \qquad (10.26)$$

and hence it is clear that $\overline{X}_f$ belongs to the contact distribution whereas X_f does not. However, these definitions are coordinate-dependent in contradiction to Definition 3.3.

From the definition of X_f and the property of the Lie derivative, $\mathcal{L}_X = i_X d + d i_X$ [9], [10], it is easy to see that X_f is a contact vector field because

$$\mathcal{L}_{X_f}\theta = d i_{X_f}\theta + i_{X_f} d\theta = df - Df = (\xi f)\theta \sim \tau\theta. \qquad (10.27)$$

This also shows that X_f is a generator of a continuous contact transformation on M with $\tau = \xi f$. On the contrary, $\overline{X}_f$ is not a contact field because

$$\mathcal{L}_{\overline{X}_f}\theta = d i_{\overline{X}_f}\theta + i_{\overline{X}_f} d\theta = -Df = -df + (\xi f)\theta \neq \tau\theta. \qquad (10.28)$$

Moreover, the vector fields X_f form a Lie algebra because

$$\mathcal{L}_{[X_f, X_g]}\theta = [\mathcal{L}_{X_f}, \mathcal{L}_{X_g}]\theta = \mathcal{L}_{X_f}((\xi g)\theta) - \mathcal{L}_{X_g}((\xi f)\theta)$$
$$= (\mathcal{L}_{X_f}(\xi g) - \mathcal{L}_{X_g}(\xi f))\theta \sim \tau\theta. \qquad (10.29)$$

The fields $\overline{X}_f$ do not form a Lie algebra. Despite that, both X_f and $\overline{X}_f$ are useful in defining the Poisson (although degenerate), Jacobi, and weaker structures in thermodynamics [13].

In local coordinates or in a coordinate independent way one can show that X_f and $\overline{X}_f$ have the following important properties [11]:

(a) $X_c = c\xi$ $(X_1 = \xi)$, $\qquad \overline{X}_c = 0$ (c = a constant)
(b) $X_{-f} = -X_f$, $\qquad \overline{X}_{-f} = -\overline{X}_f$,
(c) $X_{f+g} = X_f + X_g$, $\qquad \overline{X}_{f+g} = \overline{X}_f + \overline{X}_g$,
(d) $X_{fg} = f X_g + g X_f - fg\xi$, $\qquad \overline{X}_{fg} = f \overline{X}_g + g \overline{X}_f$,
(e) $X_f f = f(\xi f)$, $\qquad \overline{X}_f f = 0$,
(f) $X_f f^n = n f^n (\xi f)$.

Comparing now X_f with the general form of a vector field X on M,

$$X = \dot{x}^i \frac{\partial}{\partial x^i} + \dot{p}_i \frac{\partial}{\partial p_i} + \dot{x}^0 \frac{\partial}{\partial x^0}, \qquad (10.30)$$

we obtain $2n+1$ ordinary differential equations

$$\dot{x}^i = \frac{\partial f}{\partial p_i} \equiv \mathcal{P}_i f,$$
$$\dot{p}_i = p_i \frac{\partial f}{\partial x^0} - \frac{\partial f}{\partial x^i} \equiv -\mathcal{X}_i f, \qquad (10.31)$$
$$\dot{x}^0 = f - p_i \frac{\partial f}{\partial p_i} \equiv f - p_i \mathcal{P}_i f.$$

Thus the flow induced by X_f represents a sort of *contact Hamiltonian equations* with a *contact Hamiltonian f*. In Section 10.5 we shall see that these equations have a very practical meaning in thermodynamics.

From (10.31) one sees that, unlike the Hamiltonian flows considered in classical mechanics (cf. Section 10.4), the contact Hamiltonian flow depends not only on the derivatives of f but also on f. An important feature of the contact Hamiltonian flows can easily be deduced from the property (e) of X_f. The proof of (e) goes as follows: $X_f f \equiv L_{X_f}(\theta(X_f)) = (L_{X_f}\theta)(X_f) + \theta([X_f, X_f]) = (\xi f)\theta(X_f) = f \cdot \xi f$. Writing this result in the form $X_f f \equiv df(X_f) = f \cdot \xi f$ we infer that, in the general case, X_f is not tangent to all the level surfaces $f = $ const. It is tangent only to one level surface on which $f = 0$. Moreover, if it happens that a Legendre manifold $\mathcal{S}$ is contained in the zero level surface of f, $\mathcal{S} \subset f^{-1}(0)$, then X_f is tangent to $\mathcal{S}$ (see Section 10.5). In such a case the contact Hamiltonian equations can be interpreted as describing a thermodynamic process.

10.4 Bracket Structures in Thermodynamics

Before defining various types of brackets in thermodynamics let us remember first some elementary facts concerning classical mechanics and classical Poisson brackets [2]. The basic notion of Hamiltonian conservative mechanics is that of a $2n$-dimensional phase space P^{2n} endowed with a nondegenerate closed two-form ω ($\omega^n \neq 0$, $d\omega = 0$). The pair (P^{2n}, ω) is called a *symplectic manifold* and ω a *symplectic form*. According to the Darboux theorem [2] there exists on P a system of local canonical coordinates (p_i, q^i) in which $\omega = dp_i \wedge dq^i$. The dynamics of a system with a Hamiltonian function $H(p,q)$ is governed by a *Hamiltonian vector field* X_H defined by H and ω according to the formula $i_{X_H}\omega = -dH$, where i again denotes the interior product. In canonical coordinates X_H is given by

$$X_H = \frac{\partial H}{\partial p_i}\frac{\partial}{\partial q^i} - \frac{\partial H}{\partial q^i}\frac{\partial}{\partial p_i}. \qquad (10.32)$$

That means that the *Hamiltonian equations* (Hamiltonian flow) for the system are of the form

$$\dot{q}^i = \frac{\partial H}{\partial p_i}, \qquad \dot{p}_i = -\frac{\partial H}{\partial q^i}. \qquad (10.33)$$

It is easy to show that ω is invariant with respect to X_H, i.e., the Lie derivative of ω is equal to zero, $\mathcal{L}_{X_H}\omega = 0$. For any two smooth real-valued functions f and g on P one can define a *Poisson bracket* $\{f,g\}$ in a number of equivalent ways,

$$\{f,g\} = \omega(X_f, X_g) = X_f g = \mathcal{L}_{X_f} g, \qquad (10.34)$$

10.4. Bracket Structures in Thermodynamics

or in canonical coordinates

$$\{f,g\} = \frac{\partial f}{\partial p_i}\frac{\partial g}{\partial q^i} - \frac{\partial f}{\partial q^i}\frac{\partial g}{\partial p_i}. \tag{10.35}$$

Before defining the brackets of two functions in thermodynamics we first generalize the notion of a Poisson bracket and a Poisson manifold [14].

Definition 4.1. A *Poisson bracket* on a smooth manifold N is a mapping $\{\,,\,\}: C^\infty(N,\mathbf{R}) \times C^\infty(N,\mathbf{R}) \longrightarrow C^\infty(N,\mathbf{R})$ on the space of C^∞ real-valued functions on N with the following properties (the dimension of N, say r, may be arbitrary):

(1) bilinearity, $\{f, \lambda g_1 + \mu g_2\} = \lambda\{f,g_1\} + \mu\{f,g_2\}$, $\forall \lambda, \mu \in \mathbf{R}$;

(2) skew-symmetry, $\{f,g\} = -\{g,f\}$;

(3) Jacobi identity, $\{f,\{g,h\}\} + \{g,\{h,f\}\} + \{h,\{f,g\}\} = 0$; and

(4) Leibniz rule, $\{f, gh\} = g\{f,h\} + \{f,g\}h$.

It is also said that $\{\,,\,\}$ defines a *Poisson structure* on N and the pair $(N, \{\,,\,\})$ is called a *Poisson manifold*.

Following this definition one can show that the Poisson bracket in local coordinates $(y^1, \ldots, y^r)$ has the form [14]

$$\{f,g\} = \sum_{k,l=1}^{r} J^{kl}(y) \frac{\partial f}{\partial y^k}\frac{\partial g}{\partial y^l} = \sum_{k,l=1}^{r} \{y^k, y^l\} \frac{\partial f}{\partial y^k}\frac{\partial g}{\partial y^l}, \tag{10.36}$$

where $J^{kl}(y)$ are some functions on N. These functions $J^{kl}(y)$ must be such that they satisfy all the properties listed in Definition 4.1. One can also see that $\{f,g\}$ is known if we know Poisson brackets for the local coordinates. Moreover, f and g enter the Poisson bracket only through their first partial derivatives (this is due to the Leibniz rule). We shall soon see that this is not the case for other brackets where f and/or g may appear directly as well.

According to the Darboux theorem there exists on N a canonical system of local coordinates $(p^1, \ldots, p^n, q^1, \ldots, q^n, z^1 \ldots, z^h)$ in which

$$\{f,g\} = \sum_{k,l=1}^{n} \left(\frac{\partial f}{\partial p^k}\frac{\partial g}{\partial q^l} - \frac{\partial f}{\partial q^k}\frac{\partial g}{\partial p^l} \right), \quad 2n + h = r. \tag{10.37}$$

Poisson manifolds form a bigger class than that of symplectic manifolds because, in particular, N need not be even-dimensional. Besides, the Poisson bracket may be degenerate for any dimension of N; it can be seen from the two previous formulas.

Definition 4.2. A *Jacobi bracket* on a differential manifold N is a mapping $\{\,,\,\}: C^\infty(N,\mathbf{R}) \times C^\infty(N,\mathbf{R}) \to C^\infty(N,\mathbf{R})$ on the space of C^∞

real-valued functions on N having the first three properties of the Poisson bracket, but with the Leibniz rule replaced by a weaker condition,

(4') $\operatorname{Supp}\{f,g\} \subset \operatorname{Supp} f \cap \operatorname{Supp} g$.

Any Poisson structure on a contact manifold (M,θ) must be degenerate because $\dim M = 2n+1$. However, one may define on (M,θ) some less symmetric structures [13], [14] than the Poisson bracket.

Definition 4.3. On a contact manifold (M,θ) we define the following brackets:

a *Jacobi bracket*

$$\{f,g\} = \theta([X_f, X_g]) = X_f(g) - (\xi f)g$$
$$= -d\theta(X_f, X_g) + f(\xi g) - g(\xi f), \quad (10.38)$$

a *Cartan bracket*

$$[f,g] = \overline{X}_f(g) = -d\theta(X_f, X_g) = -d\theta(\overline{X}_f, \overline{X}_g), \quad (10.39)$$

and a *Lagrange bracket*

$$(f,g) = X_f(g) = \{f,g\} + g(\xi f) = [f,g] + f(\xi g). \quad (10.40)$$

In the above definitions we have also included some properties (or equivalent definitions) of these brackets. These brackets could also be defined in terms of the vector fields $\mathcal{P}_i$, $\mathcal{X}_i$, and ξ as

$$\{f,g\} = (\mathcal{P}_i f)\mathcal{X}_i g - (\mathcal{X}_i f)\mathcal{P}_i g + f\xi g - g\xi f,$$
$$[f,g] = (\mathcal{P}_i f)\mathcal{X}_i g - (\mathcal{X}_i f)\mathcal{P}_i g, \quad (10.41)$$
$$(f,g) = (\mathcal{P}_i f)\mathcal{X}_i g - (\mathcal{X}_i f)\mathcal{P}_i g + f\xi g.$$

The latter relations are particularly convenient to see the symmetry properties of the brackets.

In contact coordinates

$$\{f,g\} = \frac{\partial f}{\partial p_i}\frac{\partial g}{\partial x^i} - \frac{\partial f}{\partial x^i}\frac{\partial g}{\partial p_i} + p_i\left(\frac{\partial f}{\partial x^0}\frac{\partial g}{\partial p_i} - \frac{\partial f}{\partial p_i}\frac{\partial g}{\partial x^0}\right)$$
$$+ f\frac{\partial g}{\partial x^0} - g\frac{\partial f}{\partial x^0}, \quad (10.42)$$

$$[f,g] = \frac{\partial f}{\partial p_i}\frac{\partial g}{\partial x^i} - \frac{\partial f}{\partial x^i}\frac{\partial g}{\partial p_i} + p_i\left(\frac{\partial f}{\partial x^0}\frac{\partial g}{\partial p_i} - \frac{\partial f}{\partial p_i}\frac{\partial g}{\partial x^0}\right), \quad (10.43)$$

$$(f,g) = \frac{\partial f}{\partial p_i}\frac{\partial g}{\partial x^i} - \frac{\partial f}{\partial x^i}\frac{\partial g}{\partial p_i} + p_i\left(\frac{\partial f}{\partial x^0}\frac{\partial g}{\partial p_i} - \frac{\partial f}{\partial p_i}\frac{\partial g}{\partial x^0}\right)$$
$$+ f\frac{\partial g}{\partial x^0}. \quad (10.44)$$

Remark 4.1. In the subspace $C_0^\infty(M, \mathbf{R}) \subset C^\infty(M, \mathbf{R})$ of the first integrals of ξ, i.e., for f and g such that $\xi f \equiv \mathcal{L}_\xi f = \xi g \equiv \mathcal{L}_\xi g = 0$ all these brackets reduce to the standard Poisson bracket $\{f, g\}_{d\theta} = -d\theta(X_f, X_g)$ as can be easily seen in local coordinates,

$$\{f,g\} = [f,g] = (f,g) \equiv \{f,g\}_{d\theta} = \frac{\partial f}{\partial p_i}\frac{\partial g}{\partial x^i} - \frac{\partial f}{\partial x^i}\frac{\partial g}{\partial p_i}. \quad (10.45)$$

From the local expressions for these brackets one can see that the $\{\,,\,\}$ bracket has all the properties of the Poisson bracket but the Leibniz rule. The $[\,,\,]$ bracket is missing the Jacobi identity. The $(\,,\,)$ bracket is neither antisymmetric nor fulfills the Jacobi identity; however, it is bilinear and obeys the Leibniz rule but only in the second entry. Despite these shortcomings only the third bracket produces directly the contact Hamiltonian flow because the components of X_f are equal to

$$\dot{x}^i = (f, x^i), \qquad \dot{p}_i = (f, p_i), \qquad \dot{x}^0 = (f, x^0). \quad (10.46)$$

For more details about these brackets we refer the reader to [13].

10.5 Thermodynamic Examples of Contact Flows

In thermodynamics M is usually a subset of $\mathbf{R}^{2n+1}$ and in the energy representation [1] we have the following correspondence

$$(x^0; x^1, x^2, x^3, \ldots; p_1, p_2, p_3, \ldots) \iff (U; S, V, N_1, \ldots; -T, P, -\mu_1, \ldots), \quad (10.47)$$

and, respectively,

$$\theta = dU - T\,dS + P\,dV - \mu_k\,dN^k, \qquad k = 1, \ldots, n-2, \quad (10.48)$$

where all the symbols have their standard thermodynamic meaning [1]. It is important that unless restricted to a Legendre submanifold, all these $2n+1$ variables are treated as independent. Of course, we could have worked in the entropy or in any other representation.

The most obvious thermodynamic applications of X_f, and the brackets which come to one's mind, could be as follows:

- any X_f induces a continuous contact transformation of M, i.e., a mapping of M onto itself such that $\mathcal{L}_{X_f}\theta = (\xi f)\theta \sim \theta$;
- as already mentioned above, X_f or the $(\,,\,)$ bracket allows one to find the *thermodynamic contact Hamilton equations* $\dot{x}^i = (f, x^i)$, $\dot{p}_i = (f, p_i)$, $\dot{x}^0 = (f, x^0)$; and
- $(f, g) = X_f g = \mathcal{L}_{X_f} g$ allows one to find first integrals of the flow induced by X_f. If $(f, g) = 0$ and $(f, h) = 0$, then also $(f, gh) = 0$.

270   10. Geometrical Methods in Thermodynamics

In our previous papers [11]–[13] we have discussed some examples of the contact Hamiltonian equations. Some of them described thermodynamic processes while the others mapped (deformed) submanifolds representing one kind of system onto submanifolds representing another kind of system. For instance, the ideal gas was mapped onto the gas of hard spheres or van der Waals-like gas.

Before we go to concrete examples of X_f and their associated contact Hamiltonian equations, we shall prove a general theorem about the mutual relations between vector fields X_f and the Legendre submanifolds S.

Theorem 5.1 [11].   *Let S be a Legendre submanifold of a contact manifold (M, θ). X_f is tangent to S if and only if f vanishes on S, i.e., $S \subset f^{-1}(0)$.*

Proof. First assume that X_f is tangent to S. Then, from the definitions of X_f and S, we have $\theta(X_f) = f = 0$, and the result follows.

Conversely, if $S \subset f^{-1}(0)$, then $\theta(X_f) = 0$ on S. We know that S is n-dimensional, so let $v_1, \ldots, v_n$ be a basis of n linearly independent vector fields tangent to S. Then $\theta(v_i)|_S = 0$, $df(v_i)|_S = 0$, and $d\theta(v_i, v_j)|_S = 0$ for $i, j = 1, \ldots, n$. It suffices now to show that X_f is a linear combination of $v_1, \ldots, v_n$ at every point of $f^{-1}(0)$. Indeed,

$$d\theta(X_f, v_i) = \left(i_{X_f} d\theta\right)(v_i) = -Df(v_i) = -df(v_i) + (\xi f)\theta(v_i) = 0$$

on $f^{-1}(0)$. Hence, according to the nondegeneracy condition (10.1) and the Frobenius theorem, we see that $X_f = \alpha_i v_i$ on S (α_i are some functions on M). This means that X_f is tangent to $S \subset f^{-1}(0)$.   □

Example 5.1.   For $f = U - TS + RNT - \mu N$, according to (10.24), (10.47), and (10.48) we have

$$X_f = (S - RN)\frac{\partial}{\partial S} + N\frac{\partial}{\partial N} + P\frac{\partial}{\partial P} + RT\frac{\partial}{\partial \mu} + U\frac{\partial}{\partial U}, \quad (10.49)$$

and hence the contact Hamiltonian equations (10.31) (defined by components of X_f) have the form

$$\dot{T} = \dot{V} = 0, \quad \dot{P} = P, \quad \dot{\mu} = RT, \quad \dot{S} = S - RN, \quad \dot{N} = N, \quad \dot{U} = U. \quad (10.50)$$

Their integral curves are given by

$$T = T_0, \quad P = P_0 e^t, \quad \mu = RT_0 t + \mu_0,$$
$$S = (S_0 - RN_0 t)e^t, \quad V = V_0, \quad N = N_0 e^t, \quad U = U_0 e^t. \quad (10.51)$$

Because, for an ideal gas $f = 0$, X_f is tangent to the Legendre submanifold S representing this gas and describes a "thermodynamic process" with constant volume V_0 and temperature T_0. It is easy to check that during this

10.5. Thermodynamic Examples of Contact Flows

"process" all relations between thermodynamic parameters for an ideal gas are preserved, for instance,

$$PV = NRT, \qquad U = \tfrac{3}{2}NRT, \qquad \text{or} \qquad U = TS - PV + \mu N. \quad (10.52)$$

Example 5.2. For $f = NRT - \tfrac{2}{5}TS - \tfrac{2}{5}\mu N$ one obtains

$$X_f = \left(\tfrac{2}{5}S - RN\right)\frac{\partial}{\partial S} + \tfrac{2}{5}N\frac{\partial}{\partial N} - \tfrac{2}{5}T\frac{\partial}{\partial T} + \left(RT - \tfrac{2}{5}\mu\right)\frac{\partial}{\partial \mu}, \quad (10.53)$$

and thus the integral curves of X_f take the form

$$S = (S_0 - RN_0 t)e^{2t/5}, \quad V = V_0, \quad N = N_0 e^{2t/5},$$

$$T = T_0 e^{-2t/5}, \quad P = P_0, \quad \mu = (\mu_0 + RT_0 t)e^{-2t/5}, \quad U = U_0. \quad (10.54)$$

They describe an isobaric, isochoric, and isoenergetic "process." Again it is easy to prove that the relations (10.52) are preserved.

In the two examples above the functions f have been chosen in such a way that the Legendre submanifold S of the ideal gas was placed on the level hypersurfaces $f^{-1}(0)$. Therefore, X_f was tangent to S and could be treated as a "thermodynamic process." The situation is quite different if S is not placed on $f^{-1}(0)$. In the following examples X_f is not tangent to S and cannot be treated as a generator of a thermodynamic process; rather as a generator of a one-parameter family of thermodynamic systems.

Example 5.3. Let f be an affine function of the intensive parameters only, $f = a + b^i p_i$. Then

$$\dot{x}^i = b^i, \qquad \dot{p}_i = 0, \qquad \dot{x}^0 = a, \quad (10.55)$$

and, subsequently,

$$x^i = x_0^i + b^i t, \qquad p_i = p_{i0}, \qquad x^0 = x_0^0 + at. \quad (10.56)$$

Thus the intensive parameters are kept constant, whereas the extensive ones are linear functions of t. None of equations (10.52) is preserved in this case. Instead X_f produces a continuous one-parameter family of thermodynamic systems (one-parameter family of Legendre submanifolds S_t). An interesting situation occurs for f reduced to $f = bP$ [11]. Then $V = V_0 + bT$ while all the other parameters are fixed. For a fixed b, S_t represents a one-parameter family of gases of hard spheres (cf. also Example 5.6).

Example 5.4. If $f = a + b_i x^i$ is an affine function of the extensive parameters, then X_f also belongs to the new class of contact vector fields. The integral curves of X_f now assume the form

$$\dot{x}^i = x_0^i, \qquad \dot{p}_i = p_{i0} - b_i t, \qquad \dot{x}^0 = x_0^0 + (a + b_i x_0^i)t, \quad (10.57)$$

and they do not represent a thermodynamic process. The meaning of X_f in this case is not clear.

Example 5.5. Let us now take $f = x^0 - \phi(x^1, \ldots, x^n)$. Then

$$\dot{x}^i = 0, \qquad \dot{p}_i = p_i + \frac{\partial \phi}{\partial x^i}, \qquad \dot{x}^0 = x^0 - \phi. \tag{10.58}$$

Again X_f produces a one-parameter family of Legendre submanifolds $\mathcal{S}_t$ from a given $\mathcal{S}$. However, if it happens that $\phi(x^1, \ldots, x^n)$ is such that $x^0 = \phi(x^1, \ldots, x^n)$ represents the fundamental relation [1] for the system (cf. (10.5)), then $X_f|_\mathcal{S} = 0$ and $\mathcal{S}$ is obviously preserved.

Example 5.6. If we take $f_1 = bP$, where b is a nonnegative constant, the integral curves of $X_{f_1} = b\partial/\partial V$ are such that all parameters are preserved except the volume V which changes according to $V = V_0 + bt$. Therefore, X_{f_1} maps ideal gas into a gas of noninteracting hard spheres.

On the other hand, for $f_2 = -aV^{-1}$, $a > 0$, $X_{f_2} = (-a/V)\partial/\partial U - (a/V^2)\partial/\partial P$ is such that (notice a new parameter τ)

$$U = U_0 - \frac{a}{V_0}\tau, \qquad P = P_0 - \frac{a}{V_0^2}\tau, \tag{10.59}$$

while all the other parameters are preserved. This time one can say that X_{f_2} maps an ideal gas into a gas of interacting pointlike particles.

Now let us take $f = f_1 + f_2 = bP - aV^{-1}$. The integral curves of X_f are such that T, S, N, and μ do not change, whereas

$$V = V_0 + bt, \qquad U = U_0 - \frac{a}{b}\ln\frac{V_0 + bt}{V_0}, \qquad P = P_0 - \frac{at}{V_0(V_0 + bt)}. \tag{10.60}$$

The equation of state for the ideal gas, $P_0 V_0 = N_0 R T_0$, is no more preserved and it goes over into an equation of state

$$\left(P + \frac{at}{V(V - bt)}\right)(V - bt) = NRT, \tag{10.61}$$

which for $t = 1$ resembles the well-known van der Waals equation of state. In fact, for a fixed a and b, we have obtained a one-parameter family of van der Waals gases.

Example 5.7. Two other modifications of the van der Waals gas can be obtained if, instead of one transformation induced by $X_{f_1+f_2}$, we consider two consecutive transformations [15]: that of X_{f_1} followed by X_{f_2} and vice versa. We receive two different two-parameter transformations since the transformations induced by f_1 and f_2 do not commute. This can be seen from the Lie bracket,

$$[X_{f_1}, X_{f_2}] = \left[b\frac{\partial}{\partial V}, -\frac{a}{V}\frac{\partial}{\partial U} - \frac{a}{V^2}\frac{\partial}{\partial P}\right] = \frac{ab}{V^2}\frac{\partial}{\partial U} + \frac{2ab}{V^3}\frac{\partial}{\partial P} \neq 0, \tag{10.62}$$

or from the Jacobi bracket (10.42),

$$\{f_1, f_2\} = \left\{bP, \frac{-a}{V}\right\} = \frac{\partial f_1}{\partial P}\frac{\partial f_2}{\partial V} = \frac{ab}{V^2} \neq 0. \tag{10.63}$$

In the case when X_{f_1} is followed by X_{f_2}, instead of (10.61) we receive a two-parameter family of equations of state

$$\left(P + \frac{a}{V^2}\tau\right)(V - bt) = NRT. \tag{10.64}$$

The result is different if X_{f_2} is followed by X_{f_1}, where

$$\left(P + \frac{a}{(V - bt)^2}\tau\right)(V - bt) = NRT. \tag{10.65}$$

As a matter of fact, (10.64) exactly reproduces the standard van der Waals equation. I have been informed [15] that it is possible to find a function f which allows one to obtain the van der Waals equation of state from the ideal gas in just one step.

Other examples of X_f can be found in [11]–[13]. Recently, the inverse problem, i.e., finding f for a given transformation of equations of state, has been studied by J. Jurkowski [16].

10.6 Almost Contact and Contact Metric Structures

In the remaining part of this chapter we shall discuss metric structures in thermodynamics. First we shall consider a $(2n + 1)$-dimensional differentiable manifold denoted by M^{2n+1}, or briefly by M, but having a more general structure than the contact structure. Later on we shall return to M with a contact structure.

Definition 6.1. A differentiable $(2n + 1)$-dimensional manifold M is said to have an *almost contact structure* if it admits three tensor fields φ, ξ, and η of the type (1,1), (1,0), and (0,1), respectively, such that

$$\varphi^2 = -I + \eta \otimes \xi, \quad \eta(\xi) = 1, \tag{10.66}$$

where I is the $(2n + 1) \times (2n + 1)$ identity operator.

Note that ξ is not necessarily the same vector field as previously and η does not have to be a contact form, for instance, it may be degenerate.

From (10.66) one can derive three additional properties of φ, ξ, and η, namely

$$\varphi\xi = 0, \quad \eta \circ \varphi = 0, \quad \text{rank } \varphi = 2n. \tag{10.67}$$

An almost contact structure is also called a (φ, ξ, η)-structure.

Definition 6.2. If M with an almost contact structure admits an additional Riemannian metric G such that for any vector fields X, Y, or M one has

$$G(\varphi X, \varphi Y) = G(X, Y) - \eta(X)\eta(Y), \qquad (10.68)$$

then G is said to be an *associated Riemannian metric* to the given almost contact structure and M is called an *almost contact metric manifold*.

One can also say that M has a (φ, ξ, η, G)-structure and that G is a metric compatible with the given (φ, ξ, η)-structure. For $Y = \xi$, due to (10.66)–(10.68), we have

$$\eta(X) = G(X, \xi), \quad \text{i.e.,} \quad \eta_i = G_{ij}\xi^j, \qquad (10.69)$$

which means that η is the covariant counterpart of ξ with respect to G. Due to Sasaki [4] we have the following:

Proposition. *Let M be a differentiable manifold with a (φ, ξ, η)-structure, then there exists a positive definite metric G on M such that $G(\varphi X, \varphi Y) = G(X, Y) - \eta(X)\eta(Y)$.*

To prove this result one takes an arbitrary Riemannian metric ℓ on M which, in general, has nothing in common with the (φ, ξ, η)-structure of M. Next one defines another Riemannian metric h as

$$h(X, Y) = \ell(\varphi^2 X, \varphi^2 Y) + \eta(X)\eta(Y) \qquad (10.70)$$

for which $h(X, \xi) = \eta(X)$. Finally, one defines G by the formula

$$G(X, Y) = \tfrac{1}{2}\bigl[h(X, Y) + h(\varphi X, \varphi Y) + \eta(X)\eta(Y)\bigr]. \qquad (10.71)$$

Now, replacing X and Y by φX and φY in (10.71) one gets the required property (10.68) for G.

Remark 6.1. The metric G is not unique on M because it depends on an arbitrarily chosen metric ℓ.

The geometrical meaning of G is as follows. The equation $\eta = 0$ defines on M a $2n$-dimensional distribution $\mathcal{D}$, i.e., a field of $2n$-dimensional hyperplanes tangent to M. The equation $\eta(\xi) = 1$ means that the vector field ξ defines a one-dimensional distribution, complementary to $\mathcal{D}$. Since rank $\varphi = 2n$ and $\varphi\xi = 0$, φ acts in a nontrivial way only on the distribution $\mathcal{D}$, i.e., φ acting on an arbitrary vector X tangent to M at a point $m \in M$ produces a vector $\varphi X \in \mathcal{D}$. Hence, the metrics h and G coincide on $\mathcal{D}$, whereas ξ is orthogonal to $\mathcal{D}$ with respect to h. From (10.68) we see that G is φ-invariant on $\mathcal{D}$ and that ξ is orthogonal to $\mathcal{D}$ with respect to G. In such a way, G is an extension of a Hermitian metric in a $2n$-dimensional almost complex space to a $(2n+1)$-dimensional almost contact space [3].

It was shown by Sasaki and Hatakeyama [4] that any contact manifold admits an almost contact metric structure. The four-step construction goes as follows:

1. Again let (M, θ) be a contact manifold and let ξ and D again be two complementary distributions defined by (cf. Section 10.2)

$$\theta(\xi) = 1, \qquad i_\xi \, d\theta = 0, \qquad D = \{X : \theta(X) = 0\}. \qquad (10.72)$$

2. Having an arbitrary metric ℓ on M one can define a new Riemannian metric h by

$$h(X, Y) = \ell(-X + \theta(X)\xi, -Y + \theta(Y)\xi) + \theta(X)\theta(Y). \qquad (10.73)$$

Setting $Y = \xi$ one obtains $h(X, \xi) = \theta(X)$.

3. From (10.73) one can see that ξ is orthogonal to D with respect to h. Moreover, from (10.1) and (10.72) one sees that $d\theta$ is nondegenerate on D. Thus one can choose a metric g' and a tensor field φ' on D such that $g'(X, \varphi' Y) = d\theta(X, Y)$ and $\varphi'^2 = -I$.

4. One extends g' to a metric G on M in such a way that G is compatible with h in the direction of ξ, and one extends φ' to φ such that $\varphi \xi = 0$ and $\varphi^2 = -I + \theta \otimes \xi$.

Remark 6.2. Every contact manifold admits a (φ, ξ, η)-structure but not vice versa. For instance, for a (φ, ξ, η)-structure η may be such that $d\eta = 0$. This is not the case for contact manifolds.

10.7 Construction of a Contact Metric

Now we shall construct a Riemannian metric on a contact space. In a sense, the proposed construction seems to be arbitrary, and in particular the choice of φ seems to be arbitrary. However, the obtained metric is physically well motivated and in the next section we shall show how it may be derived in another way in statistical physics.

Again let (M, θ) be a contact manifold and let (x^0, p_i, x^i) be local contact coordinates in which θ assumes the simplest form $\theta = dx^0 + p_i \, dx^i$, $i = 1, \ldots, n$. Then ξ is given by $\xi = \partial/\partial x^0$.

Let us choose φ in the form

$$\varphi = \begin{pmatrix} 0 & p_1 \cdots p_n & 0 \\ \hline 0 & 0 & I_n \\ \hline 0 & -I_n & 0 \end{pmatrix}, \qquad (10.74)$$

where I_n is an $n \times n$ unit matrix. It is easy to show that φ, ξ, and θ satisfy the conditions

$$\varphi^2 = -I + \theta \otimes \xi, \qquad \varphi \xi = 0, \qquad \theta \circ \varphi = 0, \qquad (10.75)$$

276   10. Geometrical Methods in Thermodynamics

where I is a $(2n+1) \times (2n+1)$ unit matrix. Therefore, it is clear that the given φ, ξ, and θ define an almost contact or a (φ, ξ, θ)-structure on M. An associated Riemannian metric may be introduced in the following way.

Definition 7.1. The bilinear, nondegenerate, and symmetric form

$$G = dp_i\, dx^i + \theta \otimes \theta, \qquad i = 1, \ldots, n, \tag{10.76}$$

will be called an *associated Riemannian metric* on the contact space (M, θ). Here $dp_i\, dx^i = dp_i \otimes dx^i + dx^i \otimes dp_i$.

Remark 7.1. The first equation of (10.75) means that any vector X tangent to M may be decomposed into the "horizontal" and "vertical" parts, $\operatorname{hor} X = -\varphi^2 X$ and $\operatorname{ver} X = \theta(X)\xi$, respectively, i.e.,

$$X = \operatorname{hor} X + \operatorname{ver} X = -\varphi^2 X + \theta(X)\xi. \tag{10.77}$$

Remark 7.2. The metric defined by (10.76) fulfills the condition (10.68), i.e.,

$$G(X, Y) = G(\varphi X, \varphi Y) + \theta(X)\theta(Y) \tag{10.78}$$

for any vector fields X and Y on M.

In contact coordinates G takes the form

$$G = dx^0\, dx^0 + 2p_i\, dx^0\, dx^i + dp_i\, dx^i + p_i p_j\, dx^i\, dx^j, \tag{10.79}$$

and the components of G are given by the matrix

$$(G_{\mu\nu}) = \left(\begin{array}{c|c|c} 1 & 0 & p_1 \cdots p_n \\ \hline 0 & 0 & I_n \\ \hline \begin{matrix} p_1 \\ \vdots \\ p_n \end{matrix} & I_n & p_i p_j \end{array} \right), \tag{10.80}$$

with $i, j = 1, \ldots, n$ and $\mu, \nu = 1, \ldots, 2n+1$.

The determinant of G is equal to

$$\det(G_{\mu\nu}) = (-1)^n, \tag{10.81}$$

so G is nondegenerate.

From the construction of G one can easily see that G is a universal metric on a $(2n+1)$-dimensional thermodynamic phase space M. It is universal in the sense that G is the same for all systems having the same number of thermodynamic degrees of freedom of the same type. It does not take into account any individual properties of any concrete system.

At the moment it seems that in thermodynamics the full metric G is not as important as its reduction g to the Legendre submanifolds $\mathcal{S}$ of θ. As an example let us take a special Legendre submanifold $\mathcal{S}$ given by the following $n+1$ equations (cf. (10.5)):

$$x^0 = \Omega(x^1,\ldots,x^n), \qquad p_i = -\frac{\partial \Omega(x^1,\ldots,x^n)}{\partial x^i}. \tag{10.82}$$

It is obvious that

$$g = G|_\mathcal{S} = dp_i\, dx^i|_\mathcal{S} = -\frac{\partial^2 \Omega(x)}{\partial x^i\, \partial x^j}\, dx^i\, dx^j, \tag{10.83}$$

and thus we have obtained a metric on the n-dimensional Legendre manifold $\mathcal{S}$ given by the second derivatives of a "potential" function $\Omega(x)$. This result may be generalized in such a way that g will no longer be connected with the particular parameterization (10.82) of $\mathcal{S}$. To this end, we use Theorem 2.2 of Section 10.2 which says that any $\mathcal{S}$ can be parameterized by a subset of contact coordinates in one of 2^n possible ways as

$$x^i = \frac{\partial \phi}{\partial p_i}, \qquad p_j = -\frac{\partial \phi}{\partial x^j}, \qquad x^0 = \phi - p_i \frac{\partial \phi}{\partial p_i}, \tag{10.84}$$

where $\phi = \phi(p_I, x^J)$ is a function of n variables p_i, $i \in I$, and x^j, $j \in J$. Using (10.84) we obtain

$$g^\phi = \frac{\partial^2 \phi}{\partial p_i\, \partial p_{i'}}\, dp_i\, dp_{i'} - \frac{\partial^2 \phi}{\partial x^j\, \partial x^{j'}}\, dx^j\, dx^{j'}, \qquad i, i' \in I, \quad j, j' \in J. \tag{10.85}$$

We notice that in contact coordinates there are no terms of the type $dp_i\, dx^j$. This means that the nonzero entries of the metric matrix reside in diagonal blocks.

The metric (10.85) corresponds exactly to the metrics introduced in quite different ways in thermodynamics by Weinhold and Ruppeiner. Weinhold [17] used the energy representation and he rather arbitrarily defined the components of his metric tensor as the second derivatives of internal energy U with respect to volume V, entropy S, and so on. On the contrary, Ruppeiner [18], [19] worked in the entropy representation because he started from the Einstein fluctuation formula. Our formula (10.85) generalizes their metrics to an arbitrary potential ϕ and puts them on the same footing. However, it must be added that the energy and entropy representations are equivalent up to a discrete contact transformation (with $\rho = T^{-1}$) which is not a Legendre transformation.

All these representations proved to be very useful in the thermodynamic fluctuation theory and in the theory of stability of thermodynamic systems.

10.8 Statistical Derivation of G

The basic question to be answered is why G has been chosen in the form (10.76). In particular, one has to give some justification for the term $dp_i\, dx^i$ in G because the second term $\theta \otimes \theta$ is rather standard and generally accepted [3], [4] because it only removes degeneration of the first term in G. It turns out that in thermodynamics the contact structure and the metric structure may be derived in a systematic way from a generalized Gibbs probability distribution ρ [20].

Statistical mechanics aims at explaining the thermal properties of substances by taking into account their microscopic (discrete) structure. To this end, it uses classical and quantum mechanics as well as probability and information theory. Statistical mechanics does not investigate the detailed states of all individual micro-objects (atoms, molecules,...) in the system, but tries to describe only their collective statistical behavior. This collective behavior is described by a function (or operator in the quantum case) defining a probability density over the space of all plausible microstates of the system. Having such a probability density, it is possible to calculate mean values of various physical quantities as well as their fluctuations around these mean values.

Let us consider a physical system with Γ as the space of all of its microscopical states. These states will be labeled by $y = (y^1, \ldots, y^l)$, where l denotes the number of microscopic degrees of freedom. Let $\rho : \Gamma \to \mathbf{R}_+$ be a normalized or nonnormalized probability distribution on Γ and let $F^i : \Gamma \to \mathbf{R}$, $i = 1, \ldots, n$, be a set of stochastic variables on Γ. Following the Jaynes maximum entropy (information) principle [21] we take ρ in the form

$$\rho(y; w, p_1, \ldots, p_n) = \exp\bigl[-w + p_i F^i(y)\bigr], \qquad y \in \Gamma, \qquad (10.86)$$

where $p = (p_1, \ldots, p_n)$ are some macroscopic (nonstochastic) parameters called statistical temperatures; they characterize the state of environment [22]. For a nonnormalized probability distribution, w is a free parameter, while for ρ normalized, w is a function of $p_1, \ldots, p_n$, namely

$$Z(p) = e^w = \operatorname{Tr}\exp\bigl[p_i F^i(y)\bigr]. \qquad (10.87)$$

The mean values x^i of $F^i(y)$ are now given by

$$x^i = \langle F^i \rangle = \operatorname{Tr}(\rho F^i) = \frac{\partial \ln Z}{\partial p_i} = \frac{\partial w}{\partial p_i}, \qquad (10.88)$$

whereas their variances are equal to

$$\langle (F^i - x^i)(F^j - x^j) \rangle = \frac{\partial^2 \ln Z}{\partial p_i\, \partial p_j} = \frac{\partial^2 w}{\partial p_i\, \partial p_j} = \frac{\partial x^i}{\partial p_j} = \frac{\partial x^j}{\partial p_i}. \qquad (10.89)$$

10.8. Statistical Derivation of G

The last equation also gives

$$dx^i = \langle (F^i - x^i)(F^j - x^j)\rangle \, dp_j. \tag{10.90}$$

The relation (10.90) holds for nonnormalized ρ as well.

Now let us define the microscopic entropy

$$s = -\ln\rho = w - p_i F^i, \tag{10.91}$$

and its differential

$$ds = dw - F^i \, dp_i, \tag{10.92}$$

where differentiation is to be understood only with respect to the macroscopic parameters w and p_i. The mean value of ds,

$$\vartheta = \langle ds\rangle = dw - x^i \, dp_i, \tag{10.93}$$

leads to a contact form on the space of parameters $w, p_1, \ldots, p_n, x^1, \ldots, x^n$. To this end one has to assume that ρ is (temporarily) nonnormalized, i.e., w is a free parameter, and that x^i are independent parameters as well. Then ϑ becomes a contact form because under such assumptions one has $\vartheta \wedge (d\vartheta)^n \neq 0$. For ρ normalized, ϑ becomes zero and subsequently w and x^i become functions of p_i. These functions define a Legendre submanifold of ϑ. The full Legendre transformation transforms ϑ into another contact form θ equal to

$$\theta = dx^0 + p_i \, dx^i, \tag{10.94}$$

where $x^0 = w - p_i x^i$. The metric G discussed in Section 10.7 may be derived from the variance of ds which, due to (10.89), (10.90), and (10.93), is equal to

$$\langle (ds - \langle ds\rangle)^2\rangle = \langle (F^i - x^i)(F^j - x^j)\rangle \, dp_i \, dp_j = dp_i \, dx^j. \tag{10.95}$$

Let us again assume that p_i and x^i are independent. Then $\langle (ds - \langle ds\rangle)^2\rangle = dp_i dx^i$ becomes a bilinear, positive definite, and symmetric form on the $(2n+1)$-dimensional space M^{2n+1} of parameters w, p_i, and x^i. However, the form $dp_i \, dx^i$ is degenerate on M^{2n+1}. To remove this degeneration we simply add $\theta \times \theta$ to $dp_i \, dx^i$, i.e., we define G as

$$G := dp_i \, dx^i + \theta \otimes \theta. \tag{10.96}$$

Note that the degeneracy of $dp_i \, dx^i$ might be removed in any other way, for instance, by adding $dw\,dw$ or $dx^0\,dx^0$. However, our choice has the advantage that G, reduced to any Legendre submanifold $\mathcal{S}$,

$$g = G|_{\mathcal{S}} = \langle (ds - \langle ds\rangle)^2\rangle|_{\mathcal{S}} = dp_i \, dx^i|_{\mathcal{S}}, \tag{10.97}$$

has a very simple form not depending on the way the degeneracy has been removed. Its statistical and physical interpretation is also obvious from (10.97) and (10.89).

Remark 8.1. The metric G defined by means of (10.94)–(10.96) is equivalent (up to a Legendre transformation) to the metric

$$\mathcal{H} = \langle (ds)^2 \rangle = dw\, dw - 2x^i\, dw\, dp_i + x^i x^j\, dp_i\, dp_j + dx^i\, dp_i, \quad (10.98)$$

where ρ is nonnormalized. For ρ normalized $\mathcal{H}$ is reduced to $\mathcal{S}$ and

$$g = G|_\mathcal{S} = \langle (ds - \langle ds \rangle)^2 \rangle |_\mathcal{S} = \mathcal{H}|_\mathcal{S} = \langle (ds)^2 \rangle |_\mathcal{S} = dp_i\, dx^i |_\mathcal{S}. \quad (10.99)$$

The metric $\mathcal{H}$ has the virtue of being nondegenerate at once.

10.9 Relative Information and Riemannian Metric

In order to gain better and deeper insight into the notion of a thermodynamic metric we shall now give another construction of the Riemannian metric tensor g; we shall not discuss G here. This construction of g is based on the notion of relative information (Kullback information) or relative entropy. It turns out that g obtained in this way is identical to Fisher's information matrix introduced in mathematical statistics in the early 1920s. Initially, Fisher's matrix had been used only in pure mathematical statistics and in some biological problems. Much later it had also found some applications in thermodynamics. Nowadays it is generally recognized that differential geometry, and in particular Riemannian geometry, find numerous and important applications in mathematical statistics and in information theory [23]–[25]. Modern differential geometry proved to be a very useful and convenient tool to compare different statistical hypotheses (probability distributions).

Let us consider two probability distributions $\rho(y)$ and $\sigma(y)$ on a measurable space (Γ, B), where Γ is again a space of all microscopical states, $y \in \Gamma$, and B is a σ-algebra of subsets of Γ. If $\rho(y)$ and $\sigma(y)$ are mutually absolutely continuous then we may define their *relative information* $I(\rho|\sigma)$ (also called *directed divergence* or *directed distance*) as

$$I(\rho|\sigma) = \int \rho(y) \ln \frac{\rho(y)}{\sigma(y)}\, dy, \quad (10.100)$$

and an analogous formula for $I(\sigma|\rho)$. In information theory $I(\rho|\sigma)$ is called a *Kullback information*, a *Rényi–Kullback information*, or *information gain*. It is well known [25] that $I(\rho|\sigma) \geq 0$ and $I(\rho|\sigma) = 0$ iff $\rho(y) = \sigma(y)$. There is also considered a symmetrized counterpart of $I(\rho|\sigma)$, namely

$$J(\rho, \sigma) = I(\rho|\sigma) + I(\sigma|\rho) = \int (\rho - \sigma) \ln \frac{\rho}{\sigma}\, dy \quad (10.101)$$

called a *divergence*, *information distance*, or again *information gain*.

10.9. Relative Information and Riemannian Metric

Probability distributions in statistical physics typically depend not only on the microscopic variables y (phase space variables), but also on some macroscopic thermodynamic parameters $p_1, ..., p_n$ such as the temperature T, the pressure P, the chemical potential μ, and so on. Let us assume that the set of all admissible values of $p = (p_1, ..., p_n)$ forms a differentiable manifold $\mathcal{P}^n$ and that ρ and σ are given by two identical functions but with the different numerical values of the parameters p. In such a case the two probability distributions ρ and σ are represented by two different points of $\mathcal{P}^n$.

Let us consider now a special case of two probability densities represented by two neighboring points p and $p + \Delta p$ and let us denote

$$\rho(y,p) = \rho(p), \qquad \sigma(y, p + \Delta p) \equiv \rho(y, p + \Delta p) = \rho(p + \Delta p),$$
$$J\left(\rho(p), \rho(p+\Delta p)\right) = J(p, p + \Delta p). \tag{10.102}$$

If $\rho(p)$ satisfies some regularity conditions [25], then $J(p, p + \Delta p)$ may be expanded into the Taylor series, and taking into account only the first nonvanishing terms we have

$$J(p, p + \Delta p) = g^{ij}(p)\, \Delta p_i\, \Delta p_j, \tag{10.103}$$

where

$$g^{ij}(p) = \int \rho(y,p) \frac{\partial \ln \rho(y,p)}{\partial p_i} \frac{\partial \ln \rho(y,p)}{\partial p_j} dy = E\left(\frac{\partial \ln \rho}{\partial p_i} \frac{\partial \ln \rho}{\partial p_j}\right) \tag{10.104}$$

(E means the expectation or mean value). The same procedure applied to $I(p|p + \Delta p)$ gives

$$I(p|p + \Delta p) = \tfrac{1}{2} g^{ij}\, \Delta p_i\, \Delta p_j. \tag{10.105}$$

It is seen that both $J(p, p + \Delta p)$ and $I(p|p + \Delta p)$ may be represented by quadratic forms with coefficients identical to elements of Fisher's information matrix [25]. It is obvious that the quadratic forms (10.103) and (10.105) are symmetric and positively definite. Therefore g^{ij} may be taken as a Riemannian metric tensor on $\mathcal{P}^n$. It was C. R. Rao who first proposed using the components of the information Fisher matrix as components of a Riemannian metric.

Let us now restrict ourselves to a system described by a generalized n-parameter Gibbs distribution function [22] (an exponential family [24])

$$\rho(y,p) = Z^{-1}(p) e^{-p_i F^i(y)}, \qquad i = 1, ..., n. \tag{10.106}$$

where, as in Section 10.8, $F^i : \Gamma \longrightarrow \mathbf{R}^1$ are stochastic variables (observables) describing the system, whereas $p = (p_1, ..., p_n)$ are macroscopic parameters (statistical temperatures) that characterize the environment of the system. It is assumed that $F^1, \ldots, F^n$ and the identity observable $F^0 \equiv I$ are linearly (but not statistically) independent. $Z(p)$ is the *partition function*

$$Z(p) = \int_\Gamma e^{-p_i F^i(y)} \, dy \qquad (10.107)$$

(in the quantum case the integral should be replaced by Tr). The numerical values of $F^i(y)$ fluctuate around their mean values

$$x^i = EF^i = \int_\Gamma F^i(y) \rho(y,p) \, dy = -\frac{\partial \ln Z}{\partial p_i}, \qquad (10.108)$$

whereas p_i change only if the state of the environment is changed.

The Riemannian structure on the parameter space $\mathcal{P}^n$ is defined by means of the square infinitesimal distance between two points p and $p + dp$ according to the formula (cf. (10.105))

$$dl^2 = 2I(\rho(p+dp)|\rho(p)) = g^{ij}(p) \, dp_i \, dp_j, \qquad (10.109)$$

where $g^{ij}(p)$ play the role of the components of a Riemannian metric tensor. They may be given in a number of equivalent ways

$$g^{ij}(p) = \frac{\partial^2 I}{\partial p_i \partial p_j} = E\left(\frac{\partial \ln f}{\partial p_i} \frac{\partial \ln f}{\partial p_j}\right) = \frac{\partial^2 \ln Z(p)}{\partial p_i \partial p_j}$$
$$= -\frac{\partial x^i}{\partial p_j} = -\frac{\partial x^j}{\partial p_i} = E\big[(F^i(y) - x^i)(F^j(y) - x^j)\big]. \quad (10.110)$$

For practical calculations the best is the formula involving the partition function $Z(p)$. However, for physical interpretation we rather use the last formula which says that components of the metric tensor are given by the mixed second statistical moments (correlations, covariances) of the stochastic variables F^i. The diagonal elements g^{ii} are simply given by the second central moments of F^i. The metric tensor g^{ij} degenerates if some or all of F^i are statistically independent.

The fact that components of the metric tensor are given by second derivatives of the function $\ln Z(p)$ considerably simplifies calculation of the Christoffel symbols, the components of the Riemann curvature tensor, and the scalar curvature. The Christoffel symbols simply reduce to

$$\Gamma_{ijk} = \frac{1}{2} \frac{\partial g_{ij}}{\partial p^k} = \frac{1}{2} \frac{\partial^3 \ln Z}{\partial p^i \partial p^j \partial p^k} = -\tfrac{1}{2} E\left[(F_i - x_i)(F_j - x_j)(F_k - x_k)\right], \qquad (10.111)$$

where $p^i = g^{ij} p_j$ and so on, and g_{ij} is the inverse of the tensor g^{ij}, i.e., $g_{ij} g^{jk} = \delta_i^k$. Consequently, the components of the curvature tensor

$$R_{ijkl} = \frac{1}{2}\left[\frac{\partial^2 g_{jk}}{\partial p^i \partial p^l} - \frac{\partial^2 g_{ik}}{\partial p^j \partial p^l} + \frac{\partial^2 g_{il}}{\partial p^j \partial p^k} - \frac{\partial^2 g_{jl}}{\partial p^i \partial p^k}\right]$$
$$+ g^{mn}(\Gamma_{mil}\Gamma_{njk} - \Gamma_{mik}\Gamma_{njl}) \qquad (10.112)$$

10.9. Relative Information and Riemannian Metric

reduce to
$$R_{ijkl} = g^{mn}(\Gamma_{mil}\Gamma_{njk} - \Gamma_{mik}\Gamma_{njl}), \qquad (10.113)$$

because the second derivatives of g^{ij} cancel each other in the square brackets. It is important to note that R_{ijkl} are functions of the second and third derivatives of $\ln Z$, and therefore they are functions of the second and third correlations of the stochastic variables F^i, cf. the formulas (10.110) and (10.111). Higher-order derivatives of $\ln Z$ do not have this property, for instance, the fourth derivatives of $\ln Z$ are given by linear combinations of the fourth moments and products of the second moments. These facts considerably simplify the physical interpretation and computation of R_{ijkl}, and hence the computation and interpretation of the scalar Riemannian curvature

$$R = g^{jk} R^i{}_{jik}. \qquad (10.114)$$

In spite of this simplification, calculations are very involved and difficulties grow enormously with the growing dimension n of $\mathcal{P}^n$. Computations are relatively simple only for $n = 2$ where there exists only one independent nonvanishing component R_{1212} of the curvature tensor. The scalar curvature for $n = 2$ is given by a simple formula

$$R = \frac{2}{g} R_{1212}, \qquad (10.115)$$

where $g = \det(g_{ij})$. It may also be written in the form of a determinant

$$R = \frac{-2}{g^2} \begin{vmatrix} g_{11} & g_{12} & g_{22} \\ \dfrac{\partial g_{11}}{\partial p_1} & \dfrac{\partial g_{12}}{\partial p_1} & \dfrac{\partial g_{22}}{\partial p_1} \\ \dfrac{\partial g_{11}}{\partial p_2} & \dfrac{\partial g_{12}}{\partial p_2} & \dfrac{\partial g_{22}}{\partial p_2} \end{vmatrix}. \qquad (10.116)$$

Numerical calculations of R have been done for several classical and quantum systems described by means of the grand canonical and Boguslavski distributions (or their counterparts) [18], [26], [27]. It turned out that in all cases R was nonnegative with only one exception – the ideal fermion gas for which R was negative. Moreover, R was zero for the classical ideal gas. For nonideal gases (interacting particles), $R \to 0$ in the high-temperature low-pressure limit, and $R \to \infty$ at a critical point. It is well known that far from critical points thermodynamic fluctuations may be neglected and systems are stable. On the contrary, in the vicinity of critical points, fluctuations become very large and systems become very unstable, fluctuations may cause phase transitions. This was the reason to treat R^{-1} as a scalar measure of "nonideality" or instability of thermodynamic systems. This interpretation was backed by the analysis of two model magnetic systems [26]: a one-dimensional Ising model (short-range interactions) and a mean field model (long-range interactions). For both models R was positive and $R \to \infty$ at the Curie point. Standard thermodynamic criteria of stability use only second fluctuation moments of F^i. From the physical point of view R is a new measure of stability because it takes into account the second and third moments. As a measure of stability it is related to the second law of thermodynamics. It is hoped that R may be very useful in the description of the states in the vicinity of critical points.

A slightly different approach to the concept of thermodynamic curvature is presented in an excellent review paper by Ruppeiner [19] in which a number of important references may be also found.

A challenging task in the years to come is to extend the methods of the contact and metric geometries to nonequilibrium thermodynamics.

10.10 References

[1] H. Callen: *Thermodynamics*, Wiley, New York, 1960.

[2] V. I. Arnold: *Mathematical Methods of Classical Mechanics*, Springer-Verlag, Berlin, 1976.

[3] D. E. Blair: *Contact Manifolds in Riemannian Geometry*, Lecture Notes in Mathematics, vol. 509, Springer-Verlag, Berlin, 1976.

[4] S. Sasaki: *Almost Contact Manifolds*, Lecture Notes, Tōhoku University, **1**, 1965; **2**, 1967; **3**, 1968, Tōhoku, Japan.

[5] N. E. Hurt: *Geometric Quantization in Action*, Reidel, Dordrecht, 1983.

[6] P. Liebermann and Ch.-M. Marle: *Symplectic Geometry and Analytical Mechanics*, Reidel, Dordrecht, 1987.

[7] P. Liebermann: *Diff. Geom. Appl.* **1**, 57, 1991.

[8] A. A. Kirillov: *Uspekhi Mat. Nauk* **31**, 57, 1976 (in Russian).

[9] S. Kobayashi and K. Nomizu: *Foundations of Differential Geometry*, vol. 1, Wiley, New York, 1978.

[10] Y. Choquet-Bruhat, C. deWitt-Morette, and M. Dillard-Bleick: *Analysis, Manifolds and Physics*, Part I, North-Holland, Amsterdam, 1991.

[11] R. Mrugała, J. D. Nulton, J. Ch. Schön, and P. Salamon: *Rep. Math. Phys.* **29**, 109, 1991.

[12] R. Mrugała: *Rep. Math. Phys.* **33**, 149, 1993.

[13] R. Mrugała: *Tensor, N.S.* **56**, 37, 1995.

[14] P. J. Olver: *Applications of Lie Groups to Differential Equations*, Springer-Verlag, New York, 1986.

[15] P. Valentin and L. Benayoun: private communication.

[16] J. Jurkowski: *Rep. Math. Phys.* **41**, 351, 1998.

[17] F. Weinhold: *J. Chem. Phys.* **63**, 2479, 2484, 2488, 2496, 1975; **65**, 559, 1976.

[18] G. Ruppeiner: *Phys. Rev. A* **20**, 1608, 1979.

[19] G. Ruppeiner: Thermodynamic curvature: origin and meaning, in: *Advances in Thermodynamics Series*, vol. 3, *Nonequilibrium Theory and Extremum Principles* (eds. by S. Sieniutycz and P. Salamon), Taylor & Francis, New York, 1990, pp. 129–158.

[20] R. Mrugała, J. D. Nulton, J. Ch. Schön, and P. Salamon: *Phys. Rev. A* **41**, 3156, 1990.

[21] E. T. Jaynes: *Phys. Rev.* **106**, 620, 1957.

[22] R. K. Pathria: *Statistical Mechanics*, Pergamon, Oxford, 1972.

[23] Shun-ichi Amari: *Differential-Geometrical Methods in Statistics*, Lecture Notes in Statistics, vol. 28, Springer-Verlag, Berlin, 1985.

[24] M. K. Murray and J. W. Rice: *Differential Geometry and Statistics*, Chapman & Hall, London, 1993.

[25] S. Kullback: *Information Theory and Statistics*, Wiley, New York, 1959.

[26] H. Janyszek and R. Mrugała: *Phys. Rev. A* **39**, 6515, 1989.

[27] H. Janyszek and R. Mrugała: *J. Phys. A: Math. Gen.* **23**, 467, 1990.

11
From Statistical Distances to Minimally Dissipative Processes

L. Diósi
P. Salamon

ABSTRACT. A quantitative notion of statistical distinguishability led R. A. Fisher to his idea of statistical distance which has since been developed into Riemannian geometries on the space of statistical ensembles. Parallel to, though independently, of this progress, Riemannian geometries were being proposed on spaces of quantum states and also of thermodynamic states. Riemannian geometries in various fields have found various applications as different as population dynamics and fractional distillation, just to mention the first and the most recent ones. For decades, however, little attention was paid to the common theoretical basis of these geometric methods.

This chapter intends to fill the gap. We present an elementary introduction to the concept and mathematics of statistical distance in order to help understand the emergence of Riemannian geometrical structures. While we put more emphasis on the thermodynamical aspects, the main goal is still the interpretation of different applications on equal footing and using a unified framework.

11.1 Introduction

The Riemannian metric structure of thermodynamic theory, initiated by Weinhold [1] and Ruppeiner [2], contains important and hitherto barely tapped information concerning a physical system. The structure runs deep; its presence can be felt at all levels of physical description. The Riemannian metric of thermodynamics is, as shown first by Diósi et al. [3], in fact a realization of R. A. Fisher's concept of statistical distinguishability [4]. He had applied it in 1922 to measure genetic drift and later it became the basis for the mathematical theory of information geometry. The corresponding notion of statistical distance has since been introduced for various statistical systems. At the quantum level the distance measures the reliability of experiments designed to optimally distinguish between the two states along a one-parameter family of density operators [5]. At the statistical mechanical level, distance is the number of statistically distinguishable intermediate states as we transform one state into another [6]. This leads to a natural Riemannian metric on the space of distributions in the thermodynamic limit of Gibbs' statistical ensembles. Numerous authors have speculated about the meaning of the curvature defined by this geometry as a measure of stability or interaction strength (cf. Ruppeiner's recent review [7]). The requirement of covariance with respect to this geometry can be used to give an important correction to thermodynamic fluctuation theory [7]. Finally, at the macroscopic level, the square of this same distance between two equilibrium states of a thermodynamic system equals the minimum entropy produced in a process that transforms one state into the other, multiplied by the number of relaxations during the transformation [8], [9]. This result has become known as the horse–carrot theorem.

In this chapter, we recapitulate basic ideas and results concerning the Riemannian metric structure of thermodynamics while we attempt to shed light on the underlying concept of statistical distance used in a much broader context.

11.2 Empirical Statistical Distance

The class of continuous variables spans from typical continuous quantities of physics to approximate continuous quantities, e.g., in population statistics. Consider a continuous variable corresponding to a measurement of a real number x to a certain precision Δx. The true value of x lies in the confidence interval $(x - \Delta x, x + \Delta x)$ with a probability that amounts to 68% when Δx is, as we generally assume, the standard deviation of x.

The values Δx provide a measure of distinguishability between different values of x. Two values, say x and x', are statistically indistinguishable if $|x - x'| < \Delta x + \Delta x'$. In the opposite case they are well distinguishable. By convention, we shall say that x and x' are statistically *distinguishable* if the equality holds

$$|x - x'| = \Delta x + \Delta x'. \tag{11.1}$$

To illustrate continuous variables with nonconstant precision we consider an example taken from statistics itself. Let x be the relative frequency of a certain event from a large sample N. As is well known, its variance is inversely proportional to the square root of N. The exact expression would be $\sqrt{x(1-x)/N}$. For simplicity's sake, we restrict our considerations to small values of x and we use an approximate[1] expression

$$\Delta x = \sqrt{x/N} \equiv c/\sqrt{N}, \tag{11.2}$$

where, for later purposes, we introduce the square root c of the relative frequency x.

---

[1] See, however, footnote 3 on p. 293

288 11. From Statistical Distances to Minimally Dissipative Processes

Well distinguishable:

Not distinguishable:

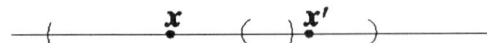

Distinguishable:

FIGURE 11.1. The two values x and x' are, by convention, distinguishable when their confidence intervals contact each other.

11.2.1 *Optimum Calibration*

An immediate application of the notion of statistical distinguishability occurs when we calibrate a scale of a measuring apparatus. We shall assume that the precision Δx of the measuring apparatus is known for all measured values of x. Let x_0 be the first point of calibration. Where shall we put the next mark x_1? Obviously, the reasonable choice is such that x_0 and x_1 be distinguishable: $x_1 = x_0 + \Delta x_0 + \Delta x_1$. The subsequent calibration marks satisfy the distinguishability condition (11.1):

$$x_{\nu+1} = x_\nu + \Delta x_\nu + \Delta x_{\nu+1} \quad (\nu = 0, 1, \ldots). \tag{11.3}$$

Hence, the resolutions of measurements and calibration marks will match.

These calibration marks are not equidistantly distributed on the scale x. We can, nevertheless, reparameterize the scale by a new variable $\tilde{x}$ such that its standard deviation is constant

$$\Delta \tilde{x} \equiv 1. \tag{11.4}$$

Then the calibration marks are thus located at the equidistant steps $\tilde{x}_\nu = \nu$ for $\nu = 0, 1, 2, \ldots$. The natural scale $\tilde{x}$ defines the so-called statistical length [4]. By construction, the statistical lengths of the confidence intervals equal 1, according to (11.4). The statistical length of a bigger interval (x_i, x_f) is equal to the maximum number of nonoverlapping confidence intervals between x_i and x_f, which turns out to be

$$\ell_{if} = |\tilde{x}_f - \tilde{x}_i|. \tag{11.5}$$

We can formulate the principle of optimum calibration in such a way: the neighboring marks should be separated by unit statistical lengths from each other.

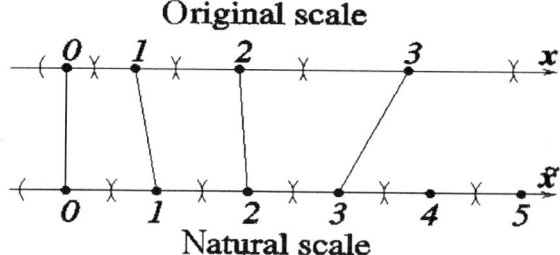

FIGURE 11.2. The natural scale $\tilde{x}$ replaces x in such a way that the statistical lengths of the confidence intervals become equal to 1. The optimum calibration marks are separated by 1 on the new scale $\tilde{x}$.

Let us derive the natural scale $\tilde{x}$ of the relative frequency x. From the (11.2), (11.4) and from the (asymptotic) relation $\Delta \tilde{x} = (d\tilde{x}/dx)\Delta x$ we obtain

$$\tilde{x} = 2\sqrt{N}c, \qquad (11.6)$$

where $c = \sqrt{x}$. The ν'th optimum calibration point is $\tilde{x}_0 + \nu$. This will correspond to $x_\nu = (x_0 + \nu/\sqrt{4N})^2$ on the scale of relative frequencies. Equation (11.5) yields the statistical length of an interval (x_i, x_f):

$$\ell_{if} = 2\sqrt{N}|c_i - c_f|. \qquad (11.7)$$

11.2.2 Naive Optimum Control

We turn to another application of the concept of statistical length in the field of optimum control in a noisy environment. A system with a single continuous parameter x is to be driven from an initial state x_i into a given final state x_f. The standard deviation Δx of the parameter may depend on the current value x. Initially, the minimum significant change is just Δx_i. So, a cautious strategy might consist of a sequence of minimum, yet significant, steps from x_i to x_f: each step is equal to the local value of Δx. Thus the steps must correspond to the optimum calibration marks (11.3) starting from $x_i = x_0$.

On the natural scale $\tilde{x}$, each step will have unit length, according to (11.4). If, furthermore, we perform a constant number v of steps per unit time, then

$$\frac{\Delta \tilde{x}}{\Delta t} \equiv v. \qquad (11.8)$$

This is the principle of *constant statistical speed* which seems to be a kind of cautious control in noisy environments.

Imagine, for instance, the inflation rate x in an economy whose financial policy is to decrease x from a higher value x_i to the lower one x_f.

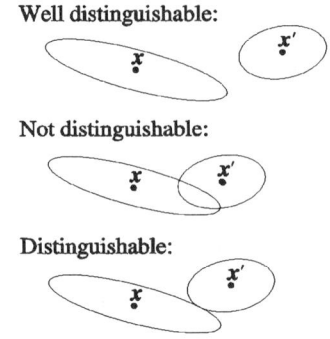

FIGURE 11.3. Two vectors **x** and **x**′ are considered statistically *distinguishable* if their confidence ellipsoids contact each other.

To minimize the risk of uncontrolled changes, one can exert minimum significant perturbations to the current economical system. For instance, the inflation rate can be decreased by Δx each year, where Δx is the variance of the given annual inflation x. This is a strategy at unit statistical speed. If necessary, the process can be made more intensive if we choose a higher constant velocity $|v| > 1$, i.e., we perform $|v|$ such steps per year.

The ad hoc principle of constant statistical speed and similar "cautious control schedules" will be formulated analytically in Section 11.5. We will see that such controls yield a powerful tool in various tasks of process optimization.

11.2.3 *More Parameters*

We can easily extend the above statistical concepts to the case with more continuous parameters $\mathbf{x} = (x_1, x_2, \ldots, x_k, \ldots)$ to characterize the given system. The simultaneous standard deviations Δx_k are correlated and the role of confidence intervals are played by multidimensional ellipsoidal confidence volumes. Two vectors **x** and **x**′ are, by our convention, statistically *distinguishable* if their confidence ellipsoids contact each other.

The optimum calibration of a curve in the multidimensional parameter space invokes considerations similar to the single parameter scale. Neighboring calibration marks should be distinguishable. Their confidence volumes, centered along the curve, will contact each other. The statistical length of the curve will be equal to the number of such confidence volumes. This length may completely depart from any apparent length of the curve. The statistical length depends largely on the variability of the ellipticities and orientations of the confidence volumes along the curve. The statistical length of a curve connecting two points $\mathbf{x}_i$ and $\mathbf{x}_f$ will depend on the curve itself. The minimum length can, as usual in geometry, be called the statistical distance between $\mathbf{x}_i$ and $\mathbf{x}_f$. The minimizing curve will correspond to a geodesic curve in the analytic theory. For the time being, however, we are discussing the empirical concepts. Of course, the naive optimum control at constant statistical speed, suggested for a single variable, applies easily to the simultaneous control of parameters $\mathbf{x} = (x_1, x_2, \ldots)$ at a constant statistical speed along a given curve. A new feature of the optimization is the choice of the shortest path (geodesic) to connect the initial and final parameters $\mathbf{x}_i, \mathbf{x}_f$.

We recall the rescaling procedure which led to uniform confidence intervals (11.4) for a single variable x. In perfect analogy, one would attempt to use new variables $\tilde{\mathbf{x}} = (\tilde{x}_1, \tilde{x}_2, \ldots)$ such that their confidence volumes be hyperspheres of unit radii. Hence, in these natural variables $\tilde{\mathbf{x}}$, the statistical lengths of curves are equal to the ordinary lengths. Consequently, statistical distances coincide with the corresponding Euclidean distances[2]:

$$\ell_{if} = \|\tilde{\mathbf{x}}_i - \tilde{\mathbf{x}}_f\|. \tag{11.9}$$

Minimum lengths curves, i.e., geodesics, are straight lines in $\tilde{\mathbf{x}}$.

The existence of the above natural parameterization is a delicate problem. Even at an empirical level we can see that sometimes the natural parameters $\tilde{\mathbf{x}}$ may *not* exist. The variation of the confidence ellipsoids from site to site may paralyze our attempts at constructing the natural parameterization. This problem, with all its complexity, is part of a paradigm. If one goes beyond the empirical considerations, it turns out [10], [11] that the parameter space $\mathbf{x}$ of statistical ensembles constitutes a manifold of Riemannian geometry. If, in particular, this geometry is Euclidean then, and only then, will the natural parameters $\tilde{\mathbf{x}}$ exist.

Previously we considered, as an example, the relative frequency x of a single event in a large sample N. Now we are going to generalize the example for the relative frequencies $x_1, x_2, \ldots$ of a number of mutually exclusive events from the same large sample N. When the frequencies are small, their variances Δx_k are independent and given by the same equation (11.2), respectively, for each x_k. The k'th axis of the confidence ellipsoid is parallel to the coordinate axis x_k and its half-length is equal to

$$\Delta x_k = 2\sqrt{N} c_k, \tag{11.10}$$

with the notation $c_k = \sqrt{x_k}$. Introducing the natural parameters $\tilde{\mathbf{x}}$ by (11.6), we achieve that all $\Delta \tilde{x}_k$ are of unit length: the confidence volumes become unit spheres in the parameter space $\tilde{\mathbf{x}}$. Consequently, one can apply (11.9) to write the statistical distance between arbitrary two parameters $\mathbf{x}_i$ and $\mathbf{x}_f$:

$$\ell_{if} = 2\sqrt{N}\|\mathbf{c}_f - \mathbf{c}_i\|. \tag{11.11}$$

---

[2] The squared Euclidean norm of a vector $\mathbf{x}$ is defined as $\|\mathbf{x}\|^2 = \sum_k x_k^2$.

11.3 Theory of Statistical Distance

The concept of statistical distinguishability, considered on intuitive grounds in the previous section, implies a natural geometry on the space of statistical ensembles. In the present section we give an insight into the general structure of this geometry for classical as well as quantum ensembles.

11.3.1 Classical Statistics

We consider discrete classical statistical ensembles. They are parameterized by normalized probability distributions $\mathbf{p} = (p_1, p_2, \ldots, p_k, \ldots)$ corresponding to a complete set of mutually exclusive events. The parameter space is a hyperplane

$$\sum_k p_k = 1, \quad p_k \geq 0. \tag{11.12}$$

We can introduce an alternative parameterization $\mathbf{c} = (c_1, c_2, \ldots)$ where the components of the vector $\mathbf{c}$ are the *square roots of the probabilities*: $c_k = \sqrt{p_k}$ for $k = 1, 2, \ldots$. Then the parameter space (11.12) becomes spherical

$$\|\mathbf{c}\| = 1, \quad c_k \geq 0, \tag{11.13}$$

i.e., a sector on the surface of the unit hypersphere [11].

We note that in practice the theoretical probabilities p_k appear as relative frequencies x_k of the corresponding events in a given sample. Regarding the statistical distinguishability of our ensembles, the probabilities (11.12) will be treated as relative frequencies in a large sample of size N. Then we can apply the equations and considerations of the previous section. We know from (11.6) that in natural parameterization $\tilde{\mathbf{x}} = 2\sqrt{N}\mathbf{c}$ the confidence volumes are just the unit spheres. In this parameterization the statistical distance is identical to the Euclidean distance (11.11). One would conclude that the geometry of the parameter space of discrete ensembles is Euclidean. This is not exactly the case! Nevertheless, the true geometry is amazing as we will see below.

There has been a loophole in our derivation of (11.11). We assumed that the events and the corresponding relative frequencies $x_1, x_2, \ldots$ were independent and this holds indeed for small values of the relative frequencies. If, however, we identify them by the normalized probabilities (11.12) then the very constraint of normalization puts constraint upon them. The Euclidean geometry (11.11), when extended deliberately[3] for all possible values of $x_k = p_k = (c_k)^2$, induces a non-Euclidean (spherical) geometry

---

[3] Our derivation is illustrative. Rigorous calculations [12] confirm the distance (11.14).

on the parameter space (11.13):

$$\ell_{if} = 2\sqrt{N} \arccos(\mathbf{c}_i \mathbf{c}_f) \qquad (11.14)$$

which is just $2\sqrt{N}$ times the angle[4] between the two unit vectors $\mathbf{c}_i$ and $\mathbf{c}_f$ [11], [6].

This result is remarkable! From purely classical considerations we have found something very similar to the quantum mechanical formalism. We found that the natural parameters of statistical ensembles are unit *vectors* $\mathbf{c} = (c_1, c_2, \ldots)$. The ensemble's usual parameters, i.e., the normalized probabilities $p_1, p_2, \ldots$ are equal to the *square* of the vector's components

$$p_1 = (c_1)^2, \quad p_2 = (c_2)^2, \quad \ldots. \qquad (11.15)$$

The resemblance to quantum mechanics is puzzling! Note, however, that the analogy is not perfect: our state vectors $\mathbf{c}$ are always real [6].

We close this section with an interesting relation between the statistical distance and the entropy function $s(\mathbf{p}) = -\sum p_k \log p_k$. Eq. (11.14), and (11.15) yield the following result for the squared statistical distance $d\ell$ between two infinitesimally close distributions $\mathbf{p}$ and $\mathbf{p} + d\mathbf{p}$:

$$d\ell^2 = N \sum_k (dc_k)^2 = N \sum_k \frac{(dp_k)^2}{p_k}. \qquad (11.16)$$

Anticipating the formalism of Riemannian geometry (Section 11.4), we mention that the metric tensor, yielding this infinitesimal statistical length, turns out to be the second derivative of the entropy function times -1:

$$g^{ik}(\mathbf{p}) = N\delta_{ik} \frac{1}{p_k} = -N \frac{\partial^2 s(\mathbf{p})}{\partial p_i \, \partial p_k}. \qquad (11.17)$$

11.3.2 Quantum Statistics

The concept of distinguishability can be interpreted for quantum ensembles as well. The same considerations that we have applied to classical ensembles extends to them. The result is a unique notion of statistical distance. We are not going to present the details of specific problems which make, as usual, the quantum case more subtle than the classical one. We only present the resulting equations.

The so-called *pure* quantum ensembles are characterized by normalized *complex* state vectors $\mathbf{c}$. Wootters [6] pointed out first that the statistical distance between pure quantum ensembles

$$\ell_{if} = 2\sqrt{N} \arccos |\mathbf{c}_i^* \mathbf{c}_f| \qquad (11.18)$$

---

[4]The angle γ between two vectors $\mathbf{a}, \mathbf{b}$ is defined via their scalar product $\mathbf{ab} = \sum_k a_k b_k = \cos\gamma \|\mathbf{a}\|\|\mathbf{b}\|$.

coincides with the statistical distance (11.14) between classical ensembles, apart from the fact that the state vector components are now complex numbers.

General quantum ensembles are described by Hermitian positive definite density matrices $\hat{\rho} \equiv \{\rho_{kl}\}$. The statistical squared distance between two ensembles $\hat{\rho}_i, \hat{\rho}_f$ has the following form:

$$\ell_{if}^2 = 8N\left(1 - tr\sqrt{\hat{\rho}_i^{1/2}\hat{\rho}_f\hat{\rho}_i^{1/2}}\right) \qquad (11.19)$$

which was, with a different prefactor, suggested by Bures [13]. (It can be proved that the distance remains the same if we interchange the two density matrices.)

It is instructive to study the distance between two ensembles whose density matrices commute with each other. In this case they have a common canonical basis where they are both diagonal. The nested square roots largely simplify

$$\ell_{if} = 2\sqrt{2N}\sqrt{1 - tr\sqrt{\hat{\rho}_i\hat{\rho}_f}}. \qquad (11.20)$$

Recall that the diagonals of the density matrices serve as classical probability distributions $\mathbf{p}_i$ and $\mathbf{p}_f$, respectively. This means that we can formally apply the classical distance (11.14) to our commuting density matrices, yielding[5]

$$\ell_{if}^{(cl)} = 2\sqrt{N}\arccos\left(tr\sqrt{\hat{\rho}_i\hat{\rho}_f}\right). \qquad (11.21)$$

Let's compare this to the quantum distance (11.20):

$$\frac{\ell_{if}}{4\sqrt{N}} = \sin\left(\frac{\ell_{if}^{(cl)}}{4\sqrt{N}}\right). \qquad (11.22)$$

The quantum and classical distances tend to coincide for neighboring ensembles. For distant ensembles the quantum distance is always smaller. This is not surprising geometrically. Quantum states are usually harder to distinguish from each other due to their overlap in Hilbert space. Furthermore, the shortest path between the two endpoints goes through commuting density matrices in the case of the classical distance while it may find a shorter way through non-commuting matrices in the quantum case.

One can calculate the distance between infinitesimally close density matrices $\hat{\rho}$ and $\hat{\rho} + d\hat{\rho}$. For convenience, we use the canonical basis where $\rho_{kl} = \delta_{kl}p_k$. It is important to note that, while ρ_{kl} is now diagonal, its increment $d\rho_{kl}$ may be nondiagonal. The squared statistical distance (11.19) yields the following:

$$d\ell^2 = 2N\sum_{k,l}\frac{|d\rho_{kl}|^2}{p_k + p_l}. \qquad (11.23)$$

---

[5] We apply the identity $\mathbf{p}_i\mathbf{p}_f = tr(\hat{\rho}_i\hat{\rho}_f)$.

The proof that this distance is in fact a measure of the Fisher statistical distinguishability (Section 11.2) has been carried out by Braunstein and Caves [5]. Their proof shows that the Bures distance (11.19) is really the Fisher statistical distance applied this time to quantum statistical ensembles.

11.4 Riemannian Geometry

We have shown in the previous section that the natural geometry, reflecting statistical distinguishability of ensembles (probability distributions) is non-Euclidean. In various fields of applications, we are dealing with certain subclasses of probability distributions. Here we restrict ourselves to the case where the probability distributions are parameterized by a finite number of parameters. Gibbs distributions in statistical physics are particularly relevant: they underlie the statistical geometry of thermodynamic parameter space.

11.4.1 *Parameterized Statistics*

Consider, for simplicity, discrete statistical ensembles whose probability distributions **p** are parameterized by a finite number of parameters[6] $\mathbf{y} = (y^1, y^2, \ldots, y^n)$. These probability distributions constitute an n-dimensional submanifold of the spherical hypersurface (11.13) and this hypersurface inherits a Riemannian geometry from the enveloping space. The resulting curvature and metric will, in general, change with **y**.

We can no longer write the statistical distance between two distant elements $\mathbf{p}(\mathbf{y}_i)$ and $\mathbf{p}(\mathbf{y}_f)$ in simple angular form as in (11.14), since now the distance must be measured along a path staying entirely within the submanifold of distributions described by our parameters **y**. Fortunately, the distance $d\ell$ between infinitesimally close elements of parameter values **y** and $\mathbf{y} + d\mathbf{y}$ remains the same. In particular, (11.11) holds:

$$d\ell = 2\sqrt{N}\|d\mathbf{c}\|, \tag{11.24}$$

where

$$d\mathbf{c} = \sum_{k=1}^{n} \frac{\partial \mathbf{c}}{\partial y^k} \, dy^k. \tag{11.25}$$

If we introduce the following *Riemann metric* $g_{ik}(\mathbf{y})$ on the parameter space **y** [10]:

$$g_{ik} = 4N \sum_{r} \frac{\partial c_r}{\partial y^i} \frac{\partial c_r}{\partial y^k}, \tag{11.26}$$

---

[6]We use coordinates with superscripts following the traditional notation in Riemannian geometry and tensor analysis.

then we can write the infinitesimal statistical distance (11.24) in the standard Riemannian form

$$d\ell = \sqrt{\sum_{i=1}^{n}\sum_{k=1}^{n} g_{ik}(\mathbf{y})\, dy^i\, dy^k}. \qquad (11.27)$$

These equations define a Riemannian geometry on the manifold of parameterized discrete distributions $\mathbf{p(y)}$. Common Riemannian expressions will yield the statistical lengths/distances between two arbitrary distributions of respective parameter values $\mathbf{y}_i$ and $\mathbf{y}_f$.

It is straightforward to extend the above considerations beyond discrete probability distributions. Invoking the relation (11.15), we write the statistical metric (11.26) as follows:

$$\begin{aligned} g_{ik}(\mathbf{y}) &= 4N \sum_r \frac{\partial c_r}{\partial y^i} \frac{\partial c_r}{\partial y^k} \\ &= N \sum_r p_r \frac{\partial \ln p_r}{\partial y^i} \frac{\partial \ln p_r}{\partial y^k}. \end{aligned} \qquad (11.28)$$

This has the following compound forms:

$$g_{ik}(\mathbf{y}) = N \left\langle \frac{\partial \ln p(\mathbf{y})}{\partial y^i} \frac{\partial \ln p(\mathbf{y})}{\partial y^k} \right\rangle = -N \left\langle \frac{\partial^2 \ln p(\mathbf{y})}{\partial y^i \partial y^k} \right\rangle, \qquad (11.29)$$

where $\langle \ldots \rangle$ denotes expectation value calculated with $\mathbf{p(y)}$. These equations apply to continuous distributions as well. Let $p(\mathbf{\Gamma}; \mathbf{y})$ be the probability distribution of the continuous random variables $\mathbf{\Gamma}$, depending on the continuous parameters $\mathbf{y}$. The probability distributions are normalized

$$\int p(\mathbf{\Gamma}; \mathbf{y})\, d\mathbf{\Gamma} = 1. \qquad (11.30)$$

Equation (11.29) takes the following form:

$$g_{ik}(\mathbf{y}) = -N \int \frac{\partial^2 \ln p(\mathbf{\Gamma}; \mathbf{y})}{\partial y^i \partial y^k}\, p(\mathbf{\Gamma}; \mathbf{y})\, d\mathbf{\Gamma}. \qquad (11.31)$$

This metric tensor is, apart from the scale factor N, identical to Fisher's information matrix [14]. It plays a particular role in mathematical statistics, namely in the theory of parameter estimation. This metric was also used by Amari [15] as the metric for his *information geometry* (see also Chentzov [16]).

11.4.2 *From Gibbs Statistics to Thermodynamics*

Phenomenological thermodynamics can, as is well known, be derived from the statistical physics of dynamical systems. Let us start with a large dynamical system consisting of M moles of molecules. If the system is in

11.4. Riemannian Geometry

equilibrium then, according to Gibbs, its phase point $\mathbf{\Gamma}$ follows the probability distribution

$$p(\mathbf{\Gamma}; \mathbf{y}) = \exp(-\phi(\mathbf{y}) - \mathbf{y}\mathbf{F}(\mathbf{\Gamma})), \qquad (11.32)$$

where $\mathbf{y} = (y^1, y^2, \ldots, y^n)$ are the entropic intensive parameters of the equilibrium state, and $\mathbf{F}(\mathbf{\Gamma}) = (F_1(\mathbf{\Gamma}), F_2(\mathbf{\Gamma}), \ldots, F_n(\mathbf{\Gamma}))$ are the conjugated conserved dynamic quantities. The function $\phi(\mathbf{y})$ assures the normalization (11.30). When the size M of the dynamic system goes to infinity, the ratio $\phi(\mathbf{y})/M$ converges to the phenomenological *thermodynamic potential* (per moles) $\varphi(\mathbf{y})$ of the system

$$\frac{\phi(\mathbf{y})}{M} \to \varphi(\mathbf{y}). \qquad (11.33)$$

This *thermodynamic limit* is at the heart of the Gibbs theory.

Let us calculate the statistical metric of the Gibbs ensembles (11.32)! On substituting (11.32) into the rightmost expression in (11.29), we get

$$g_{ik}(\mathbf{y}) = N \frac{\partial^2 \phi(\mathbf{y})}{\partial y^i\, \partial y^k}. \qquad (11.34)$$

Assuming that the system is large enough to take the thermodynamic limit (11.33), we can replace the function ϕ in (11.34) by $M\varphi$. Now, this leaves us with a factor NM on the right-hand side of the above equation. Recall that, in the theory of statistical distance, N was the sample size. In the present case it would mean the number of M-mole thermodynamic systems in the *same* equilibrium state. It makes no difference if we unify them into a single NM-mole system. Finally, we simply absorb the factor M into the number N characterizing the overall size (in moles) of the thermodynamic system. So we obtain

$$g_{ik}(\mathbf{y}) = N \frac{\partial^2 \varphi(\mathbf{y})}{\partial y^i\, \partial y^k}. \qquad (11.35)$$

This metric, generating the statistical distance between the Gibbs ensembles at various thermodynamic parameters $\mathbf{y}$, can be expressed by the second derivative of the thermodynamic potential. Note an important aspect of this result: the concept of statistical distance has induced a notion of distance between thermodynamic states. This is the thermodynamic distance anticipated by Weinhold [1] and turned by Ruppeiner [2] into the corresponding Riemannian geometry on thermodynamic state space.

It is rather instructive to derive the Riemannian metric if we use extensive parameters[7] $x_k = -\partial \varphi/\partial y^k$ for $k = 1, 2, \ldots, n$, instead of the intensives.

---

[7] In thermodynamics we alter the traditional covariant notation of Riemannian geometry in a particular way. By our convention, the intensive parameters y^k bear upper labels but the extensive parameters x_k bear lower ones. Consequently, the covariant metric tensor will have lower labels in intensive, and upper labels in extensive coordinates.

The result reads

$$g^{ik}(\mathbf{x}) = -N \frac{\partial^2 s(\mathbf{x})}{\partial x_i \, \partial x_k}. \qquad (11.36)$$

Here $s(\mathbf{x})$ is the specific entropy function, related to the specific thermodynamic potential $\varphi(\mathbf{y})$ by the Legendre transformation $s = \mathbf{xy} - \varphi$.

Note that the forms of our metric in (11.36) can be termed macroscopic since now our metric matrix is just the matrix of second partial derivatives of the macroscopic entropy with respect to the extensive variables of the macroscopic system. It is interesting to note that the metric matrix is the second derivative in both (11.35) and (11.36). We note that this holds only for the entropy and its complete Legendre transform and not for any partial Legendre transforms [17], [18].

Another surprise is the fact that the metric matrix is the second derivative of the entropy in both macroscopic description (11.36) and in a microscopic description (11.17). This is surprising in light of the fact that metric matrices and second derivative matrices transform differently under a change of coordinates [19].

For our discussion of multiphase systems of variable composition, it is convenient to have an expression for our metric in terms of the macroscopic extensive variables $X_k = Nx_k$, $k = 1\ldots, n$ and $X_{n+1} = N$ instead of the *specific* extensive parameters[8] $s, x_1, \ldots, x_n$. This transforms our thermodynamic metric (11.36) into the following form[9]:

$$g^{ik}(\mathbf{X}) = -\frac{\partial^2 S(\mathbf{X})}{\partial X_i \, \partial X_k}, \qquad (11.37)$$

where $S(\mathbf{X}) = Ns(\mathbf{x})$ is the extensive entropy function of the system. Since

$$\frac{\partial^2 S}{\partial X_{n+1}^2} = \frac{\partial^2 S}{\partial N^2} = 0, \qquad (11.38)$$

it follows that $g^{ik}(\mathbf{X})$ is degenerate, i.e., there exist directions along which the metric measures a zero distance. Such directions always correspond to scaling one phase of the system. If the scale of the system is not fixed, the structure is only semi-Riemannian. Null directions result from the linear growth of the entropy as we scale any one phase. This linearity makes the second derivative, i.e., the components of the metric tensor, vanish along such directions. As remarked in Weinhold's original papers [20], forming an

---

[8] We refer to an extensive quantity divided by the mole number as specific extensive. A related terminology is the extensive density which is the extensive quantity divided by the volume. For simple equilibrium systems the number of independent extensive quantities exceeds the number of specific extensive quantities by 1.

[9] We use the same letter g and adopt the functional notation $g(\mathbf{X})$ to indicate the fact that the actual matrix will depend on our choice of parameters.

appropriate combination of such null directions by simultaneously scaling two phases of a pure substance can represent a phase transition. We will return to a discussion of the use of such null directions in the following sections.

11.5 Relevance of Riemannian Geometry in Thermodynamics

For macroscopic thermodynamics, the use of a closely related metric

$$G^{ik}(\mathbf{Z}) = \frac{\partial^2 U(\mathbf{Z})}{\partial Z_i \, \partial Z_k} \tag{11.39}$$

introduced by Weinhold [1] preceded the use of the metric described above. Here U is the internal energy and $\mathbf{Z} = (S, V, N_1, N_2, ...) \in I\!R^{n+1}$ is the vector of extensive variables of the system in the energy representation for which $U = U(\mathbf{Z})$ constitutes complete information [21]. Weinhold's papers used the geometry only locally to conveniently express relationships among differential changes in our state variables. He also suggested its possible use as a Riemannian metric. Soon thereafter, Salamon et al. [22] recognized that distances computed for the ideal gas using the metric G represented known expressions for the changes in the kinetic energy of the molecules in a gas resulting from the passing of a shock wave. This led to the recognition of the connection between geometry and dissipation as developed below.

At about the same time, Ruppeiner introduced the use of the metric g in (11.36) for extending the scales accessible to thermodynamic fluctuation theory. While the metrics g in (11.36) and G in (11.39) are conformally equivalent[10] [18], the metric g turns out to be more fundamental. Similar to the role of the energy representation elsewhere in thermodynamics, G serves mainly as a convenient device for calculating $g(\mathbf{X})$, with $\mathbf{X} = (U, V, N_1, N_2, ...) \in I\!R^{n+1}$ the vector of extensive variables in the entropy representation for which $S = S(\mathbf{X})$ constitutes complete information [21].

We begin our treatment of macroscopic applications of the Riemannian structure with a discussion of fluctuations since this follows most closely from the arguments in the previous section. We will then turn our attention to dissipation and a discussion of several horse–carrot theorems which relate the dissipation associated with coaxing a system to traverse a given sequence of states. The applications depend on the second-order expansion of the entropy and thus on the identity between the metric and the second derivative of entropy.

---

[10]Two metrics are conformally equivalent iff the squares of the length elements differ by a (possibly position dependent) scale factor.

11.5.1 *A Covariant Fluctuation Theory*

Consider the traditional expression for a fluctuation in a system of size[11] V inside a system of infinite size $V_0 = \infty$ with densities $\mathbf{x}_0$. The likelihood of the subsystem having densities $\mathbf{x}$ is given by the Einstein–Smoluchowski theory as

$$P(\mathbf{x}, V|\mathbf{x}_0, \infty)\, d^n\mathbf{x} = C \exp(S(\mathbf{x}, \mathbf{x}_0)/k_B)\, d^n\mathbf{x}, \qquad (11.40)$$

where C is a normalization constant, k_B is Boltzmann's constant, and $S(\mathbf{x}, \mathbf{x}_0)$ is the total entropy of the reservoir at extensive densities $\mathbf{x}_0$ containing the subsystem of finite volume V at state $\mathbf{x}$. This expression is valid for small fluctuations (i.e., at large volumes V) but it turns out that, in a subtle way, it also contains the statistics of larger fluctuations (i.e., at smaller volumes V).

At infinite volume there are no fluctuations at all

$$P(\mathbf{x}, \infty|\mathbf{x}_0, \infty)\, d^n\mathbf{x} = \delta(\mathbf{x} - \mathbf{x}_0). \qquad (11.41)$$

For large finite volumes (11.40) leads to the Gaussian approximation:

$$\begin{aligned} P(\mathbf{x}, V|\mathbf{x}_0, \infty) &= C \exp\left(\frac{V}{2k_B} \sum_{i,k} \frac{\partial^2 s(\mathbf{x}_0)}{\partial x_i \partial x_k}(x - x_0)_i (x - x_0)_k\right) \\ &\equiv C \exp\left(-\frac{1}{2k_B} \sum_{i,k} g^{ik}(\mathbf{x}_0)(x - x_0)_i (x - x_0)_k\right), \end{aligned} \qquad (11.42)$$

where the exponent becomes proportional to the square of the statistical (or thermodynamic) distance measured from the equilibrium value. Recall that this length element is the natural scale for measuring the size of fluctuations.

Ruppeiner's important observation [2] was that although (11.40) depends on which parameters $\mathbf{x}$ we use to define our state, by way of the volume form $d^n\mathbf{x}$, its Gaussian approximation (11.42) is invariant under reparameterization[12] and thus avoids this unphysical dependence. Hence, we must restore this invariance when we extend its validity for smaller volumes V.

Ruppeiner [23], [24] and Diósi and Lukács [25] used the Gaussian fluctuation theory as a starting point for an improved covariant theory of fluctuations, valid also for smaller volumes.

The physical intuition leading to the improved theory comes by considering a nested sequence of systems. We begin with an equilibrium system

---

[11] Here we follow Ruppeiner [23], [24] in using the volume to set our scale.

[12] In this approximation, the Jacobian matrix of a coordinate transformation can be taken as constant for $\mathbf{x}$ sufficiently near $\mathbf{x}_0$.

11.5. Relevance of Riemannian Geometry in Thermodynamics

in the thermodynamic limit $V = \infty$ at the state $\mathbf{x}_0$. As the volume of the subsystem we consider gets smaller, its fluctuations depend on the state of its immediate surroundings. In this way we get a Markov process for fluctuations inside fluctuations inside fluctuations .... It is worth noting that the Gaussian approximation (11.42) provides the *exact* transition rates. The corresponding Chapman–Kolmogorov equation describes how a particular fluctuation in a system of size V depends on the state of the system at slightly larger size $V' = V + dV$.

Note that the role of time in the Chapman–Kolmogorov equation is played by $1/V$ which starts at 0, where the distribution is the delta-function (11.41), and then takes on small values, where the Gaussian distribution (11.42) still holds, and then tends to infinity as the system size becomes small. This is the newly explored regime where a covariant description of fluctuations emerges.

The Chapman–Kolmogorov equation takes the form of a covariant Fokker–Planck equation. Its ultimate form, assuring all conservation laws, was derived by Diósi and Lukács [25]. Here, to abandon using covariant differential calculus, we present the equation in extensive (density) parameters

$$\frac{\partial}{\partial V^{-1}} P(\mathbf{x}, V|\mathbf{x}_0, \infty) = \frac{1}{k_B} \sum_{i,k} \Big(\frac{\partial^2}{\partial x_i \, \partial x_k} g(\mathbf{x})^{ik} P(\mathbf{x}, V|\mathbf{x}_0, \infty) \Big). \quad (11.43)$$

From the initial distribution (11.41) at $V^{-1} = 0$, this Fokker–Planck equation evolves the distribution function of thermodynamic fluctuations at all finite volumes V. By construction, the equation provides initiallly the Gaussian distribution (11.42), and preserves normalization and the mean values of the extensive variables $\mathbf{x}$. This latter assures the fulfillment of the conservation laws.

For small fluctuations, Gaussian fluctuation theory does well. It seems that it also yields an overall covariant formalism which improves the match with experiment for fluctuations of moderate size [7]. This is especially useful for understanding system behavior near the critical point where fluctuations become large.

Requiring covariance of such partial differential equations along with some hypotheses connecting the Riemannian curvature and the free energy yields equations of state connecting the critical exponents [7].

11.5.2 *Entropy Production*

At the macroscopic level, the Riemannian structure introduced above is intimately connected with entropy production. Let $\mathbf{X} = (U, V, N_1, N_2, ...) \in \mathbb{R}^{n+1}$ and $\tilde{\mathbf{X}} = (\tilde{U}, \tilde{V}, \tilde{N}_1, \tilde{N}_2, ...) \in \mathbb{R}^{n+1}$ be the vectors of extensive variables of systems A and $\tilde{A}$, respectively. In an interaction between these two systems, in which the infinitesimal vector of flows $d\mathbf{X}$ moves from $\tilde{A}$

to A, the entropy production is

$$dS_u = dS_A + dS_{\tilde{A}} = \sum_{i=1}^{n}(Y_i - \tilde{Y}_i)\, dX_i \qquad (11.44)$$

where $\mathbf{Y} = \partial S/\partial \mathbf{X} = (1/T, p/T, \mu_1/T, \mu_2/T, ...)$, and we have made use of the conservation laws $dX_i = -d\tilde{X}_i$, $i = 1, ..., n$. Equation (11.44) is the familiar flow-times-force expression for entropy production and, as we will see, bears a close resemblance to the length element $d\ell^2$ in our Riemannian geometry. To emphasize this similarity, we rewrite (11.44) in the form

$$dS_u = -\sum_{i=1}^{n} \Delta Y_i\, dX_i = -\Delta \mathbf{Y} \cdot d\mathbf{X}, \qquad (11.45)$$

where the $\Delta \mathbf{Y} = \tilde{\mathbf{Y}} - \mathbf{Y}$. Note that this sum must be positive by the second law.

11.5.3 *The Metric as a Symmetric Product*

We now express our length element

$$d\ell^2 = \sum_{j=1}^{n}\sum_{i=1}^{n} g(\mathbf{X})^{ij}\, dX_i\, dX_j \qquad (11.46)$$

as a symmetric product. Since

$$dY_j = \sum_{i=1}^{n} \frac{\partial Y_j}{\partial X_i}\, dX_i \qquad (11.47)$$

$$= \sum_{i=1}^{n} \frac{\partial^2 S}{\partial X_i\, \partial X_j}\, dX_i \qquad (11.48)$$

$$= -\sum_{i=1}^{n} g(\mathbf{X})^{ij}\, dX_i, \qquad (11.49)$$

we can write the length element as

$$d\ell^2 = -d\mathbf{Y}\, d\mathbf{X}. \qquad (11.50)$$

Note the similarity between this expression and (11.45). Note also that while $d\mathbf{Y}$ and $\Delta \mathbf{Y}$ look similar, they represent very different quantities: $d\mathbf{Y}$ is an infinitesimal change in the state of system A, while $\Delta \mathbf{Y}$ is the difference $\tilde{\mathbf{Y}} - \mathbf{Y}$.

11.5. Relevance of Riemannian Geometry in Thermodynamics

From the symmetric product $-d\mathbf{Y}\,d\mathbf{X}$ in (11.50), it is easy to obtain the conformal equivalence between the metrics G and $g(\mathbf{X})$. On substituting

$$dX_1 = dU = TdS + \sum_{i=2}^{n} W_i\,dZ_i, \qquad (11.51)$$

$$W_i = \frac{\partial U}{\partial Z_i} = -TY_i, \quad i = 2,\ldots,n, \qquad (11.52)$$

$$Z_i = X_i, \qquad i = 2,\ldots,n, \qquad (11.53)$$

into (11.50) we obtain on, rearrangement,

$$d\ell^2 = -d\mathbf{Y}\,d\mathbf{X} = \frac{d\mathbf{W}\,d\mathbf{Z}}{T} \qquad (11.54)$$

$$= \frac{G^{ij}\,dZ_i\,dZ_j}{T} = \frac{dL^2}{T}. \qquad (11.55)$$

Equations (11.54) and (11.55) are the infinitesimal form of the Gouy–Stodola theorem [26] expressing the well-known relationship between loss of availability at a temperature T and the associated entropy production. This fact will become more apparent after our discussion of the discrete horse–carrot theorem.

11.5.4 The Group of Transformations

The conformal equivalence of the geometries defined by the second derivatives of U, S, and ϕ leads naturally to the question of what other potential functions $\wp$ one might consider with the property that the metric $g(\mathbf{X})$ is a multiple of the second derivative matrix of $\wp$ with respect to $\wp$'s natural variables. The question is elegantly posed using the formalism introduced by Hermann [27]. Define an n degree of freedom thermodynamic system as a maximal integral submanifold of a contact form[13] ω on a space of dimension $2n+1$. In usual coordinates this takes the form $\omega = dS - \sum_{i=1}^{n} Y_i\,dX_i$. Asking for a maximal integral submanifold is asking for an n-dimensional surface on which the differential expression of the first law, $\omega = 0$, holds. The additional structure implied by the symmetric two-form $\eta = d\mathbf{Y}\,d\mathbf{X}$ gives an interesting class of manifolds [28], [29], [30]. It turns out that there are very many potentials $\wp$. The set of such potentials can be characterized by considering the group of coordinate transformations which preserve ω and η up to scale factors. The group turns out to equal the semidirect product of the integers modulo 2, Z_2, the multiplicative group of nonzero real numbers, $\mathbb{R}^*$, the general linear group, $\mathrm{Gl}(n)$, and the Heisenberg group, $H(n)$. If we ask that η be preserved without scaling, the group shrinks to

---

[13] A nowhere vanishing differential form of maximal rank.

just the semidirect product $\mathrm{Gl}(n) \otimes_\alpha \mathrm{H}(n)$ but which excludes the transformation from S to either U or ϕ.

While this shows there are many possible such potentials, they have not been extensively applied. Diósi and Lukács showed that the action of the renormalization operator belongs to this group [25]. See also the recent work by Brody and Ritz [31]. Mrugała [28] has applied these contact transformations to find laws of corresponding states. More recently, Eu [32] has used them to study Pfaffian formulations of uncompensated heat.

11.5.5 *Dissipation in a Small Equilibration*

There is an important special case for which there is a close relationship between $\Delta \mathbf{Y}$ and $d\mathbf{Y}$ in (11.45) and (11.50). This is the case of a small equilibration with a bath. Let us take $\tilde{A}$ sufficiently large that any changes in its intensive variables $\tilde{\mathbf{Y}}$ can be neglected. Furthermore, we allow A to equilibrate to $\tilde{A}$ so the final values of $\mathbf{Y}$ equal $\tilde{\mathbf{Y}}$. Integrating to equilibrium gives

$$\Delta S_u = \int dS_u = \int -\Delta \mathbf{Y} \, d\mathbf{X} \tag{11.56}$$

which in light of (11.49) becomes

$$\Delta S_u = \int \Delta \mathbf{X}^t g(\mathbf{X}) \, d\mathbf{X} \tag{11.57}$$

to first order in $\Delta \mathbf{X} = \mathbf{X}_0 - \mathbf{X}$ where $\mathbf{X}_0$ is the vector of extensive variables of system A after equilibration with the bath $\tilde{A}$. To this order we may take the metric matrix $g(\mathbf{X})$ to be constant in which case (11.57) can be integrated to give

$$\Delta S_u = \tfrac{1}{2} \Delta \mathbf{X}^t g(\mathbf{X}) \Delta \mathbf{X} \tag{11.58}$$
$$= \tfrac{1}{2} \Delta \ell^2. \tag{11.59}$$

11.5.6 *The Discrete Horse–Carrot Theorem*

The discrete horse–carrot theorem follows at once from the general expression (11.59) expressing the relationship between the length of a small equilibration and the corresponding entropy production. Consider a path in the state space of system A, and the process whereby we select the states of k baths to match the system's intensive variables $\mathbf{Y}$ at k points along this path. The discrete horse–carrot theorem answers the question: How should the states be chosen so as to minimize the total entropy produced in the k successive equilibrations which bring the system to equilibrium with the k successive baths along the path? For large k the answer is simply that one should place the baths equidistant in the geometry given by $g(\mathbf{X})$. This

11.5. Relevance of Riemannian Geometry in Thermodynamics

follows by noting that the total entropy production is given by

$$\Delta S_u = \sum_{l=1}^{k} \Delta_l S_u = \frac{1}{2} \sum_{l=1}^{k} \Delta_l \ell^2 \qquad (11.60)$$

which is to be minimized while fixing the total length

$$\ell = \sum_{l=1}^{k} \Delta_l \ell. \qquad (11.61)$$

This minimization is easily handled using Lagrange multipliers to give

$$\Delta_l \ell = \text{constant} = \frac{\ell}{k}. \qquad (11.62)$$

Substituting this back into (11.60) gives the horse–carrot inequality

$$\Delta S_u \geq \Delta S_u^{\min} = \frac{\ell^2}{2k}. \qquad (11.63)$$

A more thorough analysis including the dynamics of incomplete relaxation is possible [9]. For large times the analysis tells us to allocate the same number of relaxation times to each equilibration. Thus the minimum entropy-producing way, to bring the system in a finite time along a given path using a fixed number k of intermediate equilibrations, is to make the steps equidistant with a constant number of relaxation times alloted for each step. This is the origin of the idea of constant thermodynamic speed $v = d\ell/d\xi$, where $d\xi = dt/\epsilon$, t is time, and ϵ is the relaxation time of the system. The optimality of constant thermodynamic speed for a k-step process is hereby established. The optimality of this control in other contexts has led to some confusion as we discuss further below.

Some Comments on Loss of Availability

An expression entirely analogous to our equation (11.59) can be derived for the loss of availability ΔA_u in a small equilibration

$$\Delta A_u = \tfrac{1}{2} \Delta \mathbf{Z}^t G \Delta \mathbf{Z} \qquad (11.64)$$
$$= \tfrac{1}{2} \Delta L^2. \qquad (11.65)$$

We can now see more clearly why we referred to the conformal equivalence of the two metrics given by the second derivatives of S and U expressed in (11.55) as the differential form of the Gouy–Stodola theorem

$$\Delta A_u = -T_a \, \Delta S_u, \qquad (11.66)$$

where T_a is the temperature of the atmosphere. For the infinitesimal process in (11.55), the role of the large bath is played by $\tilde{A}$ rather than by the atmosphere. The ambiguity of where the heat equivalent of the lost availability ends up severely limits the usefulness of the analogous horse–carrot inequality for ΔA_u,

$$\Delta A_u \geq \frac{L^2}{2k}. \tag{11.67}$$

For isothermal processes the two inequalities (11.63) and (11.67) are equivalent. For nonisothermal processes the lengths given by ℓ and L are not simply related. Although the lengths given by L can be of interest for system control in certain circumstances [33], dL^2 is more often useful as a device for calculating $d\ell^2$ using (11.55) as we illustrate below.

11.5.7 The Continuous Horse–Carrot Theorem

In this section we treat the continuum version of the discrete control considered above. The problem is now as follows: Given that system A traversed the path $\mathbf{X}(t)$, $t \in [0, \tau]$, how much entropy production had to occur? For sufficiently large τ, this question has a very similar answer to what we found for the discrete process although the optimal control turns out to be a constant entropy production rate rather than constant thermodynamic speed. We assume that we can control system $\tilde{A}$ reversibly and that system A is affected only indirectly through its contact with $\tilde{A}$. Our argument proceeds from the integral form of (11.45) for the total entropy production

$$\Delta S_u = \int_0^\tau dS_u = -\int_0^\tau \Delta \mathbf{Y} \, d\mathbf{X}. \tag{11.68}$$

Some Basic Expressions Connecting Dissipation and Geometry

There are a number of interesting rearrangements of (11.68) for the total dissipation which reveal connections between this dissipation and our geometry. Define $\mathbf{X}_e$ by the formula

$$-(\tilde{\mathbf{Y}} - \mathbf{Y}) = g(\mathbf{X})(\mathbf{X}_e - \mathbf{X}). \tag{11.69}$$

Since $g(\mathbf{X})$ is not necessarily invertible, the scale of different homogeneous phases in $\mathbf{X}_e$ must be set separately by specifying how these scales of A evolve. Then for $\mathbf{X}_e$ close to $\mathbf{X}$, we can interpret $\mathbf{X}_e$ as the state of A which would minimize the entropy production rate in contact with the current state of $\tilde{A}$ subject to the constraint of keeping A's state on the line through $\mathbf{X}$ in the direction $d\mathbf{X}$. On substituting (11.69) into (11.68) and replacing $d\mathbf{X}$ by $(d\mathbf{X}/d\ell)d\ell$ we have

$$\Delta S_u = \int_0^\tau (\mathbf{X}_e - \mathbf{X})^t g(\mathbf{X}) \frac{d\mathbf{X}}{d\ell} \, d\ell. \tag{11.70}$$

Recall that a metric on a vector space defines a dot product [34]. The integrand in (11.70) is the dot product using the metric $g(\mathbf{X})$ of the deviation

11.5. Relevance of Riemannian Geometry in Thermodynamics 307

$\mathbf{X}_e - \mathbf{X}$ and the unit tangent vector $d\mathbf{X}/d\ell$. Thus we may interpret this integrand as the distance between our current state and the state the system is trying to reach, projected onto the direction of $d\mathbf{X}$. We call this projected lag distance $D = dS_u/d\ell$ and see by the mean value theorem that we can write the total dissipation as

$$\Delta S_u = \int_0^\ell D \, d\ell = \bar{D}\ell. \tag{11.71}$$

This gives an expression for the total entropy production as the product of the mean distance to equilibrium and the total distance traversed.

A second interesting relation can be found by considering the quantity

$$\epsilon = \frac{dS_u/dt}{(d\ell/dt)^2}. \tag{11.72}$$

Note that ϵ has the units of time. In fact, for a sufficiently slow process with separable time scales, ϵ is just the relaxation time. We can see this by writing

$$\frac{d\mathbf{X}}{dt} = (\mathbf{X}_e - \mathbf{X})/\epsilon. \tag{11.73}$$

Note that by our definition of $\mathbf{X}_e$, $d\mathbf{X}/dt$ and $\mathbf{X}_e - \mathbf{X}$ must be in the same direction and hence must be proportional. If our dynamics is sufficiently slow and the time scales are separable, then all but the slowest mode of our system must equilibrate essentially instantaneously and thus $\mathbf{X}_0 - \mathbf{X}$ must be proportional to $d\mathbf{X}/dt$, i.e., $\mathbf{X}_e = \mathbf{X}_0$. With or without our assumptions of slow process and separable time scales, the definition in (11.72) allows us to express the total entropy production as

$$\Delta S_u = \int_0^\tau \epsilon \frac{d\mathbf{X}^t}{dt} g(\mathbf{X}) \frac{d\mathbf{X}}{dt} dt = \int_0^\tau \epsilon \left(\frac{d\ell}{dt}\right)^2 dt. \tag{11.74}$$

We can again apply the mean value theorem, to give the alternative form

$$\Delta S_u = \bar{\epsilon} \int_0^\tau \frac{d\mathbf{X}^t}{dt} g(\mathbf{X}) \frac{d\mathbf{X}}{dt} dt = \bar{\epsilon} \int_0^\tau \left(\frac{d\ell}{dt}\right)^2 dt. \tag{11.75}$$

A third interesting expression for ΔS_u results if we change parameters along the path $X(t)$ and express our dissipation integral in terms of the number of relaxations ξ. Recall that $d\xi = dt/\epsilon$ and so our integral becomes

$$\Delta S_u = \int_0^\Xi \frac{d\mathbf{X}^t}{d\xi} g(\mathbf{X}) \frac{d\mathbf{X}}{d\xi} d\xi = \int_0^\Xi \left(\frac{d\ell}{d\xi}\right)^2 d\xi, \tag{11.76}$$

where

$$\Xi = \int_0^\tau d\xi = \int_0^\tau \frac{dt}{\epsilon}. \tag{11.77}$$

All our expressions in this section are valid generally. Nevertheless, without our hypotheses of a sufficiently slow process with separable time scales, in which A and $\tilde{A}$ are near equilibrium with each other at each instant, the physical meaning of D and ϵ are merely formal [35].

For our fourth and final version of the integrated dissipation ΔS_u, we start from the Onsager–Prigogine-type of linearized flux–force relationship

$$\frac{d\mathbf{X}}{dt} = \gamma \Delta \mathbf{Y}, \tag{11.78}$$

where γ is the matrix of kinetic coefficients [36]. Note that γ must be symmetric and positive definite. If we solve this for $\Delta \mathbf{Y}$ and substitute into (11.68) for the entropy production, we get

$$\Delta S_u = \int_0^\tau \frac{d\mathbf{X}^t}{dt} \gamma^{-1} \frac{d\mathbf{X}}{dt}\, dt = \int_0^\tau \left(\frac{d\lambda}{dt}\right)^2 dt, \tag{11.79}$$

which can again be interpreted as the integral of a speed squared. This time, the lengths λ are given by yet another metric (γ^{-1}). There is a fundamental difference between this metric and the ones which we have so far considered. The coefficients in γ are *kinetic* as opposed to *equilibrium* quantities. Stated another way, the metric coefficients in $g(\mathbf{X})$ are covariances while the coefficients in γ are time correlations.

A Simple Lemma from Optimization

We now pause our development for a simple result which will show us how to minimize the entropy production and how to obtain a number of inequalities corresponding to the various expressions for ΔS_u derived in the previous subsection. While these inequalities can be obtained from the Cauchy–Shwartz inequality [8], [9], we use a variational argument here to emphasize their connection to optimal process control.

To minimize an integral of the form

$$\int_0^\tau f(x) \left(\frac{dx}{dt}\right)^2 dt, \tag{11.80}$$

with given values of $x(0)$ and $x(\tau)$, the first-order necessary conditions of Euler–Lagrange for our autonomous Lagrangian, K,

$$K = f(x) \left(\frac{dx}{dt}\right)^2 \tag{11.81}$$

give

$$K - \frac{dx}{dt} \frac{\partial K}{\partial (dx/dt)} = \text{constant}. \tag{11.82}$$

Substituting our expression for K in (11.81) into (11.82), we find that for optimality, the Lagrangian K should be constant.

11.5. Relevance of Riemannian Geometry in Thermodynamics

An immediate and useful corollary follows for the special case of (11.80) with $f \equiv 1$. For this case

$$K = \left(\frac{dx}{dt}\right)^2 = \text{constant} \tag{11.83}$$

implies

$$\frac{dx}{dt} = \text{constant} = \Delta x/\tau. \tag{11.84}$$

The minimum value of the integral then simplifies to the right-hand side of

$$\int_0^\tau \left(\frac{dx}{dt}\right)^2 dt \geq \Delta x^2/\tau. \tag{11.85}$$

This special case takes on particular importance when x is the arc length with respect to some metric M. In this case, letting $\mathbf{X}$ represent the coordinates in this space, the inequality becomes

$$\int_0^\tau \frac{d\mathbf{X}^t}{dt} M \frac{d\mathbf{X}}{dt} dt \geq \left(\int_0^\tau \sqrt{\frac{d\mathbf{X}^t}{dt} M \frac{d\mathbf{X}}{dt}} dt\right)^2 \Big/ \tau. \tag{11.86}$$

This general inequality leads directly to the fact that extremal curves for the speed squared coincide with extremal curves for the length (geodesics), a fact that is the starting point for Morse theory [37].

Applications of the Lemma

Applying the lemma to (11.74) or (11.79) tells us that to minimize entropy production, we should proceed at a *constant entropy production rate*

$$\dot{S}_u = \epsilon \left(\frac{d\ell}{dt}\right)^2 = \left(\frac{d\lambda}{dt}\right)^2 = \text{constant}. \tag{11.87}$$

Applying the corollary for squared speed to (11.75) and (11.76) leads to the continuous versions of the horse–carrot inequality

$$\Delta S_u \geq \bar{\epsilon}\ell^2/\tau, \tag{11.88}$$
$$\Delta S_u \geq \ell^2/\Xi. \tag{11.89}$$

These inequalities bound the dissipation by the squared thermodynamic distance ℓ^2 divided by the number of relaxations. As such, they bear a strong resemblance to our discrete horse carrot inequality (11.63). Despite some confused claims in the literature, these generally valid inequalities do not say anything useful about how to minimize the entropy production in a given time. The averaging process that goes into $\bar{\epsilon}$ and Ξ for fixed total time[14] hides a dependence. For the process with a fixed number Ξ of relaxations, inequality (11.89) becomes sharp and the minimum entropy production strategy is to drive the process with constant thermodynamic speed $d\ell/d\xi$. Thus constant thermodynamic speed is optimal *for a given number of relaxations* in both the discrete and the continuous case. This fact was only recently elucidated independently by [38] and [39].

The factor of $\frac{1}{2}$, present in the discrete but not in the continuous case, is real and comes from the fact that in the discrete case we repeatedly relax (almost all the way) to equilibrium, while in the continuous case we maintain an approximately fixed distance. It follows that in the discrete case we are on the average about half as far from equilibrium and thus by (11.71) should expect about half the dissipation.

Finally we note that our corollary about the integral of the squared speed also applies to (11.79) and gives an alternative route to the last part of our minimum entropy production condition (11.87) for fixed time. It also gives an associated inequality

$$\Delta S_u \geq \lambda^2/\tau. \tag{11.90}$$

The implications of the geometry of time correlations given by γ^{-1} is left for another chapter. For preliminary results in this direction, the interested reader is referred to [38], [39], and [40].

11.5.8 *Cooling Rates for Simulated Annealing*

The above formalism has been applied to the control of the temperature in simulated annealing–an algorithm for solving global optimization problems. Here the idea is to associate a (usually fictitious) physical system with the optimization problem by identifying the objective function with the energy of such a system. We then simulate relaxations to equilibrium at a decreasing sequence of temperatures by a random walk over states using the Metropolis algorithm [41]. Much has been written about the ideal cooling rate [42], [43], [44]. Several authors have advocated a constant statistical velocity cooling schedule which keeps $d\ell/dt$ constant based on arguments along the lines of Section 11.2.2. In fact, this schedule has been incorporated in the popular simulated annealing package known as Timberwolf [45], [46].

Although no direct connection has been established between entropy production and performance of the algorithm, such conjectures are tantalizing [47], [48]. Motivated by these conjectures, both constant thermodynamic speed and constant entropy production rate schedules have been tried. Empirically, it seems that the constant thermodynamic speed schedule outperforms others [42], [49], [50] but the difference for most systems is small.

The argument in favor of constant thermodynamic speed for these problems runs along the same lines as the argument already presented in Sec-

---

[14] Note that by (11.77) we can consider Ξ as a time weighted harmonic mean ϵ.

11.5. Relevance of Riemannian Geometry in Thermodynamics

tion 11.2.2 in favor of constant statistical velocity: We cool as fast as possible consistent with the constraint of never being too far out of equilibrium [51]. The distance to equilibrium is measured by the metric $g(\mathbf{X})$ and this argument leads to keeping the distance D between the system and the bath constant. The role of the bath is played here by the parameter T used in the Metropolis algorithm. For this (thermodynamically one degree of freedom) system, the metric, $g(\mathbf{X}) = -d^2 S/dE^2$, leads to the length element

$$d\ell = \sqrt{-\frac{d^2 S}{dE^2} dE^2} \tag{11.91}$$

$$= \sqrt{-\frac{d}{dE}(1/T) \, dE} \tag{11.92}$$

$$= \sqrt{\frac{1}{\sigma_E^2} dE} \tag{11.93}$$

$$= \frac{dE}{\sigma_E}, \tag{11.94}$$

where σ_E is the standard deviation of the energy and we have used the fact that the heat capacity dE/dT is equal to σ_E^2/T^2 [52]. Letting E_0 stand for the equilibrium energy of the system at the current temperature T, in the Metropolis algorithm, we get that $D = (E - E_0)/\sigma_E$ gives our distance measure of disequilibrium.

Keeping D constant keeps the system moving with its own time scale. If the cooling is sufficiently slow and the time scales are separable, the system moves with its own relaxation time. This gives one popular way to implement what has been called constant thermodynamic speed schedules. Alas, time scales are not separable in typical problems of simulated annealing interest; physically these systems act like glasses. Accordingly, keeping the lag distance constant is not equivalent to constant thermodynamic speed. A constant D schedule does share an attractive feature with constant thermodynamic speed annealing: both schedules measure energy and time on natural scales of the system. Their "optimality" thus follows from an old meta-theorem of applied mathematics: the more one exploits the structure of the problem the better.

There exist several other means of implementing constant thermodynamic speed. One popular technique for well-studied problems, that gets around the difficulty associated with adaptive algorithms, is to model or fit the constant speed schedule obtained by laborious adaptive analysis and then use a rescaled version of this schedule for other similar problems [53].

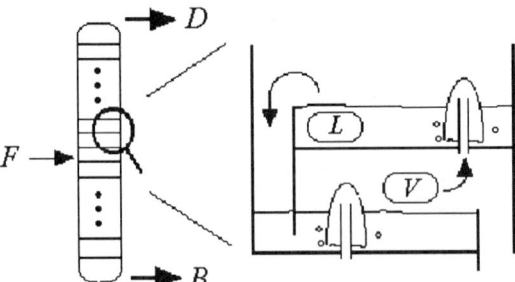

FIGURE 11.4. A schematic distillation column with flows: feed F, distillate D, and bottoms B. The close up shows two adjacent trays including overflow tubes for downward flow of liquid L and bubble caps for upward flow of vapor V.

11.6 Staged Steady Flow Processes

Most recently [54], the connection between dissipation and geometry has been extended to treat a staged steady-flow process of considerable industrial interest: fractional distillation. The example involves some surprises which hint at the existence of other applications of horse–carrot-type analyses. The first surprise is that the scale of the process is set by the flow rates rather than the states along the process. The second surprise is that the null directions for our semi-Riemannian metric turn out to be useful.

11.6.1 *Dissipation in a Distillation Column*

Fractional distillation is a process for separating a mixture of compounds based on the differences in the boiling points of the components. Fractional distillation is performed within a vertical column divided into trays that constitute the k stages for the process. The mixture to be separated is introduced near the middle of the column at the feed tray, and the separated components are removed at the top as distillate D and at the bottom as bottoms B (see Figure 11.4). Boiling occurs on each tray resulting in the formation of vapor which is then bubbled through the liquid at the next higher tray. Similarly, each tray is equipped with an overflow tube which returns excess liquid to the next lower tray.

We treat a binary mixture at constant pressure.[15] For steady-state operation, the net difference between the upward flow of vapor and the downward flow of liquid must equal D at each tray–tray interface above the feed and

---

[15] More components and a pressure differential along the column can be handled similarly, albeit at a significant cost in complexity.

B at each interface below the feed. Formally, numbering the trays from the top of the column (see Figure 11.4), and letting V_m and L_m stand for the number of moles of vapor and liquid leaving tray m, the mass balance equations at the interface between trays m and $m+1$ are

$$V_{m+1} - L_m = \begin{cases} D & \text{above feed,} \\ -B & \text{below feed,} \end{cases} \quad (11.95)$$

$$y_{m+1} V_{m+1} - x_m L_m = \begin{cases} x_D D & \text{above feed,} \\ -x_B B & \text{below feed,} \end{cases} \quad (11.96)$$

where x and y are the mole fractions of the first component in the gaseous and liquid phases, respectively. In the limit of an infinite number of trays this becomes

$$V - L = \begin{cases} D & \text{above feed,} \\ -B & \text{below feed,} \end{cases} \quad (11.97)$$

$$yV - xL = \begin{cases} x_D D & \text{above feed,} \\ -x_B B & \text{below feed,} \end{cases} \quad (11.98)$$

with x, y, V, and L now smooth functions of T except at the feed plate where we switch between the appropriate balance conditions. The whole process becomes a continuous, piecewise smooth path in the state space of the two-phase binary system by including a rescaling branch at the feed plate which contributes length zero. This path is known in the literature [55] as the minimum reflux values of V and L at each T. This is also the path we will dissect into k equal length pieces for a discrete horse–carrot process.

In the conventional operation of the column, a heat source is connected at the bottom tray and a heat sink is connected at the top tray creating a temperature gradient along the column. This results in the net upward motion of low-boiling component and downward motion of high-boiling component. We depart from the conventional design and use additional heat sources (sinks) along the column to adjust the temperature at each plate. We then ask for the sequence of temperatures which minimizes the total dissipation inside the column. We take the transport of heat and matter between the column and its surroundings as reversible.

Since we assume that each stage is in equilibrium, the losses occur as the upward flow of vapor and downward flow of liquid equilibrate at the next trays. For concreteness, consider the bubble of vapor going up–the analysis for the downward flow of liquid proceeds similarly. The losses can be counted by a conceptual rearrangement of what occurs. We consider the bubble of vapor to be isolated except for the exchange of heat and $p\,dV$ work with the two-phase fluid in the tray above. Accordingly, this fluid acts as a bath with a certain temperature and pressure. In this manner, the bubble is brought to equilibrium at the temperature and pressure of the next tray by a horse–carrot process whose entropy production is given by the distance squared, $\Delta \ell^2$. This squared distance is an extensive quantity; the scale is set by the number of moles of material moving per unit time. In the final state of each bubble, some of the vapor has condensed to liquid, but each phase is exactly at the composition in the next tray and we can reversibly mix the bubble and its surroundings. Our conceptual rearrangement of events is justified since in either case the net effect is the complete equilibration between the bubble of vapor and the equilibrium system in the next tray.

Since we assume constant pressure, the form of the metric in (11.55) is the most convenient since only one term in the sum is nonzero.

$$\Delta S_u = \tfrac{1}{2}(\Delta \ell)^2 = \frac{1}{2}\frac{\Delta T \Delta S}{T} = \frac{1}{2}\frac{C_\sigma (\Delta T)^2}{T^2}, \qquad (11.99)$$

where C_σ is the constant pressure saturation heat capacity of the two-phase mixture in equilibrium [56]. We get the same expression for the liquid, although ΔT has the opposite sign. Since the dissipation only depends on $(\Delta T)^2$, we would get the same entropy production if the liquid flow were reversed and also went up the column.

We have hereby established that the dissipation of small relaxation steps along this path equals the squared length of the corresponding displacement. Therefore, the discrete horse–carrot theorem applies and we can conclude that, to minimize total entropy production in the column, the tray temperatures should be adjusted to equalize the thermodynamic distance between trays. To find the optimal temperature profile, we need to find temperatures T_j such that

$$\int_{T_j}^{T_{j+1}} \sqrt{\frac{C_\sigma}{T}}\, dT = \frac{1}{k}\int_{T_0}^{T_k} \sqrt{\frac{C_\sigma}{T}}\, dT, \quad j = 0, ..., k-1. \qquad (11.100)$$

This derivation shows an application of the discrete horse–carrot theorem to the steady-state operation of a separation process. The results express the dissipation in terms of the length of a path in the equilibrium state space of the mixture and show how to optimally control the temperatures of the stages along such a separation. The procedure is readily adapted to any staged steady flow process in the limit of many stages. We start from the flow vectors along the process. Since these flows equilibrate at the next stage, the entropy produced by such small relaxations is the square of a length element. For the purpose of counting dissipation, all flows can be taken unidirectional and summed exactly as for distillation. The corresponding path consists of the flows for the process in the limit of infinitely many stages. This should have implications for the control of many real processes.

11.7 Conclusions

This chapter presented a review of the geometry of distinguishability in all its guises ranging from the quantum to the macroscopic. We tried to present a thorough overview of the results and applications along with the connections to related geometries.

11.8 References

[1] F. Weinhold: Metric geometry of thermodynamics I–IV, *J. Chem. Phys.* **63,** 2479, 2484, 2488, 2496, (1975).

[2] G. Ruppeiner: Thermodynamics: A Riemannian geometric model, *Phys. Rev. A* **20,** 1608–1613, (1979).

[3] L. Diósi, G. Forgács, B. Lukács, and H.L. Frisch: Metrization of thermodynamic state space and the renormalization group, *Phys. Rev. A* **29,** 3343–3345, (1984).

[4] R.A. Fisher: On the dominance ratio, *Proc. Roy. Soc. Edinburgh* **42,** 321, (1922).

[5] S. L. Braunstein and C. M. Caves: Statistical distance and the geometry of quantum states, *Phys. Rev. Lett.* **72,** 3439–3443, (1994).

[6] W. K. Wootters: Statistical distance and Hilbert space, *Phys. Rev. D* **23,** 357–362, (1981).

[7] G. Ruppeiner: Riemannian geometry in thermodynamic fluctuation theory, *Rev. Mod. Phys.* **67,** 605–659, (1995).

[8] P. Salamon and R. S. Berry: Thermodynamic length and dissipated availability, *Phys. Rev. Lett.* **51,** 1127–1130, (1983).

[9] J. Nulton, P. Salamon, B. Andresen, and Qi Anmin: Quasistatic processes as step equilibrations, *J. Chem. Phys.* **83,** 334, (1985).

[10] C. R. Rao: Information and the accuracy attainable in the estimation of statistical parameters, *Bull. Calcutta Math. Soc.* **37,** 81–91, (1945).

[11] A. Batthachariya: On a measure of divergence between two multinomial populations, *Indian J. Stat.* **7**, 401–406, (1946).

[12] C. Atkinson and A. F. S. Mitchell: Rao s distance measure, *Indian J. Stat.* **43**, 345–365, (1981).

[13] D. J. C. Bures: An extension of Kakutani's theorem on infinite product measures to the tensor product of semifinite W^*-algebras, *Trans. Amer. Math. Soc.* **135**, 199–212, (1969).

[14] R. A. Fisher: On the mathematical foundations of theoretical statistics, *Phil. Trans. Roy. Soc. A* **222**, 309–368, (1921).

[15] S. Amari: Differential-geometric methods in statistics, Lecture Notes in Statistics, vol. 28, Springer-Verlag, Berlin, 1985.

[16] N. N. Chentzov: Categories of mathematical statistics, *Dokl. Akad. Nauk. SSSR* **164**, 3, (1965) (in Russian).

[17] F. Schlögl: Thermodynamic metric and stochastic measures, *Z. Phys. B* **59**, 449, (1985).

[18] P. Salamon, J. Nulton, and E. Ihrig: On the relation between energy and entropy versions of thermodynamic length, *J. Chem. Phys.* **80**, 436, (1984).

[19] P. Salamon, J. D. Nulton, and R. S. Berry: Length in statistical thermodynamics, *J. Chem. Phys.* **82**, 2433–2436 (1985).

[20] F. Weinhold: Metric geometry of equilibrium thermodynamics V, *J. Chem. Phys.* **65**, 559, (1976). See in particular Fig. 4.

[21] H. Callen: *Thermodynamics,* Wiley, New York, 1985.

[22] P. Salamon, B. Andresen, P. Gait, and R. S. Berry: Interpretation of Weinhold's metric, *J. Chem. Phys.* **73**, 1001, (1980).

[23] G. Ruppeiner: Thermodynamic critical fluctuation theory?, *Phys. Rev. Lett.* **50**, 287, (1983).

[24] G. Ruppeiner: New thermodynamic fluctuation theory using path integrals, *Phys. Rev. A* **27**, 1116, (1983).

[25] L. Diósi and B. Lukács: Covariant evolution equation for the thermodynamic fluctuations, *Phys. Rev. A* **31**, 3415, (1985).

[26] A. Bejan: *Entropy Generation Through Heat and Fluid Flow,* Wiley, New York, 1982.

[27] R. Hermann: *Geometry, Physics and Systems,* Marcel Dekker, New York, 1981.

11.8. References

[28] R. Mrugała, J. D. Nulton, J. C. Schön, and P. Salamon: Contact transformations in thermodynamics, *Rep. Math. Phys.* **29**, 109–121, (1991).

[29] R. Mrugała, J. D. Nulton, J. C. Schön, and P. Salamon: A statistical approach to the geometric structure of thermodynamics, *Phys. Rev. A* **41**, 3156, (1990).

[30] P. Salamon, E. Ihrig, and R. S. Berry: A group of coordinate transformations preserving the metric of Weinhold, *J. Math. Phys.* **24**, 2515, (1983).

[31] D. C. Brody and A. Ritz: On the symmetry of real-space renormalization, *Nucl. Phys.* **B 522**, 588–604, (1998).

[32] B. C. Eu: Note on the nonequilibrium partition function and generalized potentials, *J. Chem. Phys.* **105**, 5525, (1996).

[33] K. H. Hoffmann, B. Andresen, and P. Salamon: Measures of dissipation, *Phys. Rev. A* **40**, 3618–3630, (1989).

[34] J. Dieudonne: *Linear Algebra and Geometry*, Houghton Mifflin, Boston, 1969.

[35] T. Feldmann, B. Andresen, Anmin Qi, and P. Salamon: Thermodynamic lengths and intrinsic time scales in molecular relaxation, *J. Chem. Phys.* **83**, 5849–5853, (1985).

[36] S. R. de Groot and P. Mazur: *Non-Equilibrium Thermodynamics*, North-Holland, Amsterdam, 1962.

[37] J. Milnor: *Morse Theory*, Princeton University Press, Princeton, 1969, pp. 70–73.

[38] L. Diósi, K. Kulacsy, B. Lukács, and A. Rácz: Thermodynamic length, time, speed, and optimum path to minimize entropy production, *J. Chem. Phys.* **105**, 11220–11225, (1996).

[39] W. Spirkl and H. Ries: Optimal finite time endoreversible processes, *Phys. Rev. E* **52**, 3485–3489, (1995).

[40] K. Oláh: The entropy production: New results, *Magyar Kemiai Folyoirat* **103**, 411, (1997).

[41] N. Metropolis, A. Rosenbluth, M. Rosenbluth, A. Teller, and E. Teller: Equation of state calculations by fast computing machines, *J. Chem. Phys.* **21**, 1087, (1953).

[42] P. Salamon, J. Nulton, J. Robinson, J. Pedersen, G. Ruppeiner, and L. Liao: Simulated annealing with constant thermodynamic speed, *Comp. Phys. Comm.* **49**, 423, (1988).

[43] K. H. Hoffmann and P. Salamon: The optimal simulated annealing schedule for a simple model, *J. Phys. A* **23**, 3511, (1990).

[44] R. Azencott: *Simulated Annealing : Parallelization Techniques*, Wiley, New York, 1992.

[45] C. Sechen and A. Sangiovanni-Vincentelli: The TimberWolf placement and routing package, *IEEE Custom Integrated Circuits Conference*, 1984.

[46] D. Mitra, F. Romeo, and A. Sangiovanni-Vincentelli: Convergence and finite-time behavior of simulated annealing, *Adv. Appl. Prob.* **18**, 747, (1986).

[47] T. Zimmermann and P. Salamon: The Demon algorithm, *Inter. J. Comp. Math.* **42**, 21, (1992).

[48] P. Salamon, K. H. Hoffmann, J. R. Harland, and J. D. Nulton: An information theoretic bound on the performance of simulated annealing algorithms, Research Report IRC. 89. 2, Interdisciplinary Research Center, San Diego State University, (1989).

[49] Y. Nourani and B. Andresen: Simulated annealing with optimal cooling strategy and the natural time scale, *J. Phys. A* **31**, 8373–8385, (1998).

[50] K. Mosegaard and P. D. Vestergaard: A simulated annealing approach to seismic model optimization with sparse prior information, *Geophys. prospecting* **39**, 599, (1991).

[51] J. Nulton and P. Salamon: Statistical mechanics of combinatorial optimization, *Phys. Rev. A* **37**, 1351–1356, (1988).

[52] R. K. Pathria: *Statistical Mechanics*, Butterworth-Heinemann, Oxford, 1996.

[53] B. Andresen and J. M. Gordon: Constant thermodynamic speed simulated annealing, *Lecture Notes Earth Sci.* **63**, 303, (1996).

[54] P. Salamon and J. D. Nulton: The geometry of separation processes: the horse–carrot theorem for steady flow systems, *Europhys. Lett.* **42**, 571–576, (1998).

[55] C. J. King: *Separation Processes*, McGraw-Hill, New York, 1971.

[56] J. S. Rowlinson: *Liquids and Liquid Mixtures*, Plenum Press, New York, 1969.

[57] A. De Vos: Some examples of thermodynamic processes in finite time, *Berlin Colloquium on Finite-Time Thermodynamics*, 13–15 Nov., 1997.

12
Distillation by Thermodynamic Geometry

B. Andresen
P. Salamon

> ABSTRACT. The thermal efficiency of a distillation column may be improved by permitting heat exchange on every tray rather than only in the reboiler and the condenser. Thermodynamic length optimizations on discrete systems specify the optimal temperature of each tray and, consequently, the amount of heat to be added or withdrawn in order to maintain that temperature.

12.1 Introduction

Geometric aspects have always played a prominent role in finite-time thermodynamic optimizations. One of the early results [1] is a general bound on the entropy produced in bringing a thermodynamic system from a given initial state to a given final state through a sequence of steps. The system is required to follow a prescribed equilibrium path specified by N intermediate equilibrations with the environment. For an arbitrary sequence of reservoirs this bound provides the lower limit on the entropy generated by this process. The derivation also shows how these N intermediate reservoirs may guide the system in the least dissipative fashion from the initial to the final state. Such a process has become known as a *horse–carrot process* and the associated bound as the *horse-carrot theorem* since the system (the "horse") is coaxed along a sequence of states by successive contacts with generalized baths (the "carrots"). The intensities of these baths (temperature, pressure, etc.) define successive states in a sequence of equilibria.

Fractional distillation [2] is a very old process used to separate a mixture of compounds into its components of specified purity by making use of differences in boiling point. The liquid and vapor phases of the mixture are brought into equilibrium with one another at successive points along a column which is heated from below and cooled at the top. In trayed columns this equilibration occurs on the trays, i.e., at a finite number of discrete points, while packed columns provide continuous contact between liquid and vapor. In this gradient of temperature the more volatile components will preferentially migrate to the top of the column and the less volatile components to the bottom. In the remainder of this chapter we consider only binary separation for simplicity and trayed columns for their correspondence to the step processes mentioned above.

It is the goal of this chapter to apply the general results for step processes to a distillation column in order to minimize its entropy production. In Section 12.2 below we define the thermodynamic geometry. Section 12.3 derives the optimal path for a staged process. Section 12.4 provides some comments on traditional distillation, as a prelude to finding the optimal interior temperature profile of the column by geometric means in Section 12.5. Finally Section 12.6 contains an example of the savings which are possible with this new temperature profile.

12.2 Thermodynamic Length

Thermodynamic geometry lives in the space of all the extensive thermodynamic variables of a system: energy, entropy, volume, amount of material 1, amount of material 2, etc. In this space Weinhold [3], [4] defined a metric which is the second derivative of one extensive quantity (usually entropy or energy) with respect to the other extensive quantities. For example, in the energy picture, the system is described by the internal energy $U(S, V, N_1, N_2, ...)$ expressed in terms of the other extensive quantities. The metric in this picture is

$$\mathbf{M}_U = -\left\{ \frac{\partial^2 U}{\partial X_i \partial X_j} \right\} \tag{12.1}$$

with the remainder of the extensive variables represented by X_i. Weinhold's purpose was the calculation of all the usual partial derivatives in traditional static thermodynamics at a particular point. However, any metric invites integration over a range of its variables, in this case leading to the definition of a thermodynamic length L [5] by the usual formula

$$L_U = \int dL_U = \int \sqrt{d\mathbf{X} \mathbf{M}_U \, d\mathbf{X}}. \tag{12.2}$$

Boldface indicates vectorial quantities. The alternative mixed form of the differential length

$$dL_U = \sqrt{d\mathbf{Y} \, d\mathbf{X}}, \tag{12.3}$$

where $\mathbf{Y} = \partial U / \partial \mathbf{X}$ is the vector of intensive quantities conjugate to $\mathbf{X}$, may be useful if some of those elements happen to be constant, e.g., temperature or pressure.

Salamon and Berry [6] found a connection between this thermodynamic length along a continuous process path and the (reversible) availability lost in the process. Specifically, if the system moves via states of local thermodynamic equilibrium from an initial equilibrium state i to a final equilibrium state f in time τ, then the dissipated availability $-\Delta A$ is bounded from below by the square of the distance (i.e. length of the shortest path) from i to f times ϵ/t, where ϵ is a mean relaxation time of the system. If the system proceeds entirely through a sequence of equilibrium states, the bound can be strengthened to

$$-A \geq \frac{L_U^2 \epsilon}{\tau}, \qquad (12.4)$$

where L_U is the length of the *traversed path* from i to f. This will be the case, for example, if the process is endoreversible, i.e. the system is reversible in its interior while all irreversibilities are associated with its coupling to the environment [7]. Equality in (12.4) is achieved at constant thermodynamic speed $v = dL/dt$, assuming that the process proceeds slowly, i.e. is close to equilibrium with the environment at all times. At higher speeds corrections may be applied [8].

Viewed in the entropy picture $S(U, V, N_1, N_2, ...)$ an analogous expression exists for the total entropy production during the process:

$$\Delta S^u \geq \frac{L_S^2 \epsilon}{\tau}, \qquad (12.5)$$

where the length L_S is then calculated relative to the entropy metric

$$\mathbf{M}_S = -\left\{ \frac{\partial^2 S}{\partial X_i \, \partial X_j} \right\}. \qquad (12.6)$$

When expressed in identical coordinates, these two metrics are related by $\mathbf{M}_U = -T\mathbf{M}_S$, where T is the temperature of the system [9].

12.3 Optimization of a Step Process

First consider a single step [1] where a system described by the extensive variables X_i is coming to equilibrium with an environment at intensities Y_i^0, the simplest possible horse–carrot process. The system intensities conjugate to X_i will, in the entropy picture, be denoted by $Y_i = \partial S/\partial X_i$. Specifically, $\mathbf{X} = (U, V, N, ...)$ and $\mathbf{Y} = (1/T, -p/T, \mu/T, ...)$. Then a second-order power series expansion of the system entropy S about equilibrium yields

$$S = S^0 + \sum_i Y_i^0 (X_i - X_i^0) - \frac{1}{2} \sum_{ij} (X_i - X_i^0) M_{ij}^0 (X_j - X_j^0), \qquad (12.7)$$

where the superscript zero denotes values at equilibrium with the environment. Then the change in entropy of the system becomes

$$\Delta S = S^0 - S = -\sum_i Y_i^0 \Delta X_i + \frac{1}{2} \sum_{ij} \Delta X_i M_{ij}^0 \Delta X_j. \qquad (12.8)$$

The corresponding change of entropy of the environment (reservoir) is

$$\Delta S^{\text{en}} = -\sum_i Y_i^{\text{en}} \Delta X_i^{\text{en}}, \tag{12.9}$$

since its intensive quantities Y_i^{en} are constant. At equilibrium $Y_i^0 = Y_i^{\text{en}}$, and conservation of energy and matter relates $\Delta X_i = -\Delta X_i^{\text{en}}$, yielding a total change of entropy of the universe as a result of this small equilibration process

$$\Delta S^u = \Delta S + \Delta S^{\text{en}} = \frac{1}{2}\sum_{ij} \Delta X_i M_{ij}^0 \Delta X_j \tag{12.10}$$

o,r in matrix notation,

$$\Delta S^u = \tfrac{1}{2}\Delta \mathbf{X} \mathbf{M}\, \Delta \mathbf{X}. \tag{12.11}$$

The superscript zero on $\mathbf{M}$ indicating an equilibrium value will be presupposed in the following.

Next consider N consecutive small steps. The nth step in this sequence produces the dissipation

$$\Delta S^{u\,n} = \tfrac{1}{2}(D_S^n)^2, \tag{12.12}$$

where

$$D_S^n = \sqrt{\Delta \mathbf{X}^n \mathbf{M}_S^n\, \Delta \mathbf{X}^n} \tag{12.13}$$

to first order in the stepsize is the thermodynamic length of the step. Summing over all N steps in the process, we find, using the Cauchy–Schwarz inequality,

$$\Delta S^u = \frac{1}{2}\sum_{n=1}^{N}(D_S^n)^2 \geq \frac{1}{2N}\left[\sum_{n=1}^{N} D_S^n\right]^2 = \frac{L_S^2}{2N}, \tag{12.14}$$

where $L_S = \sum D_S^n$ is the thermodynamic length of the full N-step process. Similar to the continuous results, (12.4) and (12.5), equality is achieved only when all D_S^n are equal.

This bound and the optimal equality of all steps (in thermodynamic geometry), valid for many small steps, are the crucial results used in the following analysis of the distillation process.

12.4 A Classical Distillation Column

A conventional binary distillation column is constructed as sketched in Figure 12.1 with feed entering at the steady state rate F around the middle of the column and being separated into distillate leaving the top at rate D and

12.4. A Classical Distillation Column

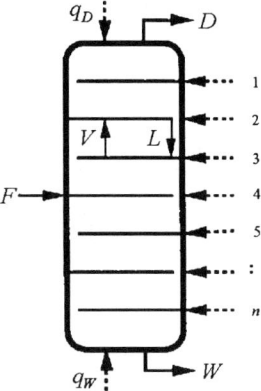

FIGURE 12.1. Sketch of a conventional distillation column with feed, distillate, and waste (bottoms) rates F, D, W, heating and cooling rates q_W and q_D, and tray numbers n.

waste W (also called bottoms B) leaving at the bottom. The mole fractions x_F, x_D, and x_W refer to the light component in those three streams. The separation is affected by a heat flow q_W entering the bottom at temperature T_W and a corresponding flow q_D leaving the top at the lower temperature T_D, thus creating vapor flows V and liquid flows L inside the column. In all large-scale distillation columns interior equilibration between vapor and liquid is achieved on trays where vapor bubbles through a thin layer of liquid.

The reversible separation (unmixing) of a feed stream F into its pure components ($x_D = 1$, $x_W = 0$) requires the power [2]

$$W_{\text{rev}} = -RT_D[x_F \ln x_F + (1 - x_F) \ln(1 - x_F)]F, \qquad (12.15)$$

where R is the gas constant. Mass and energy conservation inside a conventional distillation column, on the other hand, requires a minimum heat flow

$$q_{\min} = FR \frac{T_D T_W}{T_W - T_D} \qquad (12.16)$$

as derived in any chemical engineering textbook [2]. Using the Carnot efficiency, this is equivalent to a theoretical power

$$W_{\min} = \frac{T_W - T_D}{T_W} q_{\min} = FRT_D. \qquad (12.17)$$

Thus the thermal efficiency of a distillation column can never exceed

$$\epsilon = W_{\text{rev}}/W_{\min} = -[x_F \ln x_F + (1 - x_F) \ln(1 - x_F)] \qquad (12.18)$$

324   12. Distillation by Thermodynamic Geometry

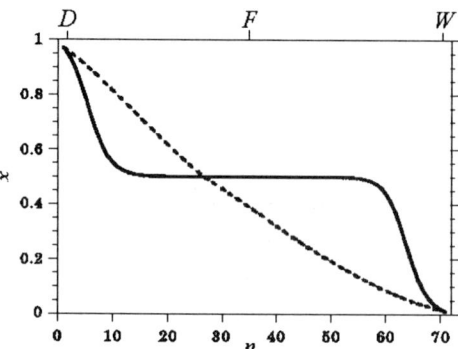

FIGURE 12.2. The liquid composition profile x as a function of tray number, counted from the condenser, for conventional (solid) and equal-thermodynamic-distance (dashed) separation of an ideal benzene–toluene system.

which has a maximum of ln(2), about 70%, for an equal mixture feed, $x_F = 0.5$. Deviation from this balanced feed as well as any nonidealities in the system will further reduce the efficiency. Considering that distillation consumes a sizable fraction of the world energy demand, this is an unfortunate situation. Of course, a completely equivalent analysis can be carried out in terms of entropy produced during the separation.

The internal distribution of temperature and mole fractions of light and heavy components in a conventional distillation column where heat is added and withdrawn only in the reboiler and the condenser, respectively, is fixed exclusively by the laws of energy and mass conservation. In most long columns that leads to an S-shaped curve of composition versus tray number with most of the variation occurring near the end points of the column, connected with a flat stretch around the feed point (see Figure 12.2). A qualitatively similar picture emerges for the temperature. This composition profile implies that the major part of the entropy production in the distillation process occurs near the ends of the column and is thus not uniformly distributed. Even more importantly, if the number of trays is increased, essentially only the middle flat section is extended while the segments of rapid variation are unchanged. This means that dissipation does not approach zero as the number of trays goes to infinity.

12.5 Optimal Temperature Profile

The principle of equal thermodynamic distance shows us how to minimize the total dissipation in a trayed column by distributing the dissipation evenly among the trays whatever their number [10], [11]. Above, a general quasistatic step process was optimized, i.e., a process composed of N discrete steps where the system equilibrates fully after each step. The standard description of a distillation column is exactly such a process where it is assumed that gas and liquid come to equilibrium at a particular temperature on each tray. Entropy is produced when the up- and down-moving flows encounter liquid on the next tray at slightly different temperature and composition. The concept of thermodynamic length defined in [1], [6] not only provides a lower bound on dissipation, it also predicts which path will achieve that bound, namely operation with equal thermodynamic distance between the trays.

Two important consequences of this general result are immediate: dissipation (here entropy production) must be equally distributed along the column; and the total dissipation approaches zero as N, the number of trays, goes to infinity (12.14), i.e., the separation becomes reversible. Neither are satisfied in ordinary distillation columns.

In binary distillation eight extensive quantities are involved on each tray. In the energy picture, these quantities are the entropy, volume, mole number of light component, and mole number of heavy component for each phase, i.e., for both vapor and liquid: $\mathbf{X} = (S_V, V_V, N_{1V}, N_{2V}, S_L, V_L, N_{1L}, N_{2L})$. This leads to an 8×8 metric matrix

$$\mathbf{M}_U = \begin{pmatrix} \mathbf{M}_U^V & 0 \\ 0 & \mathbf{M}_U^L \end{pmatrix} \begin{matrix} V, \\ L. \end{matrix} \quad \begin{matrix} V & L, \end{matrix} \qquad (12.19)$$

Fortunately a number of relations allow one to reduce the dimensionality of the problem dramatically. First of all, as already indicated by the zeros, the two physical states are usually considered energetically noninteracting, at once making $\mathbf{M}_U$ block-diagonal. After a partial Legendre transform to the mixed intensive–extensive variables $\mathbf{X} = (T, p, N_1, N_2)$ each of the submatrices $\mathbf{M}_U^V$ and $\mathbf{M}_U^L$ may be further block-diagonalized to

$$\mathbf{M}_U(T, p, N_1, N_2) = \begin{pmatrix} -G_{TT} & G_{Tp} & 0 & 0 \\ G_{pT} & -G_{pp} & 0 & 0 \\ 0 & 0 & G_{N_1 N_1} & G_{N_1 N_2} \\ 0 & 0 & G_{N_2 N_1} & G_{N_2 N_2} \end{pmatrix}, \qquad (12.20)$$

where the subscripts indicate partial derivatives with respect to the quantities specified. Note that the dependent function is now the Gibbs free energy G corresponding to the variables (T, p, N_1, N_2).

Constant pressure in the column eliminates the second row and column. Next we recombine the molar quantities N_1 and N_2 into the total amount of material, $N = N_1 + N_2$, and the molar fraction of component 1, i.e., $x = N_1/(N_1 + N_2)$ (in the liquid phase, the letter y is traditionally used for the vapor phase). The purpose is once again to effectively eliminate one variable, N, since it only describes the total scaling of the process. The metric matrix for the liquid is now down to

$$\mathbf{M}_U^L(T, x, N) = \begin{pmatrix} -G_{TT} & 0 & 0 \\ 0 & -\dfrac{G_{N_1 N_2} N^2}{x(1-x)} & 0 \\ 0 & 0 & 0 \end{pmatrix} \quad (12.21)$$

with an equivalent expression for the vapor. The square of the length element is then

$$(dL_U)^2 = -G_{TT}^V \, dT^2 - \frac{G_{N_1 N_2}^V (N^V)^2}{y(1-y)} dy^2 - G_{TT}^L \, dT^2 - \frac{G_{N_1 N_2}^L (N^L)^2}{x(1-x)} dx^2$$

$$= \left[-G_{TT}^V - \frac{G_{N_1 N_2}^V (N^V)^2}{y(1-y)} \left(\frac{dy}{dT} \right)^2 - G_{TT}^L - \frac{G_{N_1 N_2}^L (N^L)^2}{x(1-x)} \left(\frac{dx}{dT} \right)^2 \right] dT^2,$$

(12.22)

since temperatures in the two phases are equal. The vapor and liquid flows N^V and N^L are identified with the actual material flows V and L in the column at that particular tray (see Figure 12.1).

The G_{TT} terms are related to the ordinary heat capacities of the vapor and liquid mixtures, respectively,

$$G_{TT}^L = -\frac{L}{T}[x C_1^L + (1-x) C_2^L], \quad (12.23)$$

where C_i is the molar heat capacity at constant pressure of component i, with a similar equation for the vapor. The terms $G_{N_1 N_2}$ are cross derivatives of the chemical potentials,

$$G_{N_1 N_2}^L = \frac{\partial \mu_2^L}{\partial L_1} = \frac{\partial \mu_1^L}{\partial L_2}, \quad (12.24)$$

i.e., derivatives of the heavy chemical potential μ_2 with respect to the light flow $L_1 = xL$ and vice versa. The vapor–liquid equilibrium conditions $\mu_1^V = \mu_1^L$ and $\mu_2^V = \mu_2^L$ have also been invoked. For ideal mixtures, $G_{N_1 N_2} = -RT/N$, where R is the gas constant. Again analogous expressions exist for the vapor phase.

Above we have made use of the constancy of temperature and pressure to reduce the size of the problem in a fairly straightforward fashion. Mass conservation is applied in a more unusual way. In the traditional description of a distillation tray [2] the material flows of each of the two components

12.5. Optimal Temperature Profile

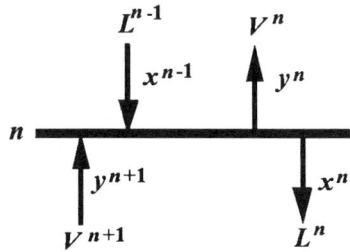

FIGURE 12.3. Definition of quantities around tray n.

entering and leaving a particular tray n are balanced:

$$V^{n+1}y_1^{n+1} + L^{n-1}x_1^{n-1} = V^n y_1^n + L^n x_1^n,$$
$$V^{n+1}y_2^{n+1} + L^{n-1}x_2^{n-1} = V^n y_2^n + L^n x_2^n. \quad (12.25)$$

As above, V and L are the vapor and liquid flow rates leaving the tray indicated in the superscript, and x_i and y_i are the molar fractions of component i in the liquid and vapor phases, respectively. These quantities are pictured in Figure 12.3.

Calculating the entropy produced in this mixing process of four streams is a bit cumbersome. Instead we consider the entropy produced in connection with the generation of a small bubble of vapor on tray $n+1$ at temperature T^{n+1} and its absorption on tray n. The vaporization creates no entropy since it is an equilibrium process; the vapor of composition (y_1^{n+1}, y_2^{n+1}) is in equilibrium with the liquid of composition (x_1^{n+1}, x_2^{n+1}) from which it is generated at that temperature. On arriving on tray n we split the process of assimilation of the bubble on that tray into two parts: first, exchange of only heat and work with the liquid–vapor system on tray n, followed by exchange of materials now at the proper temperature. In the first step the tray fluid acts as the heat and work reservoir for the equilibration, the type of simple horse–carrot process described in Section 12.3. The entropy production in this step is given to first order in the step size by (12.10) which, when expressed in the mixed intensive–extensive formulation of (12.3), has only two terms in the sum,

$$\Delta S^{u\,n} = -\frac{1}{2}\left[\Delta\left(\frac{1}{T^n}\right)\Delta U^n - \Delta\left(\frac{p}{T^n}\right)\Delta V^n\right]. \quad (12.26)$$

This is easier to calculate using the energy picture and the equivalence $\mathbf{M}_U = -T\mathbf{M}_S$ [9] so that

$$\Delta S^{u\,n} = \frac{1}{2T^n}[\Delta T^n \Delta S^n - \Delta p\, \Delta V^n] \quad (12.27)$$

since the constancy of the pressure knocks out the second term. In this thermomechanical equilibration some of the vapor in the bubble (of composition y^{n+1}) will condense so that the bubble volume now contains some liquid of composition x^n plus some vapor of composition y^n due to the new temperature T^n. Introducing the constant pressure saturation heat capacity C_r of the two-phase mixture, i.e., the effective heat capacity for the gas in equilibrium with liquid [12], so that $\Delta S = C_r\,\Delta T/T$ makes

$$\Delta S^{u\,n} = \frac{1}{2}\frac{C_r^n}{(T^n)^2}(\Delta T^n)^2. \tag{12.28}$$

Once the temperature and the pressure in our bubble have equilibrated to the surroundings, we proceed to the second step, during which each fluid phase can mix with the reservoir fluid. Since in a binary mixture T and p fix the composition, this second step occurs reversibly. The total entropy generation for this tray is thus given by (12.28) above.

On a strictly formal basis we could also recall that the energy and entropy metrics are related through $\mathbf{M}_U = -T\mathbf{M}_S$ [9] and thus quickly arrive at the dissipation between trays n and $n+1$, from (12.22),

$$\Delta S^{u\,n} = \frac{1}{2}\left(\frac{\Delta L_U^n}{T^n}\right)^2 = \frac{1}{2}\frac{C_r^n}{(T^n)^2}(\Delta T^n)^2, \tag{12.29}$$

where C_r may be expressed as

$$C_r = V[yC_1^V + (1-y)C_2^V] - G_{N_1N_2}^V\frac{TV^2}{y(1-y)}\left(\frac{dy}{dT}\right)^2$$
$$+ L[xC_1^L + (1-x)C_2^L] - G_{N_1N_2}^L\frac{TL^2}{x(1-x)}\left(\frac{dx}{dT}\right)^2, \tag{12.30}$$

and where all quantities are related to the particular tray n, and Δ indicates the difference between trays n and $n+1$.

The computational procedure is to integrate (12.29) from the distillate temperature T_D to the reboiler temperature T_W (both of course given by the required product purities) to obtain the total thermodynamic column length L_S. The distance from one tray to the next must then be fixed at $D_S = L_S/N$ for optimal performance by adjusting the tray temperatures appropriately according to (12.13).

Obviously such freedom of adjustment does not exist in a conventional adiabatic column where heat is added and removed only in the reboiler and condenser, respectively (Figure 12.1). Rather, it is necessary to allow individual heat exchange with each tray to maintain it at the desired temperature (see Figure 12.4). This heat addition/removal is of course part of the energy balance used above. The result of the whole calculation is either a graph similar to Figure 12.2 specifying the temperature of each tray in the column or a graph of the amount of heat added/removed at each tray.

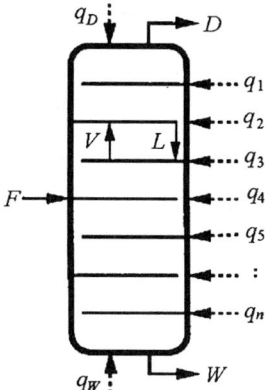

FIGURE 12.4. Sketch of an equal-thermodynamic-distance distillation column with feed, distillate, and waste (bottoms) rates F, D, W as in the traditional column Figure 12.1, but with heating or cooling on all trays n.

12.6 Example

Figure 12.5 is such a graph for a model calculation on an ideal 50/50 benzene–toluene mixture ($x_F = 0.5$). Separation of 1 mole of mixture into 99% pure products ($x_D = 0.99$ and $x_W = 0.01$) requires an exergy expenditure of 191 J in the fully optimized equal thermodynamic distance column with heat exchange on every tray. For comparison the exergy requirement is 842 J, more than a factor of 4 larger, in a conventional column heated only at the end points as indicated with the two filled circles. If for economic or purely practical reasons one does not wish to mount heat exchangers on every tray in the column, the first step of improvement will be to add just two heat exchangers at appropriate locations, making a total of four positions of heat exchange. Such a column only needs 423 J to do the job, a saving by a factor of 2 compared to the conventional column.

It should be mentioned that the total amount of *heat* used to perform a certain equal-thermodynamic-distance separation is only marginally different from that required by a conventional column, but a large part of it is used over a much smaller temperature difference than T_W to T_D, leading to a correspondingly smaller *exergy* requirement. The gradual addition and withdrawal of heat lends itself to coupling with other sources and sinks of heat at the plant which might otherwise have been lost or degraded. This distillation system is the first application of the equal thermodynamic distance principle to a chemical system. However, the entire analysis is not restricted to a distillation column but applies equally well to any staged process, be it gas separation by diffusion, staged refrigeration, or batch chemical reactions.

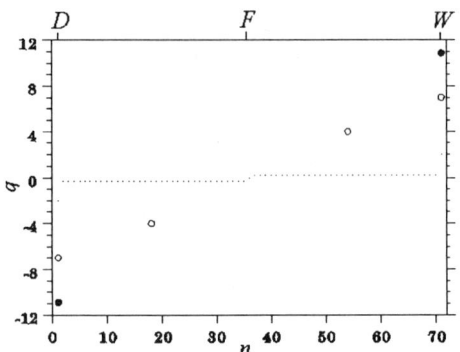

FIGURE 12.5. Amount of heat (in joules) added per mole of benzene–toluene feed on the individual trays of a 71-tray column operated in the traditional fashion with heat added only in the reboiler and withdrawn only in the condenser (•); operated according to equal thermodynamic separation specifications with heat added or withdrawn on each tray (·); and operated with just two additional heat exchange points, optimally located (o).

Acknowledgments

The work reported in this chapter has been carried out in collaboration with James Nulton and Gino Siragusa.

12.7 References

[1] J. Nulton, P. Salamon, B. Andresen, and Q. Anmin: Quasistatic processes as step equilibrations, *J. Chem. Phys.* **83**, 334, (1985).

[2] See, for example, C. J. King: *Separation Processes*, McGraw-Hill, New York, 1971.

[3] F. Weinhold: Metric geometry of equilibrium thermodynamics, *J. Chem. Phys.* **63**, 2479, (1975).

[4] F. Weinhold: Geometrical aspects of equilibrium thermodynamics, in: *Theoretical Chemistry: Advances and Perspectives* **3**, (eds. D. Henderson and H. Eyring), Academic Press, New York, 1978.

[5] P. Salamon, B. Andresen, P. Gait, and R. S. Berry: The significance of Weinhold's length, *J. Chem. Phys.* **73**, 1001, 5407E, (1980).

[6] P. Salamon and R. S. Berry: Thermodynamic length and dissipated availability, *Phys. Rev. Lett.* **51**, 1127, (1983).

[7] M. H. Rubin: Optimal configuration of a class of irreversible heat engines I, *Phys. Rev. A* **19**, 1272, (1979).

[8] B. Andresen and J. M. Gordon: On constant thermodynamic speed for minimizing entropy production in thermodynamic processes and simulated annealing, *Phys. Rev. E* **50**, 4346, (1994).

[9] P. Salamon, J. Nulton, and E. Ihrig: On the relation between energy and entropy versions of thermodynamic length, *J. Chem. Phys.* **80**, 436, (1984).

[10] P. Salamon and J. Nulton: The geometry of separation processes: A horse–carrot theorem for steady flow systems, *Europhys. Lett.* **42**, 571, (1998).

[11] P. Salamon, B. Andresen, J. Nulton, and G. Siragusa: The horse–carrot theorem and geometric optimization of distillation (in preparation), 1999.

[12] J. S. Rowlinson: *Liquids and Liquid Mixtures*, Plenum, New York, 1969.

Index

abrupt absorption model, 88
absorptance, 21, 32
absorption coefficient, 86, 87, 90
absorptivity, 95, 98
adjoint variable, 154
almost contact metric manifold, 274
almost contact structure, 273
annual control, 132
associated Riemannian metric, 274, 276
atmospheric radiation, 32, 33
atomic ion collision, 246
Auger recombination, 61, 77

backup, 134
bandgap, 51, 83, 88
bandgap shrinkage, 77
Beer's law, 91
Bellman's equation, 153
black-body radiation, 4, 8, 16, 54, 74, 95
Boltzmann statistics, 61
Bose factor, 78
Bose statistics, 60
Bouguer–Lambert law, 91
boundary conditions, 181

Carnot engine, 31, 39, 52, 100, 144
Carnot formula, 35, 52, 77, 146
Cartan bracket, 268
characteristic distribution, 262
characteristic vector field, 262
chemical potential, 60, 82, 97
cloudiness, 113
cluster ion collision, 252
cluster-surface collision, 244
combustion, 177

concentrating mirror, 38
concentration ratio, 24, 25, 32, 33
concentrator, 36, 37
conductance, 146
constitutive equations, 200
constraints, 181
contact coordinates, 259
contact diffeomorphism, 263
contact distribution, 260
contact Hamiltonian, 266
contact manifold, 259
contact vector field, 264
continuous contact transformations, 263
control theory, 106, 144, 181
control variable, 148, 181
convective heat loss coefficient, 33
cumulative power, 150
cycle of latent heat, 114

Darboux theorem, 259
dark current, 50, 92
dark saturation current, 50, 75
de Broglie wavelength, 240
detailed balance theory, 100
Diesel engine, 175
diffuse solar radiation, 26
diluted solar radiation, 15, 19, 22, 26
dilution factor, 27
dissipation, 325
distillation, 312, 322
distinguishability, 287
divergence, 280
Drude theory, 30
dynamic programming, 163
dynamical system, 208

effective mass, 88
effective temperature, 20
efficiency, 6, 8, 323
Einstein relations, 86
electroluminescence, 60
electron–phonon interaction, 80
emittance, 21
endoreversible chemical engine, 57
endoreversible engine, 41, 143, 321
endoreversible thermal engine, 52
energy balance, 200
energy gap, 74
energy transfer, 239
engine mode, 145
entropy balance, 201
entropy production, 301, 321, 327
exergy, 5, 6, 8, 329

fill factor, 82
finite-time thermodynamics, 12, 143, 319
flat-plate solar collector, 36
flux, 95, 96
Fourier's law, 200

generalized exergy, 169
generating function, 262
geodesic, 291
geometric factor, 21, 27, 91
Gibbs's law, 57
global solar radiation, 31
graded gap, 83

Hamiltonian, 152, 266
Hamiltonian equations, 266
Hamiltonian vector field, 266
heat conductional inequality, 204
heat current, 200, 210, 213
heat equation, 200, 201, 208, 231
heat leak, 178
heat pump, 145
heat radiator, 37
heat transfer coefficient, 178
Heaviside function, 79, 81, 88
heterojunction solar cell, 78

holdup time, 161
homojunction, 79
horse–carrot process, 319
horse–carrot theorem, 304, 306, 319
horse-carrot process, 321
hybrid solar energy converter, 59, 73
hyperbolic heat equation, 206

impact ionization, 62, 78, 80, 100
infinite tandem cell, 77, 93
information distance, 280
information gain, 280
information matrix, 296
information theory, 4
ion neutralization, 245
isotropic radiation, 8, 94, 95

Jacobi bracket, 267, 268
Jacobi identity, 267

Kullback information, 280

Lagrange bracket, 268
Lagrange multiplier, 162
Lambert law, 16
Lambertian surface, 94
Landsberg–Petela–Press efficiency, 16, 19, 26
Legendre submanifolds, 260
Lie derivative, 264
lifetime, 82
light current, 50
light-generated current, 82

macrostate, 5
maximum concentration, 30, 93
maximum efficiency, 26, 35, 41
maximum entropy principle, 5
maximum power, 35, 40
maximum principle, 154, 205, 209
mechanical work, 180
metric, 258, 286, 295, 320, 325
microstate, 4
mobility, 82
molecular ion collision, 246

334   Index

Monte Carlo method, 183
multistage process, 143
Müser model, 54

nonradiative recombination, 74
nondegenerate bands, 87
nonradiative recombination, 77
Novikov–Curzon–Ahlborn engine, 41, 53, 144

open-circuit point, 149
open-circuit voltage, 82, 92
optimal ignition time, 190
optimal path, 185, 320
optimal performance function, 153
optimum control, 175, 289
optimum converter temperature, 26, 30

p–n junction, 82
Page's formula, 35
pair generation, 82
parabolic band, 79
paradox of infinite velocity, 206, 209
partition function, 282
partitioning of energy, 249
penetration length, 117
perfectly forward diffuser, 27
phenomenological year, 110
photon recycling, 101
pictures, 202, 204
Planck spectrum, 56, 95
play of probabilities, 89
Poisson bracket, 266, 267
Poisson manifold, 267
Poisson structure, 267
polyatomic ion collision, 247
Pontryagin, 151
principle of superposition, 208

quantum efficiency, 50
quasi-Fermi level, 85, 86, 96

radiation, 180
radiative heat leak, 179

radiative heat transfer, 180
radiative recombination, 61
reference state, 6, 8
refractive index, 91
relative information, 280
Rényi–Kullback information, 280
Riemann metric, 295
rotational energy exchange, 243
Rutherford back scattering, 240

scattered radiation, 27
second law, 204
selective black-body radiation, 56
selective converter, 19, 21, 22
selective solar receiver, 35
selective surface, 30
shape preserving signal form, 231
shift theorem, 74
short circuit current, 99
short circuit point, 75
short-circuit current, 74
short-circuit point, 149
solar cell, 50, 59, 72
solar radiation, 72
solid angle, 94, 96
space power station, 15, 37, 41
specific humidity, 121
spectral filter, 100
spectral radiance, 94, 96
spectrum factor, 82
spontaneous emission, 89
stationary field, 203, 210
statistical distance, 291
statistical length, 288
statistical speed, 289
Stefan–Boltzmann constant, 20, 56, 97, 176
Stefan–Boltzmann law, 54
step process, 321
stimulated emission, 87, 89, 90
Stirling engine, 31, 34
stochastic optimization, 183
strict contact transformations, 263
surface collision, 239
symplectic form, 266

symplectic manifold, 266

tandem solar cell, 63, 73, 101
target problem, 108
temperature waves, 230
terrestrial solar power plant, 15
thermalization, 93
thermochemical engine, 58
thermodynamic bound, 168
thermodynamic distance, 297, 324
thermodynamic geometry, 320
thermodynamic length, 320
thermodynamic phase space, 258, 262

thermophotovoltaic conversion, 99
translational energy exchange, 241
transmittance, 25, 32
typical year, 110

ultimate efficiency, 76–78
upper bound formula, 14

variational principles, 202
vibrational energy exchange, 244
view factor, 21
voltage factor, 82

Watt governor, 107
work output, 177